BOWES and CHURCH'S
Food Values
of Portions
Commonly Used

BOWES and CHURCH'S **Food Values of Portions Commonly Used**

Thirteenth Edition, Revised by

JEAN A. T. PENNINGTON, PH.D., R.D.

Formerly Instructor in Nutrition
City College of San Francisco
San Francisco, CA

and HELEN NICHOLS CHURCH, B.S.

1817

HARPER & ROW, PUBLISHERS, New York

Cambridge, Hagerstown, Philadelphia, San Francisco,
London, Mexico City, São Paulo, Sydney

Library of Congress Cataloging in Publication Data
Bowes, Anna de Planter.
    Bowes and Church's Food values of portions commonly used.

    Bibliography: p.
    Includes index.
    1. Food—Composition—Tables. 2. Nutrition—Tables. I. Church, Charles Frederick, joint author. II. Pennington, Jean A. Thompson. III. Church, Helen Nichols. IV. Title. V. Title: Food values of portions commonly used.
TX551.B64   1980   641.1   79-27226
ISBN 0-397-54310-7
ISBN 0-06-010767-7   80 81 82 83 84 10 9 8 7 6 5 4 3 2 1
ISBN 0-06-090819-X pbk   81 82 83 84 10 9 8 7 6 5 4 3 2

DEDICATED TO THE MEMORY OF CHARLES FREDERICK CHURCH, M.D. WHOSE VISION AND EXHAUSTIVE STUDY AND DEVOTION TO THE IDEAL OF FURTHERING THE GOOD HEALTH OF ALL PEOPLE THROUGH PROPER NUTRITION MADE POSSIBLE THE FORMER EDITIONS.

Preface

The purpose of this thirteenth edition is the same as the first—to provide a quick and accurate nutrient reference. With the advent of computer nutrient data banks and the large quantity of data available from industry, the task of providing a quick and ready reference has become a matter of selectivity. We have attempted to provide a selective cross-sampling of foods commonly consumed in the U.S. plus some specialty items and a few foreign foods.

This thirteenth edition maintains the same two-lined nutrient format of the previous one. However, those familiar with previous editions will note several changes. Several food categories (e.g., fast foods, spices and flavorings, infant formulas, etc.) have been added, and many new foods, especially those from industry, are included. Thiamin and riboflavin are now expressed in milligrams rather than micrograms. Several new tables have been added (margarine, alcohol, caffeine, fluoride, etc.). Some non-nutrient information in the preceding edition has been deleted to conserve space.

Nutrients values were obtained from many sources including computer data banks, numerous nutrient tables in journal articles and in nutrition textbooks, and information supplied from the food industry. All nutrient values were assumed to be public domain. Brand names are used only to help in food identification, not to promote or endorse products. We would like to thank the food industry for making nutrient values available, and would like to request their help in correcting and updating data and providing information on new products. The authors take full responsibility for any errors in this text and invite your comments and criticisms.

Jean A.T. Pennington
Helen N. Church

June 1979

Preface to the First Edition

The purpose of this book is to supply authoritative data on the nutritional values of foods in a form for quick and easy reference.

In teaching nutrition to students of medicine, dentistry, dental hygiene and public health nursing, food values based on common measures or portions frequently served have been found most useful. This basis of calculation is particularly well suited to the practical study of comparative food values, as well as to the approximate analysis of diets from records of daily food intake. For calculations of diets from weighed portions, the actual weight of each food is given in grams or ounces.

Anna dePlanter Bowes
Charles F. Church

November, 1937

Note

After my husband's death in 1976, I continued to live on Cape Cod but retired from most of my activities in the field of nutrition.

I knew of my husband's deep and enduring desire that there should be future editions and revisions of Bowes and Church of which I had been co-author since the revision of the ninth edition.

I therefore set about to find an individual well qualified in the field of nutrition to take charge of the revision of the thirteenth edition.

I was most fortunate in securing Dr. Jean Pennington to whom I am most grateful, not only for her fine reorganization and additions, but for her long hours of precise and definitive work.

H.N.C.

Contents

FOOD AND NUTRITION BOARD, NATIONAL ACADEMY OF SCIENCES-NATIONAL RESEARCH COUNCIL
RECOMMENDED DAILY DIETARY ALLOWANCES,[a] Revised 1979
Designed for the maintenance of good nutrition of practically all healthy people in the U.S.A.

	Age (years)	Weight (kg)	Weight (lbs)	Height (cm)	Height (in)	Protein (g)	Vitamin A (µg R.E.)[b]	Vitamin D (µg)[c]	Vitamin E (mg α T.E.)[d]	Vitamin C (mg)	Thiamin (mg)	Riboflavin (mg)
Infants	0.0-0.5	6	13	60	24	kg × 2.2	420	10	3	35	0.3	0.4
	0.5-1.0	9	20	71	28	kg × 2.0	400	10	4	35	0.5	0.6
Children	1-3	13	29	90	35	23	400	10	5	45	0.7	0.8
	4-6	20	44	112	44	30	500	10	6	45	0.9	1.0
	7-10	28	62	132	52	34	700	10	7	45	1.2	1.4
Males	11-14	45	99	157	62	45	1000	10	8	50	1.4	1.6
	15-18	66	145	176	69	56	1000	10	10	60	1.4	1.7
	19-22	70	154	177	70	56	1000	7.5	10	60	1.5	1.7
	23-50	70	154	178	70	56	1000	5	10	60	1.4	1.6
	51+	70	154	178	70	56	1000	5	10	60	1.2	1.4
Females	11-14	46	101	157	62	46	800	10	8	50	1.1	1.3
	15-18	55	120	163	64	46	800	10	8	60	1.1	1.3
	19-22	55	120	163	64	44	800	7.5	8	60	1.1	1.3
	23-50	55	120	163	64	44	800	5	8	60	1.0	1.2
	51+	55	120	163	64	44	800	5	8	60	1.0	1.2
Pregnant						+30	+200	+5	+2	+20	+0.4	+0.3
Lactating						+20	+400	+5	+3	+40	+0.5	+0.5

a   The allowances are intended to provide for individual variations among most normal persons as they live in the United States under usual environmental stresses. Diets should be based on a variety of common foods in order to provide other nutrients for which human requirements have been less well defined.

b   Retinol equivalents. 1 Retinol equivalent = 1 µg retinol or 6 µg βcarotene.

c   As cholecalciferol. 10 µg cholecalciferol = 400 I.U. vitamin D.

d   α tocopherol equivalents. 1 mg d-α-tocopherol = 1 αT.E.

FOOD AND NUTRITION BOARD, NATIONAL ACADEMY OF SCIENCES-NATIONAL RESEARCH COUNCIL
RECOMMENDED DAILY DIETARY ALLOWANCES, Revised 1979
Designed for the maintenance of good nutrition of practically all healthy people in the U.S.A.

Niacin (mg N.E.)[e]	Vitamin B6 (mg)	Folacin[f] (μg)	Vitamin B12 (μg)	Calcium (mg)	Phosphorus (mg)	Magnesium (mg)	Iron (mg)	Zinc (mg)	Iodine (μg)
6	0.3	30	0.5[g]	360	240	50	10	3	40
8	0.6	45	1.5	540	360	70	15	5	50
9	0.9	100	2.0	800	800	150	15	10	70
11	1.3	200	2.5	800	800	200	10	10	90
16	1.6	300	3.0	800	800	250	10	10	120
18	1.8	400	3.0	1200	1200	350	18	15	150
18	2.0	400	3.0	1200	1200	400	18	15	150
19	2.2	400	3.0	800	800	350	10	15	150
18	2.2	400	3.0	800	800	350	10	15	150
16	2.2	400	3.0	800	800	350	10	15	150
15	1.8	400	3.0	1200	1200	300	18	15	150
14	2.0	400	3.0	1200	1200	300	18	15	150
14	2.0	400	3.0	800	800	300	18	15	150
13	2.0	400	3.0	800	800	300	18	15	150
13	2.0	400	3.0	800	800	300	10	15	150
+2	+0.6	+400	+1.0	+400	+400	+150	[h]	+5	+25
+5	+0.5	+100	+1.0	+400	+400	+150	[h]	+10	+50

e   1 NE (niacin equivalent) is equal to 1 mg of niacin or 60 mg of dietary tryptophan.

f   The folacin allowances refer to dietary sources as determined by *Lactobacillus casei* assay after treatment with enzymes ("conjugases") to make polyglutamyl forms of the vitamin available to the test organism.

g   The RDA for vitamin B12 in infants is based on average concentration of the vitamin in human milk. The allowances after weaning are based on energy intake (as recommended by the American Academy of Pediatrics) and consideration of other factors such as intestinal absorption.

h   The increased requirement during pregnancy cannot be met by the iron content of habitual American diets nor by the existing iron stores of many women; therefore the use of 30-60 mg of supplemental iron is recommended. Iron needs during lactation are not substantially different from those of non-pregnant women, but continued supplementation of the mother for 2-3 months after parturition is advisable in order to replenish stores depleted by pregnancy.

Reproduced from: Recommended Dietary Allowances, Ninth Edition (1979, in press), with the permission of the National Academy of Sciences, Washington, D.C.

Estimated Safe and Adequate Daily Dietary Intakes of Additional Selected Vitamins and Minerals[a,b]

	Age (years)	Vitamin K (µg)	Biotin (µg)	Pantothenic Acid (mg)	Copper (mg)	Manganese (mg)
Infants	0-0.5	12	35	2	0.5-0.7	0.5-0.7
	0.5-1	10-20	50	3	0.7-1.0	0.7-1.0
Children	1-3	15-30	65	3	1.0-1.5	1.0-1.5
and	4-6	20-40	85	3-4	1.5-2.0	1.5-2.0
Adolescents	7-10	30-60	120	4-5	2.0-2.5	2.0-3.0
	11+	50-100	100-200	4-7	2.0-3.0	2.5-5.0
Adults		70-140	100-200	4-7	2.0-3.0	2.5-5.0

a   Because there is less information on which to base allowances, these figures are not given in the main table of the RDA and are provided here in the form of ranges of recommended intakes.

b   Since the toxic levels for many trace elements may be only several times usual intakes, the upper levels for the trace elements given in this table should not be habitually exceeded.

Estimated Safe and Adequate Daily Dietary Intakes
of Additional Selected Vitamins and Minerals

Fluoride (mg)	Chromium (mg)	Selenium (mg)	Molybdenum (mg)	Sodium (mg)	Potassium (mg)	Chloride (mg)
0.1-0.5	0.01-0.04	0.01-0.04	0.03-0.06	115-350	350-925	275-700
0.2-1.0	0.02-0.06	0.02-0.06	0.04-0.08	250-750	425-1275	400-1200
0.5-1.5	0.02-0.08	0.02-0.08	0.05-0.1	325-975	550-1650	500-1500
1.0-2.5	0.03-0.12	0.03-0.12	0.06-0.15	450-1350	775-2325	700-2100
1.5-2.5	0.05-0.2	0.05-0.2	0.1 -0.3	600-1800	1000-3000	925-2775
1.5-2.5	0.05-0.2	0.05-0.2	0.15-0.5	900-2700	1525-4575	1400-4200
1.5-4.0	0.05-0.2	0.05-0.2	0.15-0.5	1100-3300	1875-5625	1700-5100

Reproduced from: Recommended Dietary Allowances, Ninth Edition (1979, in press), with the permission of the National Academy of Sciences, Washington, D.C.

Explanatory Notes

Each food has two lines in the main table. The heading at the top of the page indicates the nutrients and units of measurement for the numerical values listed in the two lines. The abbreviated heading names and other abbreviations and symbols used throughout the text are listed on page 18. Each food is identified by name, description, brand name (where applicable) and serving portion. The serving portion for most foods is listed in both household units and grams.

Attempts were made to conserve space and allow for the inclusion of a large number of foods. Foods are presented in their most "consumable" form. Many meats are listed only in cooked form. Raw meat values may be listed if: (1) some nutrient values were available for the raw meat, but not for the cooked one; or, (2) nutrient values for cooked meat differed significantly from the raw one due to method of preparation. For similar foods of various brand names (e.g. canned corn) the values were pooled and no brand name is listed. Brand names are used primarily to help identify a food.

Only one serving portion is offered for most foods. The portion sizes in grams and household units are taken from the previous edition of Bowes and Church, USDA Handbook No. 456, the Ohio State Nutrient Data Bank and food industry information. Due to insufficient information, some foods may have gram weights without household units and vice versa. For some foods neither gram weights nor household units were available. In these cases, the authors elected to use 50g, 100g, 200g (as in the case of combination foods), or 1 ounce (28g) portions or, whatever seemed appropriate.

Duplication of food items (repeating a food in more than one category) has been kept to a minimum. Please refer to the index if you cannot locate a food item.

The information presented here is as accurate and current as possible. There are many causes for nutrient variations in foods (season, geography, genetics, diet, processing, method of analysis, etc.), and the values listed here may have wide ranges on either side. Because of nutrient variation and the fact that the data are collected from various sources (Ohio State Data Bank, USDA Handbook No. 456, USDA Handbook No. 8, food industry, journal articles, and so forth), inconsistencies may occur.

ABBREVIATIONS & SYMBOLS

&	and	MET	methionine
amt	amount	mg	milligram(s)
AP	as purchased	Mg	magnesium
ASC	ascorbic acid	mash	mashed
avg	average	Na	sodium
bbq	barbeque	NIA	niacin
bev	beverage(s)	oz	ounce
boil	boiled	P	phosphorus
brkfst	breakfast	past	pasteurized
broil	broiled	PHE	phenylalanine
btr	butter	pkg	package
Ca	calcium	pnt	peanut
CAL	calorie(s)	pot	potato
calif	California	PRO	protein
car	carrot(s)	proc	processed
cass	casserole	PUFA	polyunsaturated fatty acids
CHO	carbohydrate	recon	reconstituted
choc	chocolate	reg	regular
cinn	cinnamon	RIB	riboflavin
ckd	cooked	roast	roasted
cnd	canned	sand	sandwich
cocnt	coconut	sce	sauce
conc	concentrate	SFA	saturated fatty acid
cond	condensed	simmer	simmered
dm	diameter	steam	steamed
(dom)	domestic	stk	steak
dress	dressing	stwd	stewed
ed	edible	stufd	stuffed
fib	fiber	s-f	sugar-free
fill	filling	sweet	sweetened
fl oz	fluid ounce	t	teaspoon
form	formula	T	tablespoon
frzn*	frozen	THI	thiamin
g	gram(s)	THR	threonine
grn	green	tom	tomato
gvy	gravy	top	topping
hmde	homemade	tot	total
hosp	hospital	tr	trace
hp	heaping	TRY	tryptophan
ISO	isoleucine	unenr	unenriched
IU	international unit	unsweet	unsweetened
jce	juice	VAL	valine
K	potassium	van	vanilla
lb	pound	veg	vegetable
LEU	leucine	vit A	vitamin A
low-cal	low-calorie	vit D	vitamin D
marb	marbled	vol	volume
marg	margarine	whip	whipped
mcg	microgram	wt	weight
med	medium	w/	with
		w/o	without
		0	none
		/	per; divided by; or

* The word frozen (frzn), when used to describe a food, refers to a commercially frozen item.

Blank spaces indicate insufficient information for nutrient values.

	WT	Proximate			Amino Acids				Minerals			Vitamins		
		CAL	CHO	FAT	TRY	LEU	LYS	MET	Na	Ca	P	THI	NIA	A
		PRO	FIB	PUFA	PHE	ISO	VAL	THR	K	Mg	Fe	RIB	ASC	D
	g	g	g	g	mg	mg	mg	mg	mg	mg	mg	mg	mg	iu

Carbonated

	WT	CAL/PRO	CHO/FIB	FAT/PUFA	TRY/PHE	LEU/ISO	LYS/VAL	MET/THR	Na/K	Ca/Mg	P/Fe	THI/RIB	NIA/ASC	A/D
ATOLE	360	76	16.9	0.4						50		.07	0.4	0
12 oz		1.4	0.7									.00	0	
BITTER LEMON	360	192	42.3	0.0					60					
12 oz		0.0												
CACTUS COOLER	360	183	45.7	0.0										
12 oz		0.0												
CHOCOLATE		166	42.1	0.0					27	0	0	.00	0.0	0
12 oz		0.0							0			.00	0	
CLUB SODA†	120	0	0.0	0.0					30	4		.00	0.0	0
4 oz		0.0							2			.00	0	
COCA COLA	360	144	36.0	0.0					1	11	61	.00	0.0	0
12 oz		0.0							4			.00	0	
COLA	360	129	36.0	0.0					22	11	2	.00	0.0	0
12 oz		0.0							0	4		.00	0	
COLA, S-F	360	2	0.1	0.0					25		39			
12 oz		0.0												
COLLINS MIXER	120	42	10.8	0.0					9					
4 oz		0.0												
CREAM SODA	360	155	39.6	0.0					24	11				
12 oz		0.0												
DIET-RITE, ROYAL CROWN	360	1	0.2	0.0					58	0	38			
12 oz		0.0							1					
FRESCA	360	4	0.1	0.0					65	36				
12 oz		0.0							0					
FRUIT-FLAVORED SODAS	360	166	43.2	0.0					27	11				
12 oz		0.0								2				
FRUIT-FLAVORED SODAS, S-F	360	3	tr	0.0					22	36				
12 oz		0.0												
GINGER ALE	360	136	33.8	0.0					4	11	54			
12 oz		0.0								4				
GINGER ALE, S-F	360	4	1.5	0.0					32	18				
12 oz		0.0												
GRAPE SODA	360	176	44.5	0.0					38	35	0			
12 oz		0.0							4	4	0.2			
LEMON-LIME	360	146	36.8	0.0					13	35	0			
12 oz		0.0							4		0.2			
LEMON-LIME, S-F	360	0	0.0	0.0					49	35				
12 oz		0.0							4	4	0.2			
MOUNTAN DEW	360	171	42.8	0.0					31	11	0			
12 oz		0.0							7		0.0			
MR. PIBB	360	140	37.5	0.0					17		42			
12 oz		0.0												
ORANGE SODA	360	167	45.4	tr					18	1				
12 oz		0.1							16		0.0			
PEPSI COLA	360	156	39.4	0.0					1	tr	52			
12 oz		0.0							10	1	0.1			
PEPSI COLA, S-F	360	1	0.2	0.0					63		61			
12 oz		0.0							12		0.1			
PEPSI LIGHT	360	71	17.6	0.0					12	0	48			
12 oz		0.0							12	tr	0.1			
PURPLE PASSION	360	188	46.8	0.0										
12 oz		0.0												
QUININE SODA‡	120	37	9.6	0.0					8	12				
4 oz		0.0												
ROOT BEER, FROSTIE	360	154	38.2	0.0					28	0	76			
12 oz		0.0							79		0.0			
ROOT BEER, FROSTIE, S-F	360	5		0.0					30					
12 oz		0.0												
ROOT BEER, HIRES	360	146	39.8	0.0					4	tr				
12 oz		0.0							19		0.0			

*Milks, milk beverages, milk flavorings, juices, and juice drinks are listed separately.

†Unsweetened carbonated water.
‡Sweetened carbonated water.

	WT	Proximate			Amino Acids				Minerals			Vitamins		
		CAL	CHO	FAT	TRY	LEU	LYS	MET	Na	Ca	P	THI	NIA	A
		PRO	FIB	PUFA	PHE	ISO	VAL	THR	K	Mg	Fe	RIB	ASC	D
	g	g	g	g	mg	mg	mg	mg	mg	mg	mg	mg	mg	iu
ROOT BEER, HIRES, S-F	360	2	0.4	0.0					53	0				
12 oz		0.0							16		0.0			
ROYAL CROWN COLA	360	156	39.0	0.0					1	0	36			
12 oz		0.0							1					
ROYAL CROWN W/A TWIST	360	147	41.0	0.0					1	0	36			
12 oz		0.0							1					
SEVEN-UP	360	144	36.0	0.0					4	0	0			
12 oz		0.0							0		0.0			
SEVEN-UP, S-F	360		0.5	0.0					82					
12 oz		0.0												
SPRITE	360	144	62.3	0.0					47	11	0			
12 oz		0.0							0					
SPRITE, S-F	360	5	0.0	0.0					48		0			
12 oz		0.0												
TAB, S-F	360	1	0.1	0.0					27		45			
12 oz		0.0												
TAHITIAN TREAT	360	204	51.1	0.0					31					
12 oz		0.0												
TEEM	360	152	38.2	0.0					31					
12 oz		0.0												
TONIC WATER	360	132	33.0	0.0					0					
12 oz		0.0												
VANILLA CREAM	360	204	51.1						27					
12 oz		0.0												
UPPER 10	360	153	38.4	0.0					19	0	0			
12 oz		0.0							0					
VERNORS	360	139							12					
12 oz		1.4												
VERNORS, S-F	360	2	tr	0.0					6					
12 oz		tr							8					
WINK	360	193	48.3	0.0					31	0	0			
12 oz		0.0							0					
Noncarbonated, sweetened														
AWAKE†, FRZN, FROM CONCENTRATE	250	139	33.3	0.0					14	0	0	0.28		2823
1 cup		0.0	0.3						125		0.0		151	
FUNNY FACE, ALL FLAVORS	240	90	22.0	0.0					0					
1 cup		0.0												
GATORADE, CITRUS	230	39	10.5	0.0					123	23	0			
1 cup		0.0							23					
GATORADE, COLA	230	39	9.2	0.0					108	0	28			
1 cup		0.0							23					
HAWAIIAN PUNCH	250	120	29.3	tr					50	20	8	.05	0.1	125
1 cup		0.1							63		0.3	.08	40	
HAWAIIAN PUNCH, LOW-CAL	250	45	11.3	0.0										
1 cup		0.1											0	
KOOL-AID, ALL FLAVORS	240	100	25.0	0.0					1	1	18	.00		150
1 cup		0.0							1		0.0	.00	10	
LEMONADE, FRZN, FROM CONCENTRATE 1 cup	250	110	28.5	0.0					0	3	3	.03	0.3	0
		0.3	0.0						40	3	0.0	.03	18	
LEMONADE FLAVORED DRINK, FROM MIX, COUNTRY TIME 1 cup	240	90	22.0	0.0					45	1	tr	.00	0.0	
		0.0	0.0						1		tr	.00	15	
LEMONADE FLAVORED DRINK, FROM MIX, WYLERS 1 cup	240	90	22.0	0.0										
		0.0												
LIMEADE, FRZN, FROM CONCENTRATE	250	103	28	0.0						3	3			
1 cup		0.0							33				5	
START‡, DRY POWDER	28	103	25.4	0.1					13	126	117	.00	0.0	2094
1 oz		0.0							1	0	tr	.00	120	
TANG‡, GRAPE, DRY POWDER	28	103	26.3	0.0					tr	123	111	.00	0.0	1960
1 oz		0.0	0.0						1	0	tr	.00	109	

*Milks, milk beverages, milk flavorings, juices, and juice drinks are listed separately.

†Imitation orange juice
‡Instant breakfast drink

	WT	Proximate			Amino Acids				Minerals			Vitamins		
		CAL	CHO	FAT	TRY	LEU	LYS	MET	Na	Ca	P	THI	NIA	A
		PRO	FIB	PUFA	PHE	ISO	VAL	THR	K	Mg	Fe	RIB	ASC	D
	g	g	g	g	mg	mg	mg	mg	mg	mg	mg	mg	mg	iu
TANG†, GRAPE, BEVERAGE	270	177	42.1	0.7					156	0	140	.00	0.0	2718
1 cup		0.1							3		0.1	.00	164	
TANG†, GRAPEFRUIT, DRY POWDER	28	101	24.7	0.1					13	115	103	.00	0.0	1907
1 oz									1	0	tr	.00	115	
TANG†, ORANGE, DRY POWDER	28	103	25.2	0.1					13	124	112	.00	0.0	1957
1 oz		0.0	0.0						75	0	0.0	.00	111	
TANG†, ORANGE, BEVERAGE	240	135	33.8	0.1					17	165	150	.00	0.0	2621
1 cup		0.0	0.0						100	0	0.0	.00		
TEA, ICED, INSTANT POWDER	3	11	2.9						tr	0	0	.00	0.0	0
W/ SUGAR & LEMON 1 t		tr							4	0		.00	1	
TEA, ICED, W/SUGAR CND	360	146	36.5	0.0					13			.00	0.0	0
12 oz		0.0	0.0						94			.00	0	
Noncarbonated, unsweetened														
COFFEE, BEVERAGE														
BREWED, W/ GRAIN, MELLOW ROAST	240	10	2.2	0.0					2					
1 cup		0.5							103					
GROUND	236	2	0.5	0.0					26	7	7	.00	1.2	0
1 cup		0.0	0.0						2	12	0.0	.00	0	
INSTANT	240	2	1.2	0.0					2	5	10	.00	0.7	0
1 cup		0.0	0.0						86	16	0.2	.00	0	
INSTANT, SANKA, DECAFFEINATED	236	5	1.2	0.0					0	0	0	.00	0.9	0
1 cup		0.0							87		0.0	.00	0	
INSTANT, ALMOND MOCHA	240	85	15.5	1.4						0		.00		0
1 cup		0.0									0.0		0	
INSTANT, BAVARIAN MINT	240	82	16.9	1.4						0		.00		0
1 cup		0.0									0.0		0	
INSTANT, CAFÉ MOCHA	240	82	16.9	1.4						0		.00		0
1 cup		0.0									0.0		0	
INSTANT, CAFÉ CAPRI	240	62	14.1	1.4						0		.00		0
1 cup		0.0									0.0		0	
INSTANT, CAFÉ VIENNESE	240	62	12.7	1.4						0		.00	0.6	0
1 cup		0.0									0.0	.05		
COFFEE, DRY														
FREEZE-DRIED	2	3	0.8						tr	3	8		0.7	
1 t									83		0.1	.03		
FREEZE-DRIED, DECAFFEINATED	2	4	0.9	0.0					tr	3	8		0.5	
1 t		0.2							82		0.1	.03		
INSTANT, REG/DECAFFEINATED	2	4	1.1	0.1					2	3	7	.00	0.6	
1 t		0.3	0.0						80	8	0.1	tr	0	
INSTANT, CAFÉ FRANCAIS	6	30	3.4	1.6					50	24	11			
1 t		0.4	tr						95					
INSTANT, CAFÉ VIENNA	7	31	5.4	0.9					48		6		0.4	
1 t		0.3	tr						74					
INSTANT, ORANGE CAPPUCCINO	7	31	1.0	5.2					48		63		0.4	
1 t		0.4							79					
INSTANT, SUISSE MOCHA	6	30	3.5	1.6					20	20			0.2	
1 t		0.4	tr						59					
INSTANT W/ GRAIN, MELLOW ROAST	3	9	1.9	tr					tr	4	8	tr	0.4	tr
1 t		0.3	tr						74	0	0.2	.01	0	
POSTUM, INSTANT BEVERAGE	236	16	3.9	0.0					7	7	61	.05	tr	
1 cup		0.5	0.0						123		0.7	.00	0	
POSTUM, INSTANT POWDER	3	10	2.5	tr					2	4	27	.02	0.1	
1 t		0.2	tr						87		0.2			
TEA, W/ ORANGE PEKOE TEA BAG	240	2	0.5	0.0								.00	0.1	0
1 cup		0.1										.05	1	
TEA, INSTANT POWDER	2	6	1.6	tr					1	tr	6		0.2	
1 t		0.2	tr						91	8	tr	.02		
TEA, INSTANT BEVERAGE	240	5	1.0	0.0					2	17	0	.00	0.0	0
1 cup		0.0							60	10	0.0	.02	0	

*Milks, milk beverages, milk flavorings, juices, and juice drinks are listed separately.

†Instant breakfast drink

	WT	CAL	CHO	FAT	TRY	LEU	LYS	MET	Na	Ca	P	THI	NIA	A
		PRO	FIB	PUFA	PHE	ISO	VAL	THR	K	Mg	Fe	RIB	ASC	D
	g	g	g	g	mg	mg	mg	mg	mg	mg	mg	mg	mg	iu
TEA, ICED, INSTANT POWDER	3	12	3.0	0.0					4	11	7	.00	0.3	0
1 t		0.0							255	21	0.1	.03		
TEA, ICED, INSTANT POWDER LEMON FLAVOR　1 t	3	5	0.5	0.0					25					
		0.0							186					
Alcoholic†														
ALE, MILD	230	98	8.0							30	41	tr	0.5	0
8 oz		1.1									0.2	.07	0	0
BEER (4.5% ALCOHOL BY VOL)	360	151	13.7	0.0					25	18	108	.00	2.2	0
12 oz		1.1	0.0						90	36	0.0	.11	0	
BEER, BIRCH	360	173	43.2	0.0					27			.00	0.0	0
12 oz		0.0										.00	0	
BEER, BUDWEISER	360	150	14.0	0.0										
12 oz		1.1												
BEER, MICHELOB	360	160	16.0	0.0										
12 oz		1.1												
BEER, NATURAL LIGHT	360	100	6.0	0.0										
12 oz		0.4												
DAIQUIRI	100	122	5.2							4	3	.01	tr	0
1 cocktail glass		0.1									0.1	tr	8	0
EGGNOG, CHRISTMAS-TYPE	123	335	18.0							44	74	.04	tr	84
1 punch cup (4 oz)		3.9									0.7	.11	tr	21
GIN, RUM, VODKA, WHISKEY	45	104		0.0					1	4	5	.00	0.0	0
1 jigger (1½ oz) 80-Proof		0.0							1	tr	0.2	.00	0	
GIN, RUM, VODKA, WHISKEY	45	112		.00					1	4	5	.00	0.0	0
1 jigger (1½ oz) 86-Proof		.00							1	tr	0.2	.00	0	
GIN, RUM, VODKA, WHISKEY	45	118		0.0					1	4	5	.00	0.0	0
1 jigger (1½ oz) 90-Proof		0.0							1	tr	0.2	.00	0	
GIN, RUM, VODKA, WHISKEY	45	124		0.0					1	4	5	.00	0.0	0
1 jigger (1½ oz) 94-Proof		0.0							1	tr	0.2	.00	0	
GIN, RUM, VODKA, WHISKEY	45	133		0.0					1	4	5	.00	0.0	0
1 jigger (1½ oz) 100-Proof		0.0							1	tr	0.2	.00	0	
GIN RICKEY	120	150	1.3							2	1	.01	tr	
1 glass		tr									tr	tr	4	0
MANHATTAN	100	164	7.9							1	1	tr	tr	35
1 cocktail (3½ oz)		tr									tr	tr	0	
MARTINI	100	140	0.3							5	1	tr	tr	4
1 cocktail (3½ oz)		0.1									0.1	tr	0	
PLANTERS PUNCH	100	175	7.9							4	3	.01	tr	0
1 glass		0.1									0.1	tr	8	0
TOM COLLINS	300	180	9.0							6	6	tr	tr	0
1 tall glass (10 oz)		0.3									tr	tr	21	0
WHISKEY SOUR	75	138	7.7	0.0					1	2	3	.02		4
1 cocktail		0.2							94			.00	8	
WHISKEY SOUR MIX	60	24	5.9	0.0					4			.00	0.0	0
2 oz		0.0										.00	0	
WINE, CHAMPAGNE (DOM)	120	84	3.0											
1 wine glass (4 oz)		0.2												
WINE, DESSERT (18.8% ALCOHOL BY VOL)	100	137	7.7	0.0					4	8	10	.01	0.2	0
1 wine glass		0.1	0.0						75	9	0.4	.02	0	
WINE, MUSCATELLE/PORT	100	158	14.0							4	8		.01	0.2
1 wine glass (3½ oz)		0.2							75			.01		
WINE, SAUTERNE, CALIFORNIA	100	84	4.0											
1 wine glass (3½ oz)		0.2												
WINE, SHERRY (DOM)	60	84	4.8						2	5		.01	0.1	
1 sherry glass (2 oz)		0.2							45			.01		
WINE, TABLE (12.2% ALCOHOL BY VOL)	100	85	4.2	0.0					5	9	10	.00	0.1	0
1 wine glass		0.1	0.0						92	9	0.4	.01	0	
WINE, VERMOUTH, FRENCH	100	105	1.0						4	8		.01	0.2	
1 wine glass									75			.01		

*Milks, milk beverages, milk flavorings, juices, and juice drinks are listed separately.

†Table I lists the caloric value and the carbohydrate and alcoholic content of alcoholic beverages. Listed here are alcoholic beverages for which other nutrients were available.

	WT	Proximate			Amino Acids				Minerals			Vitamins		
		CAL	CHO	FAT	TRY	LEU	LYS	MET	Na	Ca	P	THI	NIA	A
		PRO	FIB	PUFA	PHE	ISO	VAL	THR	K	Mg	Fe	RIB	ASC	D
	g	g	g	g	mg	mg	mg	mg	mg	mg	mg	mg	mg	iu
CANDY														
BOWLBY BITS	200	1440	40.0	132.0					120					
1 pkg		24.0							500					
BRIDGE MIX, ALMOND, KRAFT	28	154	11.3	11.9					16	66	103			
1 oz (7 pieces)		3.4	0.4						151					
BRIDGE MIX, PNT CRUNCH, KRAFT	28	125	18.4	6.3					140	10	48			
1 oz (6 pieces)		2.2	0.3						94					
BUTTERSCOTCH	5	21	4.3	0.4						1	tr	0	tr	7
1 piece		0.0	0.0								0.1	tr	0	0
BUTTERSCOTCH	28	116	24.3	2.5					19	6	2	0	tr	39
1 oz (6 pieces)		0.0							1		0.5	tr	0	
BUTTERSCOTCH MORSELS, NESTLÉ	28	148	16.8	8.7										
1 oz (28 morsels)														
CARAMELETTES, KRAFT	28	120	19.2	4.7					60	59	56			
1 oz (9 pieces)		1.7	0.0						82					
CARAMELS, PLAIN/CHOC	28	112	21.5	2.9					63	41	34	.01	0.1	3
1 oz (3 pieces)		1.1	0.1						54	1	0.4	.05		
CARAMELS, PLAIN/CHOC W/ NUTS	28	120	19.7	4.6					57	39	39	.03	0.1	6
1 oz (2 pieces)		1.3	0.1						65		0.4	.05		
CARAMELS, CHOC, KRAFT	28	112	21.1	2.5					71	48	36			
1 oz (3 pieces)		1.6	0.1						96					
CARAMELS, VANILLA, KRAFT	28	113	21.2	2.4					76	52	42			
1 oz (3 pieces)		1.6	0.0						87					
CHOC CHIPS, BAKERS	28	124	18.8	7.0					4	3	13			
1 oz		0.7	0.1						22	2	0.2		0	
CHOC CHIPS, SEMISWEET, HERSHEY	28	146	17.2	8.2					64	8	50	.01	0.1	6
1 oz		1.7	0.4	0.3					81		1.2	.01	tr	
CHOC CHIPS, SEMISWEET, NESTLÉ	28	148	17.8	7.9					3	0		.00	0.0	0
1 oz		1.0									0.0	.00	0	
CHOC MINICHIPS, HERSHEY	28	146	17.2	8.2					64	8	50	.01	0.1	6
1 oz		1.7	0.4						81		1.2	.01	tr	
CHOCOLATE-COATED														
ALMONDS	28	159	11.1	12.2					17	57	96	.03	0.5	15
1 oz		3.4	0.4						153	11	0.8	.15	tr	10
BRAZIL NUTS, KRAFT	28	162	8.9	14.1					13	57	137			
1 oz (5 pieces)		3.1	0.5						154					
CHOC FUDGE	30	129	21.9	4.8					68	30	33	.01	0.1	
1 piece		1.1	0.1						58		0.4	.04	0	
CHOC FUDGE W/ NUTS	28	127	18.8	5.8					57	28	38	.02	0.1	
1 oz		1.4	0.1						61		0.4	.04		
COCONUT CENTER	28	123	20.2	4.9					55	13	22	.01	0.1	0
1 oz		0.8	0.2						46		0.3	.02	0	
CREAM CENTER	13	51	9.4	1.8					1					
1 piece		0.5	0						15					
FONDANT	28	115	22.7	2.9					52	16	15	.01	tr	
1 oz		0.5	tr						26		0.3	.02		
FUDGE, CARAMEL, PEANUTS	28	121	17.9	5.1					57	50	52	.05	0.5	
1 oz		2.2	0.1						84		0.4	.06		
HONEYCOMBED HARD CANDY	28	130	19.8	5.5					46	22	38	.01	0.8	
W/ PNT BTR 1 oz		1.8	0.1						63		0.5	.03		
MARSHMALLOWS, KRAFT	7	23	5.4	0.2					4	1	3			
1 piece		0.2	tr						4					
MINT	22	87	15.8	3.1										
1 med		0.7												
NOUGAT & CARAMEL	28	116	20.4	3.9					48	36	34	.02	0.1	11
1 oz		1.1	0.1						59		0.5	.05		
PEANUTS	28	157	10.9	11.6					17	32	83	.10	2.1	
1 oz (4 pieces)		4.6	0.3						141		0.4	.05		
RAISINS	28	119	19.7	4.8					18	43	49	.02	0.1	42
1 oz (28 pieces)		1.5	0.2						169		0.7	.06		
VANILLA CREAM	13	57	9.1	2.2					24	17	14	.01	tr	
1 piece		0.5	tr						23		0.1	.01		
CHOC DISKS, SUGAR-COATED	28	131	20.3	5.5					20	38	39	.02	0.1	28
1 oz		1.5	0.1						70		0.4	.06		

6 CANDY, CANDY BARS (cont'd)

	WT	CAL	CHO	FAT	TRY	LEU	LYS	MET	Na	Ca	P	THI	NIA	A
	g	PRO	FIB	PUFA	PHE	ISO	VAL	THR	K	Mg	Fe	RIB	ASC	D
	g	g	g	g	mg	mg	mg	mg	mg	mg	mg	mg	mg	iu
CHOC FUDGE	28	112	21.0	3.4					53	22	24	.01	0.1	66
1 oz		0.8	0.1						41	14	0.3	.03	tr	
CHOC FUDGE W/ NUTS	28	119	19.3	4.9					48	22	32	.01	0.1	
1 oz		1.1	0.1						50		0.3	.03		
CHOC, GERMAN	28	140	16.2	9.7					7	9	35	.00	0.2	38
1 oz		1.2	0.1						92		0.8	.03	0	
CHOC KISSES, HERSHEY	28	154	14.6	9.0					22	56	87	.02	0.1	17
1 oz (6 pieces)		2.2	0.1						112		0.5	.07	tr	
CHOC STARS, KRAFT	28	145	17.0	8.1					26	64	65			
1 oz (7 pieces)		2.2	0.1						108					
CRACKER JACKS	28	114	25.5	1.0						9	46	.01	0.3	38
1 oz		0.8								22	0.3	.05	3	
FLAVORED DROPS, DIETETIC, FEATHERWEIGHT 1 oz	28	14	2.9	tr					2					
		0.6												
FONDANT	11	40	9.9	0.2					23	2	1			0
1 average		tr							1		0.1		0	
FUDGIES, KRAFT	28	114	21.1	3.6					64	29	31			
1 oz (3 pieces)		0.8	0.1						84					
GOOD AND PLENTY	43	151	374	0.1						69	tr			
1 pkg		tr												
GOOD AND FRUITY	43	160	40.0	tr						3	4			
1 pkg		0.3												
GUM DROPS	28	97	24.5	0.2					10	2	0	.00	0.0	0
1 oz (28 pieces)		tr	0.0						1		0.1	.00	0	
HARD CANDY	28	108	27.2	0.3					9	6	2		0.0	0
1 oz (6 pieces)		0.0	0.0						1		0.5	.00	0	0
JELLY BEANS	28	66	16.7	0.0					3	3	1	.00		0
1 oz (10 pieces)		0.0	0.0						tr		0.3		0	
LIFE SAVERS	10	39	9.7	0.1					3	2	1	.00	0.0	0
5 pieces		0.0	0.0						tr		0.2	.00	0	0
LIK-M-AID, SUNSHINE BRANDS	28	100	26.0	0.0										
1 serving		0.0												
LOLLIPOP	28	108	28.0	0.0					0	0	0	0	0	0
1 med		0.0	0.0						0		0	0	0	0
LOLLIPOP	57	216	56.0						0	0	.00	0	0	
1 large		0.0	0.0						0		0.0	.00	0	0
MALTED MILK BALLS, KRAFT	28	135	17.8	7.0					28	63	86			
1 oz (14 pieces)		2.3	0.1						113					
M & M'S	35	172	23.8	7.5					34					
1 pkg		2.5							137					
M & M'S PEANUT	35	177	20.1	8.9					23					
1 pkg		4.2							151					
MARSHMALLOW	8	25	6.2	0.0					4	1	1	.00		0
1 average		0.2	0.0						1		0.1		0	
MINI-MINTS, KRAFT	24	109	16.7	4.7					21	26	37			
3 pieces		0.8	0.1						45					
MINTS, BUTTER & PARTY, KRAFT	28	104	25.6	0.6					47	2	tr			
1 oz (14 pieces)		0.0	0.0						1					
PEANUT BRITTLE	25	110	18.2	3.9					8	10	31	.02	1.2	7
1 piece		2.1							38		0.5	.01	0	
PEANUT BUTTER CHIPS, REESES	28	148	12.8	7.9						20		.00	2.0	0
1 oz		5.9									0.4	.07	0	
PEANUT BUTTER CUPS, REESES	26	130	13.8	7.8					75	22	75			6
1 piece		6.9	0.2	2.0					108		0.4			
PIXY STIX, SUNLINE BRANDS	28	100	25.0	0.0										
1 serving		0.0												
SUGAR-COATED, ALMONDS	28	128	19.7	5.2					6	28	47	.01	0.3	0
1 oz (7 pieces)		2.2	0.1						71		0.5	.08	0	
VANILLA FUDGE	28	111	20.9	3.1					58	31	23	.01	tr	
1 oz		0.8	0.0						36	14	0.1	.04		
VANILLA FUDGE W/ NUTS	28	119	19.3	4.6					52	31	32	.01	tr	
1 oz		1.2	0.1						32		0.2	.04		

	WT	CAL	CHO	FAT	TRY	LEU	LYS	MET	Na	Ca	P	THI	NIA	A
		PRO	FIB	PUFA	PHE	ISO	VAL	THR	K	Mg	Fe	RIB	ASC	D
	g	g	g	g	mg	mg	mg	mg	mg	mg	mg	mg	mg	iu
CANDY BARS														
ALMOND TOFFEE, KRAFT	28	140	17.7	8.3					64	34	45			
1 bar		1.2	0.1						61					
BRKFST BAR, GENERAL MILLS	43	187	23.2	7.6										
1 bar		6.1												
BRKFST BAR, CHOC CHIP, CARNATION 1 bar	44	210	24.0	10.0	101	506	354	114	219	20	60	.30	5.0	1000
		6.0		0.9	340	326	381	251	112	60	4.5	.03	24	
BRKFST BAR, GRANOLA CINN CARNATION 1 bar	43	210	22.0	11.0	7	33	366	145	162	20	60	.30	5.0	1000
		6.0		1.1	321		404	259	84	60	4.5	.03	24	
BRKFST BAR, GRANOLA W/ PNT BTR 1 bar	43	210	20.0	12.0	87	474	298	113		20	60	.30	5.0	
		6.0		2.2	350	335	399	229		60	4.5	.03	24	
BRKFST BAR, PNT BTR CRUNCH CARNATION 1 bar	43	205	22.0	11.0	97	536	371	120	142	20	59	.29	4.9	977
		6.0		1.5	360	346	399	263	84	59	4.4	.03	24	
CHOCOLATE	28	148	16.2	9.8					9	26	40	.01	0.1	3
1 oz		1.2	0.1						75	30	0.4	.04		
CHOCOLATE	56	296	32.4	19.6					18	52	80	.02	0.2	6
2 oz		0.4	0.2						150	60	0.8	.08		
CHOC ALMOND, NESTLE	28	148	15.8	9.9					20	40		.00	0	0
1 oz		2.0									0.4	.07	0	
CHOC ALMOND, HERSHEY	32	176	15.7	11.9					28	68	80	.03	0.3	40
1 small bar		3.1							137		0.9	.16	0	23
CHOC ALMOND, HERSHEY	56	308	27.6	20.8					50	120	140	.05	0.6	70
1 bar		5.4							240		1.6	.28		40
CHOC, BITTERSWEET	28	134	13.1	11.1					1	16	80	.01	0.3	11
1 oz		2.2	0.5						172		1.4	.05	0	
CHOC, DIETETIC, CELLU	56	334	14.0	27.8										
1 bar		7.2												
CHOC, SEMISWEET	28	142	16.0	10.0					1	8	42	.00	0.1	6
1 oz		1.2	0.3						91	27	0.7	.02	0	
CHOC, SEMISWEET, BAKERS	28	130	8.8	16.4					1	8	42			6
1 oz		1.5							129		0.7			
CHOC, SEMISWEET, HERSHEY	28	147	17.5	9.2					5	9	40	.02	.11	
1 oz		1.2							95		0.9	.07		10
CHOC, SEMISWEET, HERSHEY	135	706	84.0	44.2					24	43	192	.07	0.5	
1 large bar		5.8							456		4.3	.31		48
CHOC, SPECIAL DARK, HERSHEY	41	221	25.2	12.6					6	12	66	.02	0.2	8
1 bar		2.3	0.3	0.0					137		1.2	.05		
CRUNCH, NESTLÉ	28	138	17.8	7.9										
1 bar		2.0												
ENGLISH TOFFEE, HEATH	32	220	11.1	19.1					90	0	0	.53	0.1	21
1 bar		1.0	0.1						50		0.2	.05	0	
FOREVER YOURS	32	122	26.7	1.6						22	44	.02	0.1	0
1 bar, 1¼ oz		1.1									0.2	.05	tr	
GRANOLA BAR, COCNT, NATURE VALLEY 1 bar	24	122	15.2	6.1										
		2.0												
GRANOLA BAR, CINN, NATURE VALLEY 1 bar	24	117	16.2	4.1										
		2.0												
GRANOLA BAR, HONEY 'N OATS, NATURE VALLEY 1 bar	24	117	16.2	4.1										
		2.0												
GRANOLA BAR, PNT BTR, NATURE VALLEY 1 bar	24	122	15.2	5.1										
		3.0												
KIT KAT, HERSHEY	35	180	21.5	9.3					35	53	63	.02	0.1	25
1 bar		2.5	0.1						105		0.5	.08	0	
KRACKEL, HERSHEY	40	200	21.1	11.4					64	72	96	.02	0.1	28
1 bar		2.6	0.2						146		0.6	.09	0	28
MARATHON	39	179	26.5	7.3					114					
1 bar		2.1							120					
MARS ALMOND	28	135	17.4	6.5					81					
1 oz		1.8							78					
MILK CHOCOLATE	28	146	15.9	9.0	3	19	10	2	26	64	65	.02	0.1	76
1 oz		2.2	0.	0.9	13	10	11	7	108	16	0.3	.10	0	47
MILK CHOC, HERSHEY	28	152	15.9	9.5					30	55	60	.03	0.1	40
1 oz		2.4							105		0.7	.14	0	25
MILK CHOC, HERSHEY	57	302	32.4	19.4					61	112	122	.05	0.2	82
1 bar		4.7							214		1.4	.29	1	61

	WT	Proximate			Amino Acids				Minerals			Vitamins		
		CAL	CHO	FAT	TRY	LEU	LYS	MET	Na	Ca	P	THI	NIA	A
		PRO	FIB	PUFA	PHE	ISO	VAL	THR	K	Mg	Fe	RIB	ASC	D
	g	g	g	g	mg	mg	mg	mg	mg	mg	mg	mg	mg	iu
MILK CHOC W/ ALMONDS, HERSHEY	28	155	15.1	9.3					22	59	84	.02	0.3	11
1 oz		2.6	0.4	0.9					120		0.6	.09	0	10
MILK CHOC W/ PEANUTS	50	271	22.3	19.1					33	87	147	.12	2.5	90
1 bar		7.1	0.5	2.7					244		0.7	.13	1	
MILKY WAY, MARS	52	237	35.5	9.2					137	47	57	.03	0.2	0
1 bar		2.8							139		0.4	.17	tr	
MINI-MEAL FOOD BAR, X-MAR	70	250	31.0	7.0						400		.60	11.0	2500
FOODS 1 bar		18.0								150	7.5	.85	35	200
MR GOODBAR, HERSHEY	50	285	26.0	16.5					30	65	140	.05	2.5	17
1 bar		7.2	0.6	2.7					225		1.0	.12	1	
MUNCH PEANUT	28	151	12.5	9.3					86					
1 oz		4.4							121					
PEANUT	28	144	13.2	9.0					3	12	76	.12	2.6	0
1 oz		4.9	0.3						125		0.5	.02	0	
RALLY, HERSHEY	51	260	27.1	15.4					41	46	117	.02	2.5	10
1 bar		7.0	0.3						204		0.7	.05	1	
RALLY NUT ROLL, CHOC-COVERED	42	207	22.0	11.0										
1 bar		5.0												
SEGO CHOC DIET MEAL BAR	30	155	13.5	8.4					90					
1 bar		6.2												
SNICKER, MARS	35	159	22.4	6.2					98	30	41	.01	0.1	0
1 bar		3.4							133		0.1	.11		
SNIK SNAK STICKS	28	151	16.5	8.5					26					
1 oz		2.2							108					
THREE MUSKETEERS	39	147	35.1	0.9					112	15	19	.01	0.0	0
1 bar		0.8							82		0.2	.04	tr	
TIGERS MILK BAR, CAROB-COATED	56	250	35.0	8.0						0	0	.53	0.6	0
1 bar		9.0								40	6.3	.68	0	

	WT	CAL	CHO	FAT	TRY	LEU	LYS	MET	Na	Ca	P	THI	NIA	A
	g	PRO	FIB	PUFA	PHE	ISO	VAL	THR	K	Mg	Fe	RIB	ASC	D
		g	g	g	mg	mg	mg	mg	mg	mg	mg	mg	mg	iu
ALL BRAN, KELLOGGS	56	129	40.0	1.1					567	70	494	.74	9.9	2469
1 cup		6.2	4.2						517	90	8.9	.84	30	79
ALPEN, COLGATE-PALMOLIVE	56	220	38.0	10.1										
		7.5							123					
ALPHA-BITS, POST	30	119	25.4	1.3					227	0	63	.40	5.3	1320
1 cup		1.1	0.3						35		1.1	.44	0	42
APPLE JACKS, KELLOGGS	28	109	25.5	0.3					111	4	20	.37	4.9	1235
1 cup		1.4	0.1						27	8	4.5	.42	15	40
BOO BERRY, GENERAL MILLS	28	109	23.7	1.0						20		.37	4.9	1236
1 oz		1.0									4.5	.42	15	40
BRAN, 100%, NABISCO	60	148	44.4	4.2								1.43	19.0	
1 cup		6.3								253		1.43	57	
BRAN BUDS, KELLOGGS	85	210	63.0	0					774	60	600	1.12	15.0	3748
1 cup		9.0	6.0						1079	240	13.5	1.27		120
BRAN CHEX, RALSTON	28	110	20.0	0.9					260	17	193	.38	5.0	
2/3 cup		2.9	1.3						188	72	4.5		15	
BRAN FLAKES, 40%	35	106	28.2	0.6					207	19	125	.47	6.2	1650
1 cup		3.6	1.3						137	61	5.6	.53	19	50
BUC WHEATS, GENERAL MILLS	28	108	22.7	1.0								.67	8.9	2224
1 oz		2.0									8.0	.77	27	
CAP'N CRUNCH, QUAKER	28	119	22.7	2.6		167	29	23	191	5	20	.37	4.8	0
1 oz		1.3	0.1		65	54	89	46	34	9	4.5	.25	0	0
CAP'N CRUNCHBERRIES, QUAKER	28	119	22.6	2.5					163	9	20	.37	4.9	0
1 oz		1.4	0.1						38	8	4.5	.25	0	0
CAP'N CRUNCHBERRIES PNT BTR, QUAKER 1 oz	28	125	20.6	3.8	139	236	57	39	216	5	20	.64	4.9	0
		2.2	0.1		110	90		64	52	8	4.5	.25	0	0
CAP'N CRUNCH PUNCH CRUNCH, QUAKER 1 oz	28	116	23.8	1.7							20	1.32	4.9	0
		5.2	0.1							8	4.5	.25	0	0
CAP'N CRUNCH VANILLY, QUAKER	28	116	23.4	1.8					143	21	31	.49	4.7	
1 oz		1.5	0.1						44		4.7	.60		
CHARGED BRAN, RALSTON	40	168	31.6	1.4					424	20	184	1.40	14.0	
1 cup		4.0	1.5							56	14.0	1.68		564
CHEERIOS, GENERAL MILLS	25	97	17.6	1.8					260	35	115	.33	4.4	1102
1 cup		3.5	0.3						83	34	4.0	.38	13	35
COCOA KRISPIES, KELLOGGS	28	106	25.2	0.0					199	10	20	.37	4.9	1235
1 cup		1.4	0.1						33	8	4.5	.42	15	40
COCOA PEBBLES, POST	34	139	30.0	1.9					174	1	32	.45	6.0	1500
1 cup		1.2	0.1						8		1.4	.51	0	48
COCOA PUFFS, GENERAL MILLS	28	109	24.7	1.0										
1 oz		1.0												
COOKIE CRISP, CHOC CHIP, RALSTON	28	110	25.0	1.0					198	5	8	.38	5.0	1250
1 cup		1.5	0.0						27	8	4.5	.43	15	
COOKIE CRISP, VANILLA WAFER, RALSTON 1 cup	28	110	25.0	1.0					198	5	8	.38	5.0	1250
		1.5	0.0						27	8	4.5	.43	15	
CONCENTRATE, KELLOGGS	45	188	26.1	0.2	350	2450	1500	1000	144	10	144	1.32	17.5	4409
1/2 cup		20.5	0.4		2950	1600	1800	1250	53	24	15.8	1.49	53	353
CORN CHEX, RALSTON	24	101	20.5	0.5					296		7	.28	2.8	1128
1 cup		1.6	0.1						23	2	0.6	.34		113
CORN FLAKES, GENERAL MILLS	28	109	23.7	1.0										
1 oz		2.0												
CORN FLAKES, KELLOGGS	22	84	19.1	0.0					216	1	8	.29	3.9	970
1 cup		1.5	0.1						26	3	1.4	.33	12	31
CORN FLAKES, POST TOASTIES	22	83	18.9	0.1					255	4	10	.29	3.9	970
1 cup		1.6	0.2						23		0.6	.31	8	31
CORN FLAKES, RALSTON	28	110	24.0	0.1					267	2	10	.15	1.2	156
1 cup		1.8	0.1						25	3	0.7	.03		
CORN TOTAL, GENERAL MILLS	28	109	23.7	1.0										
1 oz		2.0												
CORNY SNAPS, KELLOGGS	28	118	24.4	1.4					204	10	40	.36	4.9	1235
1 oz		1.7	0.1						71	16	1.8	.42	15	39
COUNT CHOCULA, GENERAL MILLS	28	109	23.7	1.0							0	.37	4.9	1236
1 oz		2.0									4.5	.42	15	
COUNTRY MORNING, KELLOGGS	85	390	57.0	15.0					294	120	180	.27	0.9	0
1 cup		9.0	0.9						498	120	3.3	.21	0	

	WT	CAL	CHO	FAT	TRY	LEU	LYS	MET	Na	Ca	P	THI	NIA	A
	g	PRO	FIB	PUFA	PHE	ISO	VAL	THR	K	Mg	Fe	RIB	ASC	D
		g	g	g	mg	mg	mg	mg	mg	mg	mg	mg	mg	iu
COUNTRY MORNING, RAISINS-DATES, KELLOGGS 1 cup	90	412	60.2	15.8					228	63	190	.29	1.0	0
		9.5	1.0						523	101	2.2	.22	0	
C.W. POST, GENERAL FOODS ½ cup	60	296	40.1	10.6					116	42	85	.79	10.6	2640
		4.2							222		9.5	.90	0	85
C.W. POST W/ RAISINS, GENERAL FOODS 1 oz	28	130	19.0	5.0					51	22	42	.37	5.0	1235
		2.6	0.3						113		4.5	.42	0	40
FRANKENBERRY, GENERAL MILLS 1 oz	28	308	23.7	1.0						20		.37	4.9	1236
		2.0									4.5	.42	15	40
FROOT LOOPS, KELLOGGS 1 cup	30	120	26.7	0.6					133	4	21	.40	5.3	1323
		1.5	0.1						33	8	4.8	.45	16	42
FROSTED MINI-WHEATS, KELLOGGS 4 pieces	28	103	22.8	0.3					5	10	79	.32	3.3	0
		2.5	0.5						43	39	0.6	.02	0	
FROSTED RICE, KELLOGGS 1 oz	28	109	25.7	0					225	5	20	.37	4.9	1235
		1.0	0.1						46	79	1.8	.42	15	40
FROSTED RICE KRINKLES, POST 1 cup	34	132	31.2	0.1					270	5	29	.45	6.0	1500
		1.2	0.1						48		2.5	.51	0	48
FROSTY O'S, GENERAL MILLS 1 oz	28	109	23.7	1.0						20	5	.37	4.9	1236
		2.0										.42	15	40
FRUIT BRUTE, GENERAL MILLS 1 oz	28	109	23.7	1.0						40	5	.37	4.9	1236
		2.0										.42	15	40
FRUITY PEBBLES, POST 1 cup	34	138	30.0	1.8					192	5	24	.45	6.0	1500
		0.9	0.2						48		0.9	.51	0	48
GOLDEN GRAHAM, GENERAL MILLS 1 cup	27	105	22.9	1.0						19		.36	4.8	1191
		1.9									4.0	.41	14	38
GRANOLA, CINN RAISIN, NATURE VALLEY 1 oz	28	128	18.9	4.9										
		3.0												
GRANOLA, COCONUT 'N HONEY, NATURE VALLEY 1 oz	28		17.8	4.9										
		3.0												
GRANOLA, FRUIT 'N NUT, NATURE VALLEY 1 oz	28	128	20.7	4.0										
		3.0												
GRANOLA, HONEY 'N OATS, NATURE VALLEY 1 oz	28	128	18.8	4.0										
		3.0												
GRANOLA, PILLSBURY ½ cup	56	240	38.0	8.0					80					
		6.0												
GRANOLA W/ COCNT-CASHEWS, PILLSBURY ½ cup	56	280	34.0	12.0					20					
		8.0												
GRANOLA W/ RAISINS-ALMONDS, PILLSBURY ½ cup	56	240	38.0	7.1					80					
		6.0												
GRAPE-NUTS, POST 1 oz	28	100	23.0	0.3					174	10	17	.37	5.0	1235
		3.7	0.4						75		0.5	.42	0	40
GRAPENUT FLAKES, POST 1 oz	28	100	23.0	3.6					201					
		2.0							90					
HEARTLAND GRANOLA PUFFS, PET 1 cup	60	253	42.2	8.5										
		4.2												
HEARTLAND NATURAL, PET 1 oz	28	119	18.8	4.0										
		3.0												
HEARTLAND NATURAL W/ RAISINS, PET 1 oz	28	120	18.8	4.0										
		3.0												
HONEYCOMB, POST 1 cup	23	89	20.3	0.8					186	0	16	.31	4.1	1015
		0.8							33		0.3	.35	0	32
JEAN LAFOOTTEE, CINN CRUNCH, QUAKER 1 oz	28	119	22.3	2.6		182	47	30	208	8	20	.37	4.9	0
		1.7	0.2		86	78	115	57	47	8	4.5	.25	0	0
KABOOM, GENERAL MILLS 1 oz	28	108	23.7	1.0						20		.67	8.9	2224
		2.0									8.0	.76	27	
KING VITAMAN, QUAKER 1 oz	28	119	23.0	2.4					213	5	23	.89	11.9	2471
		1.2	0.1							8	10.7	1.01	36	237
KIX, GENERAL MILLS 1 oz	28	109	23.7	1.0										
		2.0												
LIFE, QUAKER 1 cup	42	159	29.1	0.8		195	129	46	237	147	221	.55	7.4	0
		8.1	0.6		119	109	139	75	134	53	6.6	.63	0	0
LUCKY CHARMS, GENERAL MILLS 1 oz	28	110	23.7	1.0										
		2.0												
MR. WONDERFULL'S SUPRIZE, CHOC, GENERAL MILLS 1 oz	28	119	20.8	4.0										
		1.0												

Food	WT g	CAL / PRO g	CHO / FIB g	FAT / PUFA g	TRY / PHE mg	LEU / ISO mg	LYS / VAL mg	MET / THR mg	Na / K mg	Ca / Mg mg	P / Fe mg	THI / RIB mg	NIA / ASC mg	A / D iu
MR. WONDERFULL'S SUPRIZE, VAN, GENERAL MILLS 1 oz	28	119	20.8	4.0										
		1.0												
NATURAL, 100%, QUAKER 1 oz	28	137	16.8	6.2		269	160	68	18	47	98	.09	0.7	0
		3.6	0.4		152	151	212	127	146	34	0.8	.15	0	0
NATURAL, 100%, W/ APPLES-CINN, QUAKER 1 oz	28	133	17.8	5.5						10		.07	0.4	0
		3.2	0.4								0.7	.10	0	0
NATURAL, 100%, W/ RAISINS-DATES, QUAKER 1 oz	28	133	17.5	5.6					15	45	98	.08	0.7	0
		3.2	0.4						162	32	0.8	.17	0	0
OAT FLAKES, POST 1 cup	45	175	31.8	1.6					500	64	159	.60	7.9	
		7.9							159		7.2	.68	0	64
PEP, KELLOGGS 1 cup	28	101	23.5	0.3					351	98	59	.37	4.9	1235
		2.2	0.3						132	32	4.5	.42	15	40
PRODUCT 19, KELLOGGS 1 cup	28	106	23.8	0.3					413	5	40	1.48	19.7	4938
		2.5	0.1						59	16	17.8	1.68	59	395
PUFFA PUFFA RICE, KELLOGGS 1 cup	28	118	23.7	3.0					25	5	16	.37	4.8	1235
		0.8	0.1						23	5	1.8	.42	15	49
PUFFED RICE, QUAKER 1 cup	14	54	12.6	0.1					1	3	25	.10	1.2	0
		0.9	0.1						15	6	0.4	.02	0	0
PUFFED WHEAT, KELLOGGS 1 cup	14	50	10.4	0.6					1	4	32	.02	1.3	
		2.1	0.5						77	13	0.4	.01		
PUFFED WHEAT, QUAKER 1 cup	14	53	10.7	0.2					1	3	48	.08	2.0	0
		2.2	0.3						46	13	0.7	.10	0	0
QUISP, QUAKER 1 cup	24	102	19.5	2.2					160	7	20	.25	2.5	0
		1.1	0.1						30	9	2.3	.29	0	0
RAISIN BRAN, KELLOGGS 1 cup	50	130	40.0	0.5					293	35	193	.28	6.4	1603
		3.5	1.7						242	77	5.7	.55		52
RAISIN BRAN, POST 1 cup	57	173	44.8	0.8					443	30	323	.76	10.1	2514
		4.8	1.7						141		9.1	.80	0	81
RAISIN BRAN, RALSTON 1 cup	56	200	44.0	0.4					426	25	237	.10	6.2	56
		4.0	1.4						294	79	12.8	.28	1	
RICE CHEX, RALSTON 1 oz	28	110	25.0	0.5					240	4	28	.38	5.0	
		1.5	0.1						32	7	0.7		15	
RICE KRISPIES, KELLOGGS 1 cup	28	106	24.6	0.0					251	3	.39	.37	4.9	1235
		2.0	0.1						30	8	1.8	.42	15	39
RICE TOASTIES, POST 1 cup	28	111	24.2	0.1					374	13	53	.31	4.2	0
		1.8	0.1								1.0	.0	0	0
SHREDDED WHEAT 1 biscuit	25	89	20.0	0.5	21	171	83	35	1	11	97	.06	1.1	0
		2.5	0.6		120	112	144	101	87	27	0.9	.03	0	0
SHREDDED WHEAT, SPOON-SIZE 1 cup	45	174	36.4	1.6					5	22	158	.09	2.5	0
		4.8							174	51	1.7	.06	0	
SIR GRAPEFELLOW, GENERAL MILLS 1 oz	28	109	23.7	1.0						20		.37	4.9	1236
		2.0									4.5	.42	15	40
SPECIAL K, KELLOGGS 1 cup	20	76	14.6	0.0					154	7	28	.26	3.5	882
		4.2	0.1						40	11	3.2	.30	11	28
SUGAR FROSTED FLAKES, KELLOGGS 1 cup	37	141	33.7	0.0					242	1	10	.49	6.5	1631
		1.9	0.1						30	3	2.4	.56	20	52
SUGAR POPS, KELLOGGS 1 cup	28	109	25.8	0.0					63	2	8	.37	4.9	1235
		1.4	0.1						20	2	4.5	.42	15	40
SUGAR SMACKS, KELLOGGS 1 cup	28	109	24.4	0.3					34	5	40	.37	4.9	1235
		2.2	0.2						45	16	1.8	.42	15	40
SUPER SUGAR CRISPS, POST 1 cup	32	121	28.6	0.4					46	3	42	.37	5.7	1411
		2.3	0.4						56		0.7	.45	0	45
TEAM FLAKES, NABISCO 1 cup	30	109	24.0	1.0						0	40	.37	5.0	1249
		2.0								8	0.4	.37	15	
TOTAL, GENERAL MILLS 1 oz	28	109	22.7	1.0					409	39		1.48	19.7	4938
		3.0									17.8	1.68	59	
TRIX, GENERAL MILLS 1 oz	28	109	25.7	1.0										
		1.0												
WHEAT CHEX, RALSTON 1 cup	40	168	28.0	1.4					388	8	113	.46	4.6	1880
		4.0	0.8						162	38	14.0	.56	14	188
WHEAT FLAKES 1 cup	30	114	24.2	0.5	36	267	108	38	310	12	93	.19	14.7	0
		3.2	0.5		143	149	172	107	96	43	1.3	.04	0	
WHEATIES, GENERAL MILLS 1 cup	28	109	22.6	1.0					315	14	103	.38	4.9	1235
		3.0	0.5							31	4.5	.43	15	0

12 CHEESE AND CHEESE PRODUCTS

	WT	CAL / PRO	CHO / FIB	FAT / PUFA	TRY / PHE	LEU / ISO	LYS / VAL	MET / THR	Na / K	Ca / Mg	P / Fe	THI / RIB	NIA / ASC	A / D
	g	g	g	g	mg	mg	mg	mg	mg	mg	mg	mg	mg	iu

CHEESE (NATURAL AND PROCESSED)

Food	WT	CAL/PRO	CHO/FIB	FAT/PUFA	TRY/PHE	LEU/ISO	LYS/VAL	MET/THR	Na/K	Ca/Mg	P/Fe	THI/RIB	NIA/ASC	A/D
AMERICAN, PAST, PROC	28	107	0.5	8.4	91	631	475	169	318	195	216	.01	tr	340
1 oz		6.5	0.0	0.3	351	436	468	240	22	13	0.3	.11	0	
AMERICAN PIMENTO, PAST, PROC	28	106	0.5	8.8	91	555	622	162	405	174	211	.01	tr	358
1 oz		6.3	tr	0.3	319	290	376	204	46	6	0.1	.10		
BLUE	28	103	0.6	8.5	78	582	438	156	390	88	95	.01	0.4	347
1 oz		6.0	0.0	0.3	318	402	426	222	72	6	0.1	.17	0	
BRICK	28	103	0.6	8.5	87	608	453	161	157	204	127	tr	0.3	347
1 oz		6.2	0.0	0.3	335	415	446	229	38	7	0.3	.13	0	
BRIE	28	94	0.1	7.8	90	540	518	166	176	52	53	.02	0.1	187
1 oz		5.8	0.0		324	284	365	210	20		0.1	.15		
CAMEMBERT (DOM)	28	84	0.5	6.9	69	475	358	127	236	29	52	.01	0.2	283
1 oz		4.9	0.0	0.3	265	328	353	181	31	6	0.1	.21	0	
CARAWAY	28	105	0.9	8.2					193	188	137	.01	0.1	295
1 oz		7.0	0.0							6		.13	0	
CHEDDAR, AMERICAN	28	112	0.6	9.1	98	686	511	182	197	211	134	.01	tr	370
1 oz		7.0	0.0	0.3	378	469	504	259	23	13	0.3	.13	0	
CHEDDAR, AMERICAN	100	398	2.1	32.2	350	2450	1820	650	700	750	478	.03	0.1	1310
3½ oz		25.0	0.0	1.0	1350	1680	1800	920	82	45	1.0	.46	0	
CHEDDAR, AMERICAN	115	458	2.4	37.0	403	2822	2102	749	805	862	550	.03	0.1	1510
1 cup, grated		28.8	0.0	1.2	1555	1930	2074	1066	94	52	1.2	.53	0	
CHEEZOLA, PAST, PROC, FISHER	28	89	1.1	6.4					448	202				
1 oz		7.0		3.8					92					
CHESHIRE	28	108	1.3	8.6	84	627	545	171	196	180	130	.01		276
1 oz		6.5	0.0		345	406	437	233	27	6	0.1	.08	0	
COLBY	28	110	0.7	9.0	85	637	554	174	169	192	128	tr	tr	290
1 oz		6.7	0.0	0.3	350	413	444	236	35	8	0.2	.11	0	
COTTAGE, UNCREAMED (1% FAT)	226	163	6.1	2.3	312	2879	2265	843	918	138	303	.05	0.3	84
1 cup		28.0	0.0	0.1	1510	1645	1733	1243	194	11	0.3	.37	0	5
COTTAGE, UNCREAMED (2% FAT)	226	203	8.2	4.4	346	3193	2511	933	918	154	339	.05	0.3	158
1 cup		31.1	0.0	0.1	1675	1826	1923	1376	217	14	0.4	.42	0	5
COTTAGE, CREAMED	28	30	0.8	1.2	42	410	319	106	64	26	43	.01	tr	48
1 oz 1 rounded T		3.8	0.0	tr	205	220	220	179	24	1	0.1	.07	0	01
COTTAGE, CREAMED	100	106	2.9	4.2	150	1468	1142	381	229	94	152	.03	0.1	170
3½ oz 6 rounded T		13.6	0.0	tr	734	789	789	639	85	5	0.3	.25	0	2
COTTAGE, CREAMED	225	239	6.5	9.5	336	3294	2562	854	516	211	342	.07	0.2	380
1 cup		30.5	0.0	tr	1647	1769	1769	1434	191	11	0.7	.56	0	5
COTTAGE, CREAMED W/ FRUIT	226	280	30.1	7.7	249	2301	1810	673	915	108	235	.04	0.2	278
1 cup		22.4	0.0	0.2	1207	1315	1385	992	151	9	0.2	.29		5
COUNT DOWN, PAST, PROC, FISHER	28	39	2.8	0.3					434	246		.02	0.1	28
1 oz		7.6							36			.00	1	
CREAM	28	99	0.8	9.9	19	207	192	51	84	23	30	.01	tr	405
1 oz/ 2 T		2.1	0.0	0.4	119	113	125	91	34	2	0.3	.06	0	
EDAM	28	87	1.1	5.7	108	755	562	200	270	225	136	.01	0.1	510
1 oz		7.7	0.0	0.2	416	516	555	285	53	12	0.2	.14	0	
FETA	28	74	1.1	6.0					312	138	94			
1 oz		4.0	0.0	0.2					17	5	0.2		0	
FONTINA	28	109	0.4	8.7						154		.01	tr	329
1 oz		7.2		0.5						4	0.1	.06	0	
GJETOST	28	130	11.9	8.3	38	278	228	89	168	112	124			
1 oz		2.7	0.0	0.3	151	145	214	110				.23	0	
GOUDA	28	100	0.6	7.7		718	743	201	229	196	153	.01	tr	180
1 oz		7.0	0.0	0.2	401	366	506	260	34	8	0.1	.09	0	
GRUYERE	28	115	0.5	8.9	105	778	591	211	94	308	230	tr	0.1	560
1 oz		8.1	0.0	0.2	429	543	575	300	23	12	0.3	.14	0	
LIEDERKRANZ	28	85		7.4	64	451	336	120		178	120			
1 oz		4.6	0.0	0.2	248	308	331	170					0	
LIMBURGER	28	93	0.1	7.7	82	593	475	176	227	141	111	.02	0.1	363
1 oz		5.7	0.0		316	346	408	209	36	6	tr	.14	0	
LITE-LINE, PAST, PROC, BORDEN	28	52	0.6	2.2										
1 oz		7.3												
MONTERY	28	105	0.2	8.5	88	656	570	179	150	209	124			266
1 oz		6.8	0.0		361	425	458	244	23	8	0.2	.11	0	
MYSOST*	28	70	15.4	0.4	85	131	97	24		297	173	.01	0.1	
1 oz		1.6	0.0		41	92	81	85			0.2	.14	0	

*Scandinavian cheese made from whey

	WT	CAL	CHO	FAT	TRY	LEU	LYS	MET	Na	Ca	P	THI	NIA	A
		PRO	FIB	PUFA	PHE	ISO	VAL	THR	K	Mg	Fe	RIB	ASC	D
	g	g	g	g	mg	mg	mg	mg	mg	mg	mg	mg	mg	iu
MOZZARELLA 1 oz	28	79	0.6	6.1		530	552	152	104	145	104	tr	tr	222
		5.4	0.0	0.2	284	261	340	207	19	5	0.1	.07	0	
MOZZARELLA, LOW-MOISTURE 1 oz	28	89	0.7	6.9		590	614	169	116	161	115	tr	tr	253
		6.1	0.0	0.2	316	290	350	230	21	59	0.1	.08	0	
MOZZARELLA, LOW-MOISTURE, PART SKIM 1 oz	28	78	0.9	4.8		750	781	214	148	205	147	.01	tr	176
		7.7	0.0	0.1	402	369	481	293	3	7	0.1	.10	0	
MOZZARELLA, PART SKIM 1 oz	28	72	0.8	4.5		671	699	192	132	183	131	.01	tr	166
		6.9	0.0	0.1	359	330	430	262	24	7	0.1	.09	0	
MUENSTER 1 oz	28	104	0.3	8.5	93	641	606	161	178	203	133	tr	tr	318
		6.6	0.0	0.2	352	325	420	252	38	8	0.1	.09	0	
NEUFCHATEL 1 oz	28	73	0.8	6.6	25	270	249	67	112	21	38	tr	tr	318
		2.8	0.0	0.2	155	147	164	118	32	2	0.1	.05	0	
PARMESAN, GRATED 1 T	5	23	0.2	1.5	28	201	192	56	93	69	40	tr	tr	35
		2.1	0.0	tr	112	100	143	77	5	3	tr	.02	0	
PARMESAN, HARD 1 oz	28	111	0.9	7.3	137	979	937	272	454	336	197	.01	0.1	171
		10.1	0.0	0.2	545	537	696	373	26	12	0.2	.09	0	
PORT DU SALUT 1 oz	28	99	0.3	7.9	96	695	556	206	150	182	101		tr	373
		6.7	0.0		370	405	478	245			0.2	.07	0	
PROVOLONE 1 oz	28	98	0.6	7.3		643	741	192	245	212	139	.01	tr	228
		7.4	0.0	0.2	360	305	459	275	39	8	0.1	.03	0	
RICOTTA ½ cup	124	216	3.8	16.1		1514	1659	348	104	257	196	.02	0.1	608
		14.5	0.0	0.5	689	730	858	641	130	14	0.5	.24	0	
RICOTTA, PART SKIM ½ cup	124	171	6.4	9.8		1531	1678	352	155	337	227	.03	0.1	536
		13.8	0.0	0.3	660	739		649	155	19	0.5	.23	0	
ROMANO 1 oz	28	110	1.0	7.6					340	302	215		tr	162
		9.0	0.0									.11	0	
ROQUEFORT 1 oz	28	105	0.6	8.7					513	188	111	.01	0.2	297
		6.1	0.0	0.4					26	8	0.2	.17	0	
SLIM LINE, PAST, PROC, SWIFT 1 oz	28	63	2.4	2.4					606	186	119	.03	0.1	86
		7.6							94	8	0.1	.14		
SWISS, AMERICAN 1 oz	28	106	0.5	7.8					199	259	158	tr	tr	320
		7.7	0.0	0.3	485	609	650	336	29	12	0.3	.11	0	
SWISS, SWITZERLAND 1 oz	28	104	0.5	7.8	100	739	562	200	74	259	158	tr	tr	406
		7.7	0.0	0.3	408	516	547	285	31	12	0.3	.11	0	
SWISS, PAST, PROC 1 oz	28	95	0.6	7.1	102	620	696	181	388	219	216	tr	tr	229
		7.0	0.0	0.2	356	324	420	227	61	8	0.2	.08	0	
TILSIT 1 oz	28	95	0.5	7.3	99	713	571	211	211	196	140	.02	0.1	293
		6.8	0.0	0.2	380	416	491	252	18	4	0.1	.10	0	
CHEESE FONDUE, HOMEMADE ¼ cup	64	170	6.4	11.7					347	203	188	.04	0.1	563
		9.5							106		0.8	.22		

CHEESE FOOD

	WT	CAL	CHO	FAT	TRY	LEU	LYS	MET	Na	Ca	P	THI	NIA	A
AMERICAN, PAST, PROC 1 oz	28	93	2.1	7.0					337	163	130	.01	tr	259
		5.6	0.0	0.2					79	9	0.2	.13	0	
AMERICAN, PAST, PROC, COLD PACK 1 oz	28	93	2.3	6.8					270	139	112	.01	tr	197
		5.5	0.0	0.2					102	8	0.2	.12	0	
SWISS, PAST, PROC 1 oz	28	90	1.3	6.8					435	202	147	tr	tr	240
		6.1	0.0						80	8	0.2	.11	0	
CHEESE SAUCE, HOMEMADE 2 T	38	66	2.4	4.9	42	289	217	76	197	89	65	.01	tr	209
		3.0	0.0		156	194	213	110	40	6	0.1	.08	0	
CHEESE SAUCE, HOMEMADE ¼ cup	76	132	4.8	9.8	82	578	434	152	394	178	130	.02	tr	418
		6.0	0.0		312	388	426	220	80	12	0.2	.16	0	
CHEESE SAUCE, CND 1 oz	28	42	2.7	2.7					179	64	60	.01	tr	
		1.8							31		0.1	.06		

CHEESE SPREAD

	WT	CAL	CHO	FAT	TRY	LEU	LYS	MET	Na	Ca	P	THI	NIA	A
AMERICAN, PAST, PROC 1 oz	28	82	2.5	6.0		505	427	152	381	159	202	.01	tr	223
		4.7	0.0	0.2	264	236	387	178	69	8	0.1	.12	0	
CHEEZ WHIZ, KRAFT 1 T	16	43	1.0	3.3					262	75	144	tr		134
		2.5										.06		
SNACK MATE, NABISCO 1 T	15	45	0.8	3.0					186					
		2.2												

Food	WT g	CAL	CHO	FAT	PRO	FIB	PUFA	TRY	LEU	LYS	MET	PHE	ISO	VAL	THR	Na	K	Ca	Mg	P	Fe	THI	RIB	NIA	ASC	A	D
BEANS & BEEF IN TOM SCE, CND, CAMPBELLS	200	250	34.0	8.0	12.0	1.6										1020	642	92		162	3.0	.10	.10	1.8	4	386	
BEANS & FRANKFURTERS, CND 1 cup	255	367	32.1	18.1	19.4	2.6										1374	668	94		303	4.9	.18	.15	3.3	0	332	
BEANS & FRANKS IN TOM & MOLASSES SCE, CAMPBELLS	200	320	36.0	14.0	14.0	1.8										930	524	100		228	3.4	.12	.10	2.2	3	406	
BEANS & HAM	200	260	36.5	5.4	17.3	2.3	0.2									228	588	85	8	295	4.3	.43	.18	2.6	9	83	4
BEEF GOULASH, CND, BOUNTY	200	170	16.0	6.0	12.0	0.4										920	104	26		126	2.6	.06	.14	2.0	2	510	
BEEF GOULASH W/ NOODLES 1 serving	150	186	9.3	8.0	17.9	0.2										350	177	7		105	2.6	.08	.16	6.5	2	1398	
BEEFARONI, CND, CHEF BOYARDEE	200	229	29.9	7.9	8.8	0.5	2.8									1044	210	22	25	141	2.5	.13	.17	3.5	33	849	0
BEEF & MACARONI, FRZN, GREEN GIANT 1 serving	170	141	18.0	4.6	6.8	1.7										719	308	34		71	1.5	.07	.19	1.7	9	765	
BEEF, MACARONI & TOM, FRZN STOUFFERS 1 serving	180	227	21.6	10.4	11.5											278											
BEEF POTPIE, HOMEMADE 1 pie	227	558	42.7	32.9	22.7	0.9										645	361	32		161	4.1	.25	.27	4.5	7	1861	
BEEF POTPIE, COMMERCIAL, FRZN 1 pie	227	443	37.0	25.4	16.6	0.5										1008		14		117	3.4	.20	.27	4.7		1401	
BEEF POTPIE, FRZN, BANQUET 1 pie	240	409	40.8	20.0	16.3											908	14	27		125	5.2	.16	.06	1.2	2	749	
BEEF STEW W/ POT-CAR-ONION	200	190	10.9	5.3	23.5	0.7	3.2									420	452	15	22	143	3.3	.13	.20	9.4	8	2812	0
BEEF STEW, CND, BOUNTY	200	160	16.0	6.0	12.0	1.0										790	382	26		62	2.0	.04	.10	1.8	5	2830	
BEEF STEW, CND, STOKELY VAN CAMP 1 cup	258	204	18.3	8.0	15.0	0.8												5			2.3	.08	.13	2.6	8	2503	
BEEF STEW, FRZN, CAMPBELLS 1 pkg	228	205	16.0	8.0	17.1	0.5										955	391	11		163	2.3	.16	.02	4.2	7	662	
BEEF STEW, FRZN, GREEN GIANT 1 serving	170	146	13.3	5.4	4.6											564	408	26		131	4.1	.05	.22	2.9	14	2026	
BEEF STEW, FRZN, INVISIBLE CHEF	200	178	12.5	8.7	13.0											1257	185	24		150	2.4	.14	.14	3.8	14	1984	
BEEF STEW, FRZN, SARA LEE	200	304	8.0	8.6	45.0	0.6										396	690	44		420	6.0	.18	.36	8.8	5	1242	
BEEF-VEG STEW, CND 1 cup	235	186	16.7	7.3	13.6	0.7		157	1003	1079	296	552	663	724	566	966	409	28		106	2.1	.07	.12	2.4	7	228	0
BEEF-VEG STEW, HOMEMADE 1 cup	245	218	15.2	10.5	15.7	1.0	4.9	196	1446	1152	368	858	1152	931	637	91	613	29	49	184	2.9	.15	.17	4.7	17	2401	
BEEF STROGANOFF, FRZN	200	86	0.5	6.9	5.0											28	24	8		48	0.7			1.3	1	70	
BLINTZES, CHEESE 1 serving	28	54.9		3.9																							
BROCCOLI-NOODLE CASS, FRZN, GREEN GIANT	100	100	9.0	6.0	4.0											450	220	55		120	0.6	.10	.15	0.5	35	300	
BURRITO, BEAN† 1 avg	169	350	48.3	10.8	15.1	1.1										1047	479	208	78	200	1.2						
CABBAGE ROLLS, STFD, FRZN	200	193	15.7	10.5	8.7	0.7										649	404	48		111	1.7	.14	.12	2.5	42	562	
CHEESE-MACARONI, FRZN, SARA LEE	100	245	30.3	9.8	10.1	0.6										429	140	184	14	216	1.2	.15	.35	tr	tr	291	
CHEESE-MACARONI-HAM, FRZN, SARA LEE	100	253	24.5	12.3	11.3	0.7										364	114	153	14	212	1.5	.28	.32	0.9	tr	238	
CHICKEN A LA KING, HOMEMADE 1 cup	245	468	12.3	34.3	27.4											760	404	127		358	2.5	.10	.42	5.4	12	1127	
CHICKEN A LA KING, FRZN, CAMPBELLS 1 pkg	228	255	8.0	14.0	25.0	0.3										895	307	46		182	1.2	.16	.16	5.5	0	0	

*Combination foods contain two or more items from different food categories.

†From national franchisee; includes pinto beans, shredded cheese, onion, and mild sauce folded in a soft flour tortilla.

	WT g	CAL / PRO	CHO / FIB	FAT / PUFA	TRY / PHE mg	LEU / ISO mg	LYS / VAL mg	MET / THR mg	Na / K mg	Ca / Mg mg	P / Fe mg	THI / RIB mg	NIA / ASC mg	A / D iu
CHICKEN A LA KING, FRZN,	180	164	9.0	9.0					522	70	140	.07	3.6	281
CHECKERBOARD 1 serving		10.8							212		0.9	.18	5	
CHICKEN & DUMPLINGS, FRZN,	100	147	7.5	10.2					476	87	120	.05	1.9	268
SARA LEE		9.5	0.2						96	15	1.9	.09	1	
CHICKEN & NOODLES, HOMEMADE	240	367	25.7	18.5					600	26	247	.05	4.3	432
1 cup		22.3							149		2.2	.17		
CHICKEN & NOODLES, FRZN,	228	210	18.0	10.0					705	30	142	.08	3.0	569
CAMPBELLS 1 pkg		14.0							132		1.5	.16	0	
CHICKEN & NOODLES, FRZN, GREEN	170	151	15.5	4.9					816	187	105	.05	3.4	107
GIANT 1 serving		4.6	0.1						150		1.2	.09	3	
CHICKEN & NOODLES, FRZN,	200	196	13.9	5.3					78					
IDLE WILD		23.0							160					
CHICKEN & NOODLES, FRZN,	180	310	21.2	15.8					662					
STOUFFERS 1 serving		17.9												
CHICKEN & NOODLES, AU GRATIN,	200	232	16.6	10.2					970	114	262	.16	4.6	1310
SARA LEE		18.4	0.6						374	80	3.8	.20	1	
CHICKEN & NOODLES, ESCAL, FRZN,	200	306	19.6	18.4					820	74		.16	2.8	390
STOUFFERS		15.0									1.4	.32	0	
CHICKEN & RICE	200	234	14.6	12.6						122	44	.06	3.4	34
		15.0									1.4	.12	0	
CHICKEN CACCIATORE, FRZN,	200	158	9.2	5.4					592	22	162	.06	2.8	762
SARA LEEE		16.6	0.6						348	20	1.4	.14	11	
CHICKEN, CREAMED JARDINIERE,	200	180	8.2	9.4					928	118	230	.12	5.0	2548
FRZN, SARA LEE		16.6	0.6						362		3.6	.24	2	
CHICKEN FRICASSEE	200	328	6.5	18.6	1580	2680	3260	960	308	12	226	.42	1.3	142
		30.9	0.0	0.	980	2000	1840	460	280	22	1.8	.14	0	
CHICKEN PARMIGIANA	200	308	22.1						358	174	293	.16	4.8	1388
		21.6							342		6.3	.18	16	
CHICKEN POTPIE, HOMEMADE	116	273	21.2	156.6					297	35	116	.13	2.1	1543
1 pie		11.7	0.5						172		1.5	.13	2	
CHICKEN POTPIE, COMMERCIAL, FRZN	227	503	52.9	25.4					863	25	114	.23	4.3	2082
1 pie		15.7	0.7								4.3	.23		
CHICKEN POTPIE, FRZN, BANQUET	240	427	39.0	23.2					999	50	147	.11	3.4	531
1 pie		15.4							16		3.2	.11	2	
CHICKEN POTPIE, FRZN, MORTON	226	362	31.6	20.3					859	16	129	.16	4.1	791
1 pie		13.6									1.8	.16	1	
CHICKEN STEW, CND, BOUNTY	200	170	14.0	8.0					860	30	102	.04	3.8	3770
		10.0	0.4						300		1.2	.10		
CHICKEN STEW, FRZN, IDLE WILD	200	168	11.8	3.6					92					
		21.6							332					
CHICKEN TETRAZZINI, INVISIBLE	200	207	21.2	5.6					83	85	214	.36	5.0	135
CHEF		17.8							378		1.5	.21	0	
CHILI, HOMEMADE	100	200	5.8	14.8						38	152	.02	2.2	150
		10.3	0.2								1.4	.12		
CHILI, CND	142	203	6.7	15.8					615	26	64	.03	3.1	0
5 oz		8.9									2.3	.17		0
CHILI, CND, STOKELY VAN CAMP	230	460	13.3	34.0						87		.05	5.1	345
1 cup		23.7	0.5								3.2	.28		
CHILI CON CARNE W/ BEANS,	100	133	12.2	6.1					531	32	126	.03	1.3	60
HOMEMADE		7.5	0.6						233		1.7	.07		
CHILI CON CARNE W/ BEANS, CND	142	182	13.5	9.9	300	630	620	140	728	61	102	.04	1.7	213
5 oz		9.5		3.0	360	410	430	70		26	2.3	.11	0	0
CHILI CON CARNE W/ BEANS, CND	115	167	9.8	10.9						44	175	.04	2.5	
½ cup		7.5									1.6	.09		
CHILI CON CARNE W/ BEANS, CND,	100	140	13.0	7.0					505	32	91	.03	1.0	751
BOUNTY		6.0	0.7						367		1.6	.06	3	
CHILI CON CARNE W/ BEANS, WENDYS	250	250	23.0	7.0						80		.23	3.0	1000
1 serving		22.0									3.6	.17	3	
CHILI MAC, BOUNTY	100	110	13.0	4.0					550	25	67	.05	1.3	330
		5.0	0.5						218		1.2	.07	3	
CHOW MEIN, CHICKEN, HOMEMADE	220	224	8.8	8.8	264	1760	1980	594	631	51	259	.07	3.7	242
1 cup		27.3	0.7	4.0	572	1188	1100	968	416	40	2.2	.20	9	

*Combination foods contain two or more items from different food categories.

	WT	Proximate			Amino Acids				Minerals			Vitamins		
		CAL	CHO	FAT	TRY	LEU	LYS	MET	Na	Ca	P	THI	NIA	A
		PRO	FIB	PUFA	PHE	ISO	VAL	THR	K	Mg	Fe	RIB	ASC	D
	g	g	g	g	mg	mg	mg	mg	mg	mg	mg	mg	mg	iu
CHOW MEIN, CHICKEN, CND	220	84	15.6	0.2	66	374	418	132	638	40	75	.04	0.9	132
1 cup		5.7	0.7	0.0	1078	242	220	198	367	40	1.1	.09	11	
CHOW MEIN, CHICKEN, FRZN, CAMPBELLS 1 pkg	228	130	9.0	5.0					1160	41	118	.08	3.2	0
		14.0	0.7						289		0.9	.00	0	
CHOW MEIN, VEG, FRZN, LA CHOY	240	69	16.9	0.3					1273					
1 cup		2.4												
CHOP SUEY W/ MEAT, HOMEMADE	250	300	12.8	17.0					1053	60	248	.28	5.0	600
1 cup		26.0	1.3						425		4.8	.38	33	
CHOP SUEY W/ MEAT, CND	360	223	15.1	11.5					1984	126	418	.18	2.5	108
1 cup		15.8	2.9						497		6.8	.18	7	
CRAB CAKE	28	46	0.5	2.5						15	59	.02	0.5	82
1 oz		5.4								10	0.3	.03	0	
CRAB CASSEROLE	150	305	11.9	21.6										
1 serving		17.0												
CRAB, DEVILED†	100	188	13.3	9.9					867	47	137	.08	1.5	540
		11.4							166	26	1.2	.11	6	
CRAB IMPERIAL‡	200	294	7.8	15.2					1456	120	332	.12	2.2	
		29.2							262		1.8	.24	10	
DUMPLINGS, CHEESE, FRZN, SARA LEE	28	39	2.5	2.8					96	39	34	.01	0.1	96
1 oz		1.7	0.0						31	4	0.1	.04	tr	
EGG FOO YOUNG, SKILLET DINNER, FRZN, LA CHOY	200	224	11.0	17.4					1006					
		7.9												
EGG ROLL, PLAIN	100	294		236										
EGG ROLL, CHICKEN, FRZN, LA CHOY	100	242	29.4	10.2					546	27	61			
		8.1	0.9	1.1					108	19	1.0			
EGG ROLL, LOBSTER, FRZN, LA CHOY	100	217	31.0	7.3					517	23	58			
		6.9	0.4	0.8					154	21	1.2			
EGG ROLL, MEAT W/ SHRIMP, FRZN, LA CHOY	100	218	28.4	8.4					670	25	63			
		7.2	0.5	0.6					167	21	1.0			
EGG ROLL, SHRIMP, FRZN, LA CHOY	100	210	30.1	6.7					530	27	58			
		7.3	0.4	0.7					133	22	0.8			
ENCHIRITOS§	207	373	35.6	16.9					1304	203	238			
1 avg		19.2	1.4						626	76	1.9			
FISH CAKES, CND	100	111	0.0	0.6	247	2027	2398	668		49	232			
		24.7	0.0		940	1385	1286	1434			0.8			
FISH CAKES, FRIED¶	100	172	9.3	8.0										
		14.7												
FISH CAKES, FRIED, FRZN	100	270	17.2	17.9										
		9.2												
FISH CREOLE	100	86	3.4	2.7					84	12	131	.06	1.6	327
		11.6		0.4					62	24	0.5	.06	10.5	0
FISH LOAF**	100	124	7.3	3.7										
		14.1												
FRITTER, CLAM	100	311	30.9	15.0						76	195	.03	0.9	
1 serving		11.4							147		3.5	.12		
FRITTER, CORN	100	223	37.6	5.1					425	187	323	.18	2.5	365
1 serving		6.7	0.3						448		1.5	.17	2	
FRITTER, PORK	100	300	2.0	27.0					1403	15	142	.26	1.3	11
1 serving		12.0							302		0.8	.07		
FROZEN DINNERS††														
BEANS, FRANKS & SCE, MORTON	340	555	81.3	18.0					881	137	387	.26	2.5	217
		19.0									4.6	.21	17	
BEEF, SWANSON	312	412	32.5	16.8					998	22	296	.34	6.6	156
		33.1	1.6						655		4.7	.22		
BEEF, MORTON	311	292	21.8	9.3					938	38	281	.21	6.4	3133
		28.7									4.0	.26	12	

*Combination foods contain two or more items from different food categories.
†Prepared w/bread, butter, parsley, egg, lemon juice and catsup
‡Prepared w/butter, flour, milk, onion, green pepper, egg and lemon
§From national franchisee; ground beef, cheese, black olives, and mild red sauce in a soft flour tortilla.
¶Made w/cnd fish, potatoes, and egg
**Made w/cnd flaked fish, bread cubes, eggs, tomatoes, onion, and fat
††Values are for the complete dinners which usually includes the main entree plus two vegetables (e.g., fried chicken, mashed potatoes & peas).

	WT g	CAL / PRO g	CHO / FIB g	FAT / PUFA g	TRY / PHE mg	LEU / ISO mg	LYS / VAL mg	MET / THR mg	Na / K mg	Ca / Mg mg	P / Fe mg	THI / RIB mg	NIA / ASC mg	A / D iu
CHICKEN, MORTON	311	482	37.2	24.9					1153	42	247	.20	7.9	3236
		26.2									2.4	.01	12	
CHICKEN, FRIED, MASH POT, MIX VEG	312	540	35.3	26.5					1073	128	452	.22	16.2	1841
		39.9	1.2						349		3.7	.56	12	
CHICKEN, FRIED, SWANSON	340	1420	46.6	29.2					1173	136	360	.44	10.9	2893
		37.4	1.4						731		3.4	.20		
CHICKEN 'N DUMPLINGS, MORTON	340	372	30.7	17.4					1506	43	248	.22	7.1	3232
		23.0									2.7	.21	12	
CHOPPED SIRLOIN, SWANSON	284	446	39.8	19.9					977	54	301	.40	7.4	312
		26.9	1.7						667		3.4	.20		
HADDOCK, CAMPBELLS	340	326	21.4	14.3					1319	65	316	.27	5.7	313
		38.6	1.7								1.7	.24		
FISH, MORTON	248	376	42.4	13.5					556	61	268	.29	4.6	3228
		21.1									2.9	.19	20	
HAM, MORTON	283	429	49.1	18.1					278	33	189	.54	3.9	2086
		18.5									3.8	.20	25	
HAM, SWANSON	291	367	42.1	11.9					1257	70	303	.76	4.4	7845
		22.7	1.7						416		3.5	.17		
MACARONI & BEEF, MORTON	311	365	56.3	12.1					905	40	177	.26	3.9	1321
		11.5									3.2	.19	39	
MACARONI & CHEESE, MORTON	226	296	31.5	13.6					934	278	233	.33	2.1	536
		13.3									1.4	.34	0	
MEATLOAF, TOM SCE, MASH POT, PEAS	312	419	30.6	20.9					1226	59	365	.31	5.3	1342
		25.0	0.9						359		4.1	.44	12	
MEATLOAF, MORTON	311	397	27.7	21.5						138	264	.26	5.7	3600
		24.3									4.3	.30	21	
PORK LOIN, SWANSON	284	460	40.5	21.9					821	26	276	.71	5.1	
		25.3	0.6						613		2.0	.23		
POT ROAST, POT, PEAS, CORN	205	217	12.5	6.6					531	21	156	.12	4.3	226
		26.9	0.6						500		3.3	.21	10	
SALISBURY STK, MORTON	311	373	43.5	15.6					1213	93	223	.22	4.4	2472
		18.7									3.1	.28	17	
SHRIMP, MORTON	219	379	37.3	16.4					480	79	260	.29	5.1	753
		21.0									3.7	.17	27	
SPAGHETTI & MEATBALLS, MORTON	311	60	55.9	10.6					1329	90	151	.39	4.4	1278
		13.4									3.7	.27	30	
SPAGHETTI CASS, MORTON	226	289	28.9	13.1					937	29	124	.22	3.6	1305
		13.9									3.4	.19	20	
SWISS STEAK, SWANSON	284	315	34.9	15.6					682	37	173	.34	0.3	2451
		20.2	1.1						540		4.8	.20		
TURKEY, MASH POT, PEAS	312	349	39.6	9.4					1248	81	271	.22	7.2	406
		26.2	0.9						549		3.4	.28	12.5	
TURKEY, MORTON	340	445	38.6	20.1					1397	56	256	.26	6.8	3137
		26.5									3.0	.22	12	
HAMBURGER HELPER, ckd ½ cup	113	360	24.2	18.8						28	244	.18	5.7	88
		22.9								43	3.4	.26	1	
HAM CROQUETTE, pan fried 1 piece	65	163	7.6	9.8					222	45	104	.18	1.6	169
		10.6	0.1						54		1.4	.14		
HAM & LIMA BEAN CASS 1 serving	280	476	46.2	18.8					709	213	459	.48	6.6	565
		30.4							1313		6.7	.30	6	
HASH, BEEF, HOMEMADE 1 cup	225	290	28.7	99	251	1681	1795	502	40	28	171	.15	4.6	30
		21.7	0.6	0.5	902	1111	1199	941			2.9	.21	21	0
HASH, BEEF, CARNATION 3½ oz	100		11.7	15.0					793	7	63	1.70	0.1	21
		10.0							264		1.6	.08	13	
HASH, BEEF, SEXTON 3½ oz	100	133	8.0	7.2					517					
		11.1	0.6						190					
HASH, CORNED BEEF ½ cup	110	199	11.8	12.4	429	792	858	242	594	14	74	.01	2.3	0
		9.7	0.6	5.5	396	506	539	110	220	21	2.2	.10	4	0

*Combination foods contain two or more items from different food categories.

	WT	Proximate			Amino Acids				Minerals			Vitamins		
		CAL	CHO	FAT	TRY	LEU	LYS	MET	Na	Ca	P	THI	NIA	A
		PRO	FIB	PUFA	PHE	ISO	VAL	THR	K	Mg	Fe	RIB	ASC	D
	g	g	g	g	mg	mg	mg	mg	mg	mg	mg	mg	mg	iu
KOSHER DINNERS†														
BEEF, GVY, POT, CAR, GRN BEANS	336	473	24.6	26.5					333					
		34.3							2053					
CHICKEN BRST, GVY, RICE, CAR, PEAS	392	380	21.9	10.5					378					
		49.4							1243					
CHICKEN LEG, NOODLES, PEAS	392	399	30.0	13.6					400					
		39.0							1121					
HALIBUT, NOODLES, PEAS	336	287	23.9	5.6					514					
		35.2							1865					
SALISBURY STK, MACARONI, GRN BEANS	336	405	22.5	24.0					296					
		24.6							2083					
SOLE, WHIPPED POT, SQUASH	336	263	22.1	8.2					306					
		25.9							2218					
TURKEY, GVY, SWEET POT, GRN BEANS	336	318	28.9	5.3					313					
		38.8							1996					
VEAL, GVY, POT, GRN BEANS	336	388·	22.1	15.2					262					
		40.7							2144					
LAMB STEW, CND, FEATHERWEIGHT	200	223	22.3	10.7					50					
		4.7							330					
LASAGNA, CND 1 serving	170	310	44.0	9.0					825					
		14.0												
LASAGNA, CELESTE	200	280	25.4	11.4		2086	828	750	774	254	296	.30	4.5	570
		19.2	0.6		822	1098	1206	788	464	52	2.8	.42		
LASAGNA, FRZN, CAMPBELLS	228	285	4.2	14.0					1175		250			
		14.0							331				0	
LASAGNA, GREEN GIANT	200	250	23.8	10.0					964	184	202	.10	3.2	1552
		16.2	2.2						584		2.6	.34	12	
LASAGNA, INVISIBLE CHEF	200	278	21.0	14.5					900	258	264	.14	2.7	1034
		15.3							370		2.5	.28	9	
LASAGNA, SARA LEE	200	430	55.4	10.0					764	252	376	.44	4.4	1102
		24.8	0.6						868	174	3.6	.48	124	
LASAGNA, FRZN, STOUFFERS	200	284	20.0	12.0					1020	288		.22	2.2	740
		19.4									2.0	.26	5	
LOBSTER NEWBURG‡	200	388	10.2	21.2					458	174	384	.14		
		37.0							342		1.8	.22		
LOBSTER THERMIDOR 1 lobster in shell	415	405	14.8	26.6						290	451	150	4.8	984
		28.5	0.5								1.9	510	0	
LUMACHE W/ CHEESE FILL, INVISIBLE CHEF 1 serving	142	490	77.7	12.3					1125	5	184	1.04	7.0	929
		17.3							273		4.6	1.03	7	
MACARONI & BEEF, FRZN, STOUFFERS	200	·232	21.0	10.6					1080	46		.14	2.8	900
		12.0									2.6	.18	294	
MACARONI & BEEF IN TOM SCE, FRZN, CAMPBELLS 1 serving	228	215	25.0	7.1					1310	66	170	.24	3.8	2257
		13.0							500		3.9	.16		
MACARONI, BEEF & TOM CASS, FRZN, PRONTO FOODS	200	171	31.7	4.8					1128	16	93	.14	1.0	0
		18.5							428		1.8	.78	11	
MACARONI 'N BEEF IN TOM SCE, CND, FRANCO-AMERICAN	200	200	26.0	8.0					1150	34	106	.18	3.2	518
		8.0	0.4						352		2.8	.16	5	
MACARONI & CHEESE, HOMEMADE 1 cup	200	430	40.2	22.2	340	2220	1460	560	1086	362	322	.20	1.8	860
		16.8	0.2	12.0	880	1500	1720	1040	240	52	1.8	.40		
MACARONI & CHEESE, CND, FRANCO AMERICAN	200	180	24.0	8.0					900	108	132	.20	2.2	604
		6.0							54		2.0	.20	0	
MACARONI & CHEESE, FRZN, BANQUET	200	259	26.0	12.0					187	156	1188			1887
		11.5							454		83			
MACARONI & CHEESE, FRZN, GREEN GIANT	200	250	26.0	11.8					734	206	168	.10	1.0	132
		10.0	0.2						134		1.0	.36	2	
MANICOTTI, CELESTE	200	238	29.4	7.6		510	508	412	766	216	230	.30	3.2	600
		13.2	0.6		1032	736	856	514	406	46	2.6	.52		
MANICOTTI, FRZN, CAMPBELLS 1 serving	249	305	25.0	14.9					825	333	298	.18	2.1	1313
		17.0	0.6						289		1.8	.26	6	
PEPPER STK SKILLET DINNER, LA CHOY	200	182	6.8	12.3					668					
		17.6												

*Combination foods contain two or more items from different food categories.

†Values are for the complete dinner; all are made by Schrieber.

‡Prepared w/ butter, egg yolks, sherry and cream.

	WT g	CAL / PRO	CHO / FIB	FAT / PUFA	TRY / PHE	LEU / ISO	LYS / VAL	MET / THR	Na / K	Ca / Mg	P / Fe	THI / RIB	NIA / ASC	A / D
PEPPER, GRN, STFD (BEEF & CRUMBS) 1 avg	185	315	31.1	10.2					581	78	224	.17	4.6	520
		24.1							477		3.9	.81	74	
PEPPER, GRN, STFD (BEEF & SCE), FRZN, CAMPBELLS 1 avg	185	190	18.0	9.1					995	19	96	.13	2.9	24
		10.0	0.7						454		2.0	.13	30	
PEPPER, GRN, STFD, FRZN, GREEN GIANT 1 avg	226	269	18.5	16.1					1001	34	118	.07	2.7	1024
		12.9	1.1						509		2.3	.23	36	
PEPPER, GRN, STFD (BEEF & SCE), HORMEL 1 avg	191	181	18.5	7.8					882	17	99	.10	2.6	827
		9.8	1.1							31	0.7	.11	32	
PEPPER, GRN, STFD, FRZN, INVISIBLE CHEF 1 avg	181	185	15.7	7.9					1140	14	219	.10	1.2	333
		12.6							467		6.7	.12	76	
PEPPER, GRN, STFD, FRZN, PRONTO FOODS 1 avg	180	131	15.8	7.9					1145	14	220	.10	1.2	334
		12.6							469		6.7	.12	77	
PIZZA														
CHEESE, HOMEMADE 1 piece	65	145	23.7	4.4					456	144	127	.08	0.6	410
		2.7	0.2						85		0.5	.05	5	
CHEESE† 12-inch pizza	564	1307	192.3	24.8					2693	1162	1057			
		77.9	1.1						947	156	5.2			
CHEESE, THICK CRUST‡ 10-inch pizza	417	919	155.7	12.4					2265	579	637			
		46.1	1.6						947	156	5.2			
CHEESE, THIN CRUST‡ 10-inch pizza	336	718	97.8	19.6					2232	710	669			
		37.6	1.7						570	79	2.2			
CHEESE, FROM MIX 1 piece	10	236	37.9	7.3										
		4.8												
CHEESE, FRZN, CELESTE	100	225	24.9	9.2		1135	624	327	585	225	215	.16	1.3	650
		10.6	0.5		596	617	748	384	233	33	1.3	.24	0	
CHEESE, FRZN, JENOS	100	239	28.4	8.2							386	.03	1.7	141
		12.9									2.9	.23	tr	
DELUX, FRZN, CELESTE	100	224	21.7	10.8					497	153	131	.17	1.4	648
		9.4	0.6						235	34	1.3	.16		
GROUND BEEF, FRZN 1 piece	100	248	41.0	6.4										
		6.8												
MUSHROOM, FRZN, CELESTE	100	205	21.8	8.6						190		.23	1.4	658
		10.1	0.6								1.8	.27	4	
PEPPERONI, FRZN 1 piece	100	234	29.6	9.3					729	17	92	.09	1.5	560
		7.8	0.3						168		1.2	.12	9	
PEPPERONI, FRZN, CELESTE	100	253	23.7	12.4					758	181	218	.19	1.4	593
		11.5	0.5						244	36	1.4	.26		
PEPPERONI, FRZN, JENOS	100	259	29.7	10.1						195		.03	1.2	142
		12.4									3.7	.22	tr	
SAUSAGE, HOMEMADE 1 piece	67	157	19.8	6.2					488	11	62	.08	0.6	375
		5.2	0.2						113		0.5	.05	6	
SAUSAGE, FRZN 1 piece	100	250	39.2	7.4										
		6.5												
SAUSAGE, FRZN, CELESTE	100	243	23.2	12.1		1083	405	352	581	170	188	.25	1.4	666
		10.3	0.5		670	592	481	396	246	34	1.5	.18	0	
SAUSAGE, CHEF BOYARDEE	100	223	29.1	7.8					534					
		10.7												
SAUSAGE, FRZN, JENOS	100	236	28.4	8.5						190		.4	3.7	121
		11.5									4.8	.21	tr	
SAUSAGE W/ MUSHROOM, FRZN, CELESTE	100	225	21.5	11.0						147		.30	1.8	754
		10.0	0.6								1.7	.28	0	
PIZZA BURGER	100	236	11.6	12.9					382	162	207	.11	6.5	363
		17.2	0.1						263		2.0	.23	4	
PIZZA ROLLS, PEPPERONI, FRZN, JENOS 1 piece	14	42	3.8	2.3						14		.03	0.5	80
		1.4									0.5	.02	tr	
PIZZA ROLLS, SAUSAGE, FRZN, JENOS 1 piece	14	41	3.9	2.3						13		.04	0.3	196
		1.4									0.9	.02	tr	
PIZZA ROLLS, SHRIMP, FRZN, JENOS 1 piece	14	35	3.9	1.6						11		.04	0.6	49
		1.3									0.3	.02	tr	

*Combination foods contain two or more items from different food categories.

†From restaurant.
‡From national franchisee.

		Proximate			Amino Acids				Minerals			Vitamins		
	WT	CAL	CHO	FAT	TRY	LEU	LYS	MET	Na	Ca	P	THI	NIA	A
		PRO	FIB	PUFA	PHE	ISO	VAL	THR	K	Mg	Fe	RIB	ASC	D
	g	g	g	g	mg	mg	mg	mg	mg	mg	mg	mg	mg	iu
PORK & SAUERKRAUT	185	244	7.9	1.9					946	45	159	.33	7.3	53
1 serving		14.6	0.8						385		2.3	.18	15	
PORK LOIN W/ GVY & DRESSING,	228	280	17.0	14.0					830	50	193	.48	5.9	0
FRZN, CAMPBELLS 1 pkg		22.0	2.2						300		1.5	.16	1	
PORK, SWEET & SOUR, HOMEMADE	200	386	28.7	21.7					856	25	209	.71	8.8	49
1 serving		19.1	0.5						400		2.9	.24	10	
PORK, SWEET & SOUR, FRZN,	100	99	22.0	0.2					716					
LA CHOY 3½ oz		4.4												
PORK W/ DRESS & GVY, FRZN,	207	180	20.0	6.0					192	11		.44	2.9	72
INVISIBLE CHEF 1 serving		13.6							259		1.6	.15	14	
RAVIOLI, BEEF, CND, CHEF BOYARDEE	213	220	34.0	6.0					1065					
1 serving		8.1												
RAVIOLI, BEEF, FRZN, CELESTE	200	222	32.2	5.0		1024	336	286	678	118	182	.36	4.3	652
		12.2	0.8		530	586	690	376	510	54	2.7	.54		
RAVIOLI, BEEF, FRZN, INVISIBLE	170	354	39.5	15.2					160	40	235	.33	3.6	
CHEF 1 serving		10.8							404		3.0	.13	15	
RAVIOLI, BEEF, FRZN, SARA LEE	200	242	27.2	10.2					416	44	470	.26	3.4	6130
		10.4	1.2						684	154	2.6	.18	3	
RAVIOLI, CHEESE, CND, CHEF	213	220	36.0	5.0					540					
BOYARDEE 1 serving		10.0												
RAVIOLI, CHEESE, FRZN, CELESTE	200	228	33.6	5.4	658	1104	192	344	690	154	190	.36	3.8	666
		11.2	0.8		460	584	658	392	504	tr	2.8	.54		
RAVIOLI, CHEESE, FRZN, INVISIBLE	170	258	28.2	12.9					248	2	192	.23	2.1	886
CHEF 1 serving		9.2							370		2.0	.16	15	
RAVIOLI, CHEESE, FRZN, SARA LEE	200	214	28.2	7.4					860	124		.32	1.8	4610
		8.6	0.8						414	8	2.4	.18	4	
RICE, FRIED W/ MEAT, FRZN,	150	195	32.9	0.8					1280					
LA CHOY 1 cup		8.9												
RICE W/ PEAS, MUSHROOMS, FRZN,	100	148	34.3	0.4					485	12	63	.28	3.1	640
BIRDS-EYE		5.3	1.0						109	14	1.4	.17	13	
RICE W/ PEAS, MUSHROOMS, FRZN,	100	86	15.2	1.7					385	10	28	.34	1.9	143
GREEN GIANT		2.0	0.4						34		1.5	.02	2	
SALMON RICE LOAF	100	122	7.3	4.5										
		12.0												
SALT COD, CREAMED	135	267	8.1	11.7	304	2317	2690	893	10900	115	378	56	3.3	436
1 serving		30.7			1138	1556	1634	1330	216		1.4	274	0	
SHAD CREOLE†	100	152	1.6	8.7					73	19	190	.09	5.1	450
		15.0							280		0.6	.16	8	
SOUFFLE, ASPARAGUS, FRZN,	140	161	10.4	9.8					552	119		.13	0.7	923
STOUFFERS 1 serving		7.7									0.7	.28	10	
SOUFFLE, CHEESE, HOMEMADE	150	327	9.3	25.7					546	302	293	.08	0.3	1200
1 serving		14.9							182		1.5	.36		
SOUFFLE, CHEESE, FRZN, CHEF	142	359	13.6	27.1					738	300	285	.11	0.6	1380
FRANCISCO 1 serving		15.5							199		1.6	.40		
SOUFFLE, CHEESE, FRZN, STOUFFERS	150	324	12.6	23.3					795	317	302	.18	0.5	1979
		1.1							210		1.5	.42		
SOUFFLE, CORN, FRZN, STOUFFERS	150	203	23.3	9.0					630	68		.09	1.4	627
		5.9									0.9	.24	2	
SOUFFLE, SPINACH, FRZN, STOUFFERS	150	192	12.5	12.2					735	149		.15	0.5	2237
		7.4									2.3	.27	4	
SPAGHETTI W/ MEAT SCE,	292	396	39.4	20.7						27	148	.12	3.0	901
HOMEMADE 1 serving		12.7	0.7								2.1	.12	24	
SPAGHETTI W/ MEAT SCE, IDLE WILD	180	175	19.4	4.2					108					
1 serving		14.9							268					
SPAGHETTI W/ TOM SCE, HOMEMADE	220	179	34.3	1.5	46	462	240	92	840	35	95	.31	4.0	807
1 serving		7.0	0.4		304	308	266	372			2.4	.24	0	
SPAGHETTI W/ TOM SCE, CND,	221	168	34.0	1.3						35		.31	4.0	818
STOKELY VAN CAMP 1 cup		4.9	0.4								2.4	.24	9	
SPAGHETTI W/ TOM SCE & CHEESE,	302	315	44.8	10.6	63	634	329	127	1150	97	163	.31	2.7	1300
HOMEMADE 1 serving		10.6	0.6		417	423	365	510	493		2.7	.21	15	

*Combination foods contain two or more items from different food categories.

†Prepared w/ tomatoes, onion, green pepper, butter, and flour.

	WT	Proximate			Amino Acids				Minerals			Vitamins		
		CAL	CHO	FAT	TRY	LEU	LYS	MET	Na	Ca	P	THI	NIA	A
		PRO	FIB	PUFA	PHE	ISO	VAL	THR	K	Mg	Fe	RIB	ASC	D
	g	g	g	g	mg	mg	mg	mg	mg	mg	mg	mg	mg	iu
SPAGHETTI W/ TOM SCE & CHEESE, CND 1 serving	250	190	38.5	1.5	75	375	175	75	955	40	88	.35	4.5	925
		5.5	0.5	0.0	275	275	325	225	303	28	2.8	.28	10	
SPAGHETTI W/ TOM SCE & GROUND BEEF, CND, CAMPBELLS	200	230	26.0	10.0					1100	36	106	.20	3.4	950
		8.0	0.6						444		2.8	.16	2	
SPAGHETTI-TOM SCE-GROUND BEEF, CND, CHEF BOYARDEE 1 serving	213	210	28.0	9.0					1054					
		7.0												
SPAGHETTI W/ TOM SCE & MEAT BALLS, HOMEMADE 1 serving	220	295	34.4	10.3	176	1166	1144	352	895	110	208	.22	3.5	1410
		16.5	0.7		682	770	770	660	590	37	3.3	.26	20	
SPAGHETTI W/ TOM SCE & MEAT BALLS, CND 1 serving	250	258	28.5	10.3					1220	53	113	.15	2.3	1000
		12.3	0.3						245	28	3.3	.18	5	
SPAGHETTI-TOM SCE-MEAT BALLS, CND, CHEF BOYARDEE 1 serving	213	210	29.0	6.0					941					
		9.0												
SPAGHETTI-TOM SCE-MEAT BALLS, FRZN, CAMPBELLS 1 serving	228	270	25.0	12.1					1350	73	173	.16	3.7	1020
		15.0							567		2.7	.24	7	
SPAGHETTI-TOM SCE-MUSHROOMS, CND, CHEF BOYARDEE 1 serving	138	230	42.0	3.0					545					
		9.0												
SPAGHETTI SAUCE	100	120	6.9	7.1					237	16	82	.08	2.5	1055
		7.8	0.3	3.6					457	21	1.9	.10	21	2
SPAGHETTI SCE W/ MEAT 6 T	100	96	7.8	5.1					732	12	51	.04	1.2	1512
		4.5	0.6								1.8	.06	0	
SPAGHETTI SCE W/ MEAT, CND, CHEF BOYARDEE 1 serving	113	130	14.0	7.0					600					
		3.1												
SPAGHETTI SCE W/ MEAT, FRZN, BANQUET	100	116	15.0	4.2					72	76	546	.21	0.1	410
		4.9							462			.29	tr	
SPAGHETTI SCE W/ MEAT BALLS, FRZN, CAMPBELLS 1 serving	228	240	18.9	13.0					1050	64	176	.08	2.7	1760
		13.0	0.7						320		2.4	.16	8	
SPAGHETTI SCE W/ MUSHROOMS 6 T	93	81	9.0	4.5					744	12	30	.06	0.9	1812
		1.2	0.6								1.2	.06		
SPAGHETTI SCE W/ MUSHROOMS, CND, CHEF BOYARDEE 1 serving	106	60	13.0	1.0					565					
		1.0												
SPANISH RICE, HOMEMADE 1 cup	150	130	24.9	2.6					475	21	58	.06	1.1	990
		2.7	0.8						347		0.9	.04	2.2	
SPANISH RICE, FRZN, GREEN GIANT	200	200	36.0	4.0					840	20	40	.08	1.0	1300
		4.0							190		0.6	.10	6	
SPANISH RICE, STOKELY VAN CAMP	200	174	33.2	3.4					1240	28		.08	1.4	1320
		3.6	1.0						208		1.2	.06	30	
TACO† 1 avg	75	162	8.9	8.6					401	111	115			
		11.8	0.4						203	37	1.0			
TAMALES, CND 2 avg	155	217	22.0	11.0	108	356	372	108	1031	31	60			0
		7.0	0.0		201	232	232	186		14	1.9		0	
TERIYAKI, SKILLET DINNER, LA CHOY	200	266	6.3	19.1					676					
		21.5												
TUNA HELPER, CKD ½ cup	113	140	14.9	5.8						11	75	.09	2.5	54
		7.0								15	1.0	.08	0	
TUNA & NOODLES, FRZN, GREEN GIANT 1 serving	170	150	16.0	5.4					627	73	122	.07	3.9	173
		10.4	0.3						196		0.9	.14	10	
TUNA & NOODLES AU GRATIN, FRZN, SARA LEE	200	294	40.2	10.0					766	194	332	.26	5.4	666
		17.0	0.6						424	70	2.2	.38	16	
TUNA NOODLE CASS 1 cup	200	280	25.0	11.8										
		17.8												
TUNA NOODLE CASS, FRZN, STOUFFERS	200	234	20.6	10.6					740	124		.20	4.6	330
		13.2									1.8	.26	0	
TUNA POTPIE, FRZN, MORTON 1 pie	226	385	35.8	19.8					715	72	153	.20	5.4	286
		14.8									1.7	.20	5	
TURKEY A LA KING 1 serving	180	212	9.5	11.7					497	15	142	.07	4.2	486
		16.6	0.2						28		1.3	.14	6	
TURKEY, CREAMED (W/ SLICE WHOLE WHEAT TOAST) 1 cup	233	379	13.5	22.2						180	451	.25	8.2	655
		30.0		0.7							4.4	.35	tr	
TURKEY POTPIE, HOMEMADE 1 pie	227	538	42.0	30.6					620	61	229	.25	5.7	3019
		23.6	0.9						449		3.2	.30	5	

*Combination foods contain two or more items from different food categories.

†From national franchisee; consists of crisp folded corn tortilla w/ground beef, lettuce, shredded cheese, and red sauce.

	WT	Proximate			Amino Acids				Minerals			Vitamins		
		CAL	CHO	FAT	TRY	LEU	LYS	MET	Na	Ca	P	THI	NIA	A
		PRO	FIB	PUFA	PHE	ISO	VAL	THR	K	Mg	Fe	RIB	ASC	D
	g	g	g	g	mg	mg	mg	mg	mg	mg	mg	mg	mg	iu
TURKEY POTPIE, COMMERCIAL, FRZN	227	417	35.9	23.6					864	22	107	.23	3.9	2043
1 pie		15.2	0.5						259		1.6	.23	5	
TURKEY POTPIE, FRZN, BANQUET	240	415	40.6	21.6					1017	107	168	.11	3.7	788
1 pie		14.4							18		2.6	.11	2	
TURKEY TETRAZZINI, FRZN, STOUFFERS	100	141	9.3	8.3					370	45		.06	1.6	120
		6.8									0.6	.13	0	
TURKEY W/ GVY & DRESS, FRZN, CAMPBELLS 1 pkg	214	260	16.0	12.0					855	32	242	.15	5.3	0
		24.0	0.0						325		1.5	.15	0	
VEAL & PEPPERS, FRZN, SARA LEE	200	164	9.4	1.0					464	24	196	.16	6.0	1064
		29.8	0.4						614	28	3.2	.26	34	
VEAL PARMIGIANI, FRZN, CAMPBELLS 1 serving	214	295	17.0	13.9					1825	99	289	.30	0.4	618
		25.0	0.6						468		2.4	.38	12	
VEAL SCALLOPINI 1 serving	116	199	2.7	11.3					918	106	228	.05	10.1	94
		20.4	0.1						385		2.1	.29	tr	
VEAL STEW W/ CAR & ONIONS 1 cup	238	242	7.1	15.6						32	191	.07	2.9	3253
		17.6	0.6								2.9	.17	0	0
VEGETARIAN MAIN DISHES														
BEEF STYLE ROLL, WORTHINGTON FOODS	100	200	5.1	13.8								1.35	10.2	
		15.2									2.8	.36		
CHICKEN STYLE, WORTHINGTON FOODS	100	243	2.9	16.9					580	60		.59	5.2	
		15.5	0.6	11.3							2.0	.22		
CHIC-KETTS, WORTHINGTON FOODS	100	214	7.1	11.9								.63	3.6	
		20.2									1.3	.12		
CHILI, WORTHINGTON FOODS	100	136	14.3	5.2						43		.21	4.3	
		7.6									1.9	.05		
CORNED BEEF STYLE, WORTHINGTON FOODS	100	271	9.0	18.7						42			6.5	
		17.0		4.6							1.9	.27		
CROQUETTES, WORTHINGTON FOODS 1 piece	61	167	11.1	8.9					403	22		.24	2.2	
		10.0	0.1	0.0							2.0	.07		
DINNER CUTS, LOMA LINDA	100	102	4.4	1.8					572	39	101	.01	0.9	
		16.9		1.3					65	21	1.4	.23		
FRY STICKS, WORTHINGTON FOODS 1 piece	65	103	5.2	3.4					1255					
		12.9	0.5											
GRAN BURGER, WORTHINGTON FOODS	100	371	34.3	2.8						17		.86	14.3	
		51.4										1.00		
MEATLOAF MIX, WORTHINGTON FOODS	100	415	3.9	14.0					15	20				
		33.0	3.4	0.3							0.9			
NUTEENA, LOMA LINDA	100	254	9.5	18.2	164	880	576		460	32	166	.02	0.3	
		13.2		6.3		613	693	418	285	60	1.3	.15		
PEANUTS & SOY, CND	100	237	13.4	16.9										
		11.7	0.9											
PROSAGE, WORTHINGTON FOODS	100	265	7.4	17.6					1000	22			.44	8.8
		17.0	0.1	5.8							2.6	.38		
PROTEENA, LOMA LINDA	100	198	5.6	10.0	144	1517	775	776	520	36	140	.07	2.0	360
		21.4		3.3	1920	541	1002	368	144	44	1.6	.55		
REDIBURGER, LOMA LINDA	100	223	5.5	13.3	229	728	372	751	370	24	83			
		20.3		7.9	1740	870	481	621	90	19	1.5			
SMOKED BEEF STYLE, WORTHINGTON FOODS	100	243	10.0	14.3						40		2.20	8.0	
		17.1		2.5							2.4	.27		
STAKELETS, WORTHINGTON FOODS 1 piece	20	42	1.6	2.6						20		.76	0.9	
		2.8		0.0							0.6	.04		
STRIPPLES, WORTHINGTON FOODS 3 pieces	21	49	0.9	2.9					268					
		4.8												
TASTEECUTS, LOMA LINDA	100	98	3.4	1.7					420	40	89	.01	0.8	
		17.4		0.8					47	31	1.4	.20		
TURKEY STYLE, SMOKED, WORTHINGTON FOODS	100	257	4.3	19.0						43			6.6	
		15.4		7.9							2.5	.20		
VEGA-LINKS, WORTHINGTON FOODS 1 piece	34	73	1.6	6.0					146	11		.08	0.9	
		3.3	tr	1.7							0.4	.07		

*Combination foods contain two or more items from different food categories.

	WT	Proximate			Amino Acids				Minerals			Vitamins		
		CAL	CHO	FAT	TRY	LEU	LYS	MET	Na	Ca	P	THI	NIA	A
		PRO	FIB	PUFA	PHE	ISO	VAL	THR	K	Mg	Fe	RIB	ASC	D
	g	g	g	g	mg	mg	mg	mg	mg	mg	mg	mg	mg	iu
VEGEBURGER, LOMA LINDA	100	115	4.2	2.5	221	1474	972	695	350	42	97	.17	2.3	
		19.2		1.2	1753	921	944	604	75	22	1.7	.19		
VEG SKALLOPS, WORTHINGTON FOODS 1 piece	21	17.3	0.7	0.5										
		2.3									0.2			
VEGETARIAN BURGER, WORTHINGTON FOODS	100	160	9.6	6.4					900	21		.16	5.3	
		14.2	0.1	0.0							2.9	.11		
VITABURGER, LOMA LINDA	100	280	30.0	1.0	343	4075	3164	596	700	220	570	.60	16.0	0
		50.0		3.0	2603	2464	2617	2170	2200	260	10.0	.60	0	0
WHAM, WORTHINGTON FOODS 1 piece	39	78	2.2	4.7					117	16		2.34	3.2	
		5.9	tr								0.7	.14		
WHEAT PROTEIN, CND	100	109	8.8	0.8										
		16.3	0.1											
WHEAT PROTEIN W/ PEANUTS, CND	100	212	17.7	7.1										
		20.3	0.4											
WHEAT & SOY PROTEIN, CND	100	104	7.6	1.2										
		16.1	0.3											
WHEAT & SOY PROTEIN W/ VEG OIL, CND	100	150	9.5	5.6										
		16.1	0.6											
WHEAT PROTEIN & VEG OIL, CND	100	189	5.2	10.4										
		19.1												
WELSH RAREBIT 1 cup	232	415	14.6	31.6					770	582	432	.09	2.3	1230
		18.8	0.0						320		0.7	.53		

*Combination foods contain two or more items from different food categories.

	WT	Proximate			Amino Acids				Minerals			Vitamins		
		CAL	CHO	FAT	TRY	LEU	LYS	MET	Na	Ca	P	THI	NIA	A
		PRO	FIB	PUFA	PHE	ISO	VAL	THR	K	Mg	Fe	RIB	ASC	D
	g	g	g	g	mg	mg	mg	mg	mg	mg	mg	mg	mg	iu
A-1 SCE, HEUBLEIN	5	4	1.0	0.0					82	1				
1 t		0.1							17		0.1			
ANGOSTURA AROMATIC BITTERS, A-W	5	15	0.5	0.0					tr	tr		.00	tr	
1 t		tr							2		tr	.00	0	
BARBEQUE SCE	16	15	1.3	1.1					130	3	3	tr	0.1	58
1 T		0.2	0.1						28		0.1	tr	1	
BBQ SCE, HICKORY SMOKE FLAVOR, OPEN PIT 1 T	16	15	3.4	0.2					213	2	1	tr	tr	8
		0.1	0.1						6	1	tr	tr		
BBQ SCE W/ MINCED ONIONS, OPEN PIT 1 T	16	16	3.5	.2					231	3	2	tr	tr	8
		0.1	0.1						9	1	0.1	tr	tr	
CATSUP	15	16	3.8	0.1					156	3	8	.01	0.2	210
1 T		0.3	0.1	0.0					55	3	0.1	.01	2	
CHILI SCE	17	17	4.0	0.1					228	2	3	.02	0.4	320
1 T		0.5	0.1						63		0.1	.01	2	
HOLLANDAISE SCE	50	180	0.4	18.5						23	78	.03	tr	1027
¼ cup		2.2									0.9	.04	tr	
HOLLANDAISE SCE, MOCK	85	191	6.3	16.4						88	112	45	0.2	834
¼ cup		4.1									0.7	130	1	
HORSERADISH, RAW	20	17	3.9	0.1					2	28	13	.01		
		0.6	0.5						113	7	0.3		16	
HORSERADISH, PREPARED	18	7	1.7	tr					17	11	6			
1 T		0.2	0.2						52		0.2			
ITALIAN SCE	28	13	2.1	0.3					141	7	10	.02	0.4	190
1 oz		0.5	0.3						113		0.3	.07		
MANWICH SANDWICH SCE	73	40	9.6	0.3					176	4	9	.02	0.3	463
1 serving		1.2							92		0.4	.01	10	
MUSTARD, BROWN	5	4	0.3	0.4					65	5	6			
1 t		0.3	0.1						7	2	0.1			
MUSTARD, YELLOW	5	4	0.3	0.2					63	4	4			
1 t		0.2	tr						7	2	0.1			
MUSTARD W/ HORSERADISH	16	12	1.1	0.8					282	17	14			
1 T		0.9	0.1						26		0.3			
PICKLE RELISH, SOUR	15	3	0.4	0.1						4	3			
1 T		0.1	0.2								0.2			
PICKLE RELISH, SWEET	15	21	5.1	0.1					107	3	2			
1 T		0.1	0.1								0.1			
PICKLES														
CHOW CHOW, SOUR	28	8	1.2	0.4					375	9	15			
1 oz		0.4	0.2								0.7			
CHOW CHOW, SWEET	28	32	7.6	0.3					148	6	6			
1 oz		0.4	0.3								0.4			
CUCUMBER, BREAD & BUTTER	25	18	4.5	0.1	1	8	10	2	168	8	7	tr	tr	35
4 slices		0.2	0.1		5	7	8	6	28		0.4	7	2	
CUCUMBER, DILL	100	11	2.2	0.2	5	30	31	7	1428	26	21	tr	0.1	100
1 large		0.7	0.5		16	22	24	19	200	12	1.0	.02	6	0
CUCUMBER, SOUR	105	10	2.0	0.2	4	22	22	5	1353	17	15	tr	tr	100
1 large		0.5	0.5		12	16	17	14			3.2	.02	7	
CUCUMBER, SWEET	100	146	36.5	0.4	5	30	31	7	572	12	16	.00		90
1 large		0.7	0.5		16	22	24	19	1		1.2	.02	6	0
SOUR CREAM SCE†	53	141	1.5	10.8						58	85	.04	tr	478
¼ cup		2.6	0.0								0.5	.10	0	9
SOYSAUCE, KIKKOMAN	15	11	1.2	0.0	0	81	72	21	858	3	32	.01	0.5	0
1 T		1.3		0.0	47	50	57	42	54	64	0.4	.02	0	0
SPANISH SCE	95	89	5.6	7.3						15	48	.05	1.0	1282
½ cup		1.4									0.8	.07	11	
STEAK SCE, LEA & PERRINS	15	18	2.5	tr					149	6	1	.01	tr	51
1 T		tr	0.1						64		0.4	.07	11	
STEAK SCE W/ MUSHROOMS, DAWN FRESH 1 oz	30	9	1.5	0.1					158	2	5	tr	0.1	3
		0.3	0.1						17		0.1	.03	1	

*Condiment—something usually pungent, acid, salty, or spicy added to or served w/food to enhance its flavor or to give added flavor.

†Made w/2 egg yolks, ¾ cup sour cream, 1 t tarragon, vinegar; makes 1 cup (4 portions).

	WT	Proximate			Amino Acids				Minerals			Vitamins		
		CAL	CHO	FAT	TRY	LEU	LYS	MET	Na	Ca	P	THI	NIA	A
		PRO	FIB	PUFA	PHE	ISO	VAL	THR	K	Mg	Fe	RIB	ASC	D
	g	g	g	g	mg	mg	mg	mg	mg	mg	mg	mg	mg	iu
SWEET & SOUR SCE, LA CHOY	30	54	13.3	0.1					235	2	1			
1 oz		0.1							7	1				
TARTAR SCE	20	95	1.5	10.0					141	4	10	.01	tr	55
1 T		0.2							16		0.2	.01	1	
TERIYAKI SCE, KIKKOMAN	30	30	4.8	0.0	0	108	100	0	1150	7	46	.01	0.4	0
1 oz		1.8		0.0	64	69	77	49	68	18	0.5	.02	0	
TABASCO SCE	5	tr	0.1	0.0					22			tr	tr	
1 t		0.1	0.0						3			.01		
WHITE SCE, MEDIUM	33	54	3.1	4.1					125	37	31	.01	0.1	164
2 T		1.3							46	5	0.1	.05	0	
WHITE SCE, MEDIUM	66	107	6.1	8.1					250	74	61	.02	0.2	328
¼ cup		2.6							92	9	0.2	.10	tr	
WHITE SCE, THICK	33	65	3.6	5.2					132	35	30	.02	0.1	188
2 T		1.3							44		0.1	.05		
WHITE SCE, THICK	66	131	7.3	10.3					263	71	59	.03	0.2	376
¼ cup		2.6							88		0.2	.11		
WHITE SCE, THIN	30	37	2.3	2.6					105	37	30	.01	tr	107
2 T		1.2							44		tr	.05	0	
WHITE SCE, THIN	61	74	4.6	5.2					214	73	59	.02	0.1	213
¼ cup		2.4							89		0.1	.10	tr	
WORCESTERSHIRE SCE	5	4	0.9	0.0					49	5	3	tr	tr	17
1 t		0.1							40		0.3	.01	9	

*Condiment—something usually pungent, acid, salty, or spicy added to
or served w/food to enhance its flavor or to give added flavor.

	WT g	CAL g	CHO g	FAT g	PRO g	FIB g	PUFA g	TRY mg	LEU mg	LYS mg	MET mg	PHE mg	ISO mg	VAL mg	THR mg	Na mg	Ca mg	P mg	K mg	Mg mg	Fe mg	THI mg	NIA mg	A iu	RIB mg	ASC mg	D iu
		Proximate						**Amino Acids**								**Minerals**						**Vitamins**					
APPLE BROWN BETTY	140	211	41.6	4.9												214	25	31				.08	0.6	140			
½ cup					2.2	0.7													140	7	0.8				.06	1	
APPLE CRISP	145	302	58.0	8.1												228	10	15				.05	0.4	357			
1 serving					0.9	0.4													79		0.7				.03	2	
APPLE DUMPLING, FRZN	78	254	27.3	14.8												152	1	16				.16	0.5	8			
PEPPERIDGE FARM 1 dumpling					3.1	0.3													45	6	1.1				.05	2	
APPLE STRUDEL, FRZN	100	290	39.0	14.3												130	3	19				.15	0.6	10			
PEPPERIDGE FARM 1 serving					4.0	0.4													74	10	1.2				.09	3	
BAVARIAN CREAM, ORANGE	240	289	39.8	14.0													62	48				.10	0.3	790			
1 serving					4.1	tr															0.4				.06	61	
BLUEBERRY CRISP	100	149	36.3	tr												2	12	4				.02	0.2	0			
					0.5	tr	tr												26	5	0.7				.04	3	0
BREAKFAST PASTRY—COFFEECAKE																											
COFFEECAKE, FROM MIX	72	232	37.7	6.9												310	44	125				.13	1.0	115			
1 piece					4.5	0.1													78		1.2				.12		
COFFEECAKE, APPLE, FRZN	60	177	26.1	7.4												195	33	45				.02	0.2	299			
1 serving					2.1	0.2													57		.5				.05	1	
COFFEECAKE, CINN, PILLSBURY	60	225	33.2	8.6												493											
1 serving					3.2																						
COFFEECAKE, HONEY NUT, FRZN	55	218	30.9	9.7												110	46	67				.06	0.4	46			
CHEF FRANCISCO 1 serving					3.2	0.2													131		1.0				.06	tr	
BREAKFAST PASTRY—DANISH																											
ALMOND DANISH, FRZN, SARA LEE	83	344	44.7	15.1												313	22	56				.15	1.3	66			
1 roll					5.1	0.3													64	17	1.3				.12	1	
ALMOND DANISH, ICED, PILLSBURY	39	140	20.0	5.5												358											
1 roll					2.5																						
APPLE DANISH, FRZN, SARA LEE	105	360	56.1	13.5												408	14	47				.15	1.2	53			
1 roll					3.9	0.3													63	12	1.7				.11	11	
CARAMEL DANISH, PILLSBURY	43	157	20.7	7.6												324											
1 roll					2.0																						
CHEESE DANISH, FRZN, SARA LEE	83	308	31.8	16.8												330	33	62				.13	1.1	169			
1 roll					7.6	0.1													71	29	1.4				.15	tr	
CHERRY DANISH, FRZN, SARA LEE	82	274	35.2	12.7												291	31	66				.15	1.3	174			
1 roll					4.9	0.3													115	21	2.9				.15	19	
CINN DANISH, FRZN, SARA LEE	97	389	52.8	17.7												353	24	62				.18	0.1	62			
1 roll					5.1	0.3													110	17	1.7				.14	tr	
CINN DANISH W/ RAISINS	39	135	21.0	5.0												315											
1 roll					2.0																						
DANISH ROLL	35	148	16.0	8.2				33	205	77	42					128	18	38				.02	0.3	109			
1 small					2.6	tr		144	128	125	84								39	8	0.3				.05		
ORANGE DANISH, ICED, PILLSBURY	39	132	21.3	4.6												329											
1 roll					2.0																						
BREAKFAST PASTRY—SWEET ROLLS																											
CARAMEL ROLL, PILLSBURY	39	145	29.1	2.3												203											
1 roll					2.3																						
CINN ROLL	55	174	27.1	5.0				60	384	170	78					214	47	59				.04	0.4	39			
1 avg					4.7	0.1		255	240	237	160								68	18	0.4				.08	tr	
CINN ROLL, PILLSBURY	39	145	19.5	6.5												333											
1 roll					2.0																						
CINN ROLL, ICED, PILLSBURY	34	107	6.2	3.0												358											
1 roll					2.0																						
SWEET ROLL	55	174	27.1	5.0				60	384	170	78					214	47	59				.04	0.4	39			
1 avg					4.7	0.1		255	240	237	160								68	18	0.4				.08		
BROWNIES																											
BUTTERSCOTCH	30	115	16.3					5.0								98	31	45				.01	tr	242			
1 avg					1.3	tr													88		0.5				.02	0	
CHOC, HOMEMADE	30	146	15.3	9.4												75	12	44				.05	0.2	60			
1 piece					2.0	0.2													57		0.6				.03	tr	
CHOC, FROM MIX, PILLSBURY	36	140	22.0	5.0												140											
1 piece					2.0																						

Column key (each food has two rows — top values / bottom values):

- WT (g)
- Proximate: CAL, CHO (g), FAT (g) / PRO (g), FIB (g), PUFA (g)
- Amino Acids: TRY, LEU, LYS, MET / PHE, ISO, VAL, THR (mg)
- Minerals: Na, Ca, P / K, Mg, Fe (mg)
- Vitamins: THI, NIA, A / RIB, ASC, D (mg; A and D in iu)

Food	WT	CAL / PRO	CHO / FIB	FAT / PUFA	TRY / PHE	LEU / ISO	LYS / VAL	MET / THR	Na / K	Ca / Mg	P / Fe	THI / RIB	NIA / ASC	A / D
CHOC, HOSTESS	25	100	15.5	4.0					41	12	27	.03	0.2	
1 bar		1.5							60		0.5	.04	tr	
CHOC, FRZN, SARA LEE	56	244	34.1	11.0					122	12	71	.05	0.4	90
1 brownie		2.9	0.2						104	12	1.0	.04	0	
CHOC, ICED, FRZN	30	126	18.2	6.2					60	12	37	.03	0.1	66
1 piece		1.5	0.2						54		0.5	.02	tr	
CHOC W/ NUTS, HOMEMADE	30	146	15.3	9.4					75	12	44	.06	0.2	60
1 avg		2.0	0.2						57	5	0.6	.04		
CHOC W/ NUTS, FROM MIX	30	128	18.9	6.0					50	14	41	.04	0.2	30
1 avg		1.5	0.2						50		0.6	tr		
CHOC W/ NUTS, FRZN, CAMPBELLS	50	255	24.0	12.6					72	15	71	.06	0.3	5
		3.0	0.3						81		1.1	.05	1	
CHOC W/ NUTS, ICED, LANCE	42	171	28.3	5.8					97	18		.01	0.3	
1 pkg		1.5									0.5	.06		
CHOC W/ NUTS, ICED, FRZN	30	126	18.2	6.2					60	12	38	.04	0.2	66
1 bar		1.5	0.2						54		0.5	.03		
PECAN FUDGE, KEEBLER	30	130	15.7	6.4					53	8	286			
		2.2							76		0.4			
CAKE														
ANGEL FOOD, HOMEMADE	45	121	27.1	0.1	39	247	74	42	127	4	10	tr	0.1	0
1 piece		3.2	0.0		176	147	138	93	40		0.1	.06	0	
APPLE COBBLER, FRZN, CAMPBELLS	50	97	18.5	2.5					67	6	42	.05	0.3	5
1 piece		1.0	0.2						25		0.3	.02	0	
BANANA, FRZN, SARA LEE	50	158	26.5	5.4					147	12	47	.04	0.4	75
1 piece		1.6	0.1						69	5	0.4	.03	1	
BANANA/APPLESCE W/O ICING	50	170	28.9	6.5						12	22	.06	0.5	25
1 piece		1.4								7	1.1	.05	0	
BANANA W/ BUTTERCREAM ICING	50	181	29.4	6.7					122	32	47	tr	0.4	288
1 piece		1.2	tr						77		tr	.02	tr	
BOSTON CREAM PIE	110	332	54.9	10.3					205	74	111	.03	0.2	230
1 serving		5.5	0.0						98		0.5	.12	tr	
CARAMEL W/O ICING	45	173	24.2	7.8					137	35	48	.01	0.1	81
1 piece		2.0	0.1						31		0.6	.04	tr	
CARAMEL W/ CARAMEL ICING	55	208	32.5	8.1					139	46	52	.01	0.1	110
1 piece		2.0	0.0						35		0.8	.04	tr	
CARROT, FRZN	50	173	22.5	9.4					216	18	50	.05	0.3	1392
1 piece		2.0	0.3						63		0.5	.04	1	
CARROT W/ CREAM CHEESE ICING	50	178	20.8	10.2					123	12	26	.06	0.6	1070
1 piece		1.8							58		0.4	.04	1	
CHEESECAKE, FRZN, MILANI	85	160	23.1	4.8					122					
1 piece		6.3							171					
CHEESECAKE, FRZN, SARA LEE	85	255	26.0	13.9					173	88	96	.06	0.3	298
1 piece		7.1	0.0						114	32	0.6	.16		
CHEESECAKE, FRENCH, FRZN, SARA LEE	85	265	22.9	17.3					140	51	46	.03	0.2	277
1 piece		4.3	0.0						65	22	0.9	.10	tr	
CHEESECAKE, CHERRY, FRZN	85	222	29.0	11.0					98	35	32	.03	0.3	269
1 piece		2.9	0.1						72	17	0.8	.08	8	
CHEESECAKE, STRAWBERRY, FRZN	85	238	28.6	12.8					110	40	37	.03	0.3	207
1 piece		3.1	0.1						63	19	0.9	.09	18	
CHOC, FRZN, SARA LEE	50	200	27.2	9.6					195	14	81	.04	0.4	89
1 piece		2.1	0.2						54	9	0.7	.04	0	
CHOC LAYER, FRZN, SARA LEE	85	277	32.2	15.2					213	60	138	.07	0.6	200
1 piece		3.0	0.2						79	15	0.8	.11	tr	
CHOC DOUBLE LAYER, FRZN, SARA LEE	85	281	33.3	15.2					170	58	121	.05	0.6	136
1 piece		3.3	0.3						90	14	0.8	.12	tr	
CHOC FUDGE, FRZN, PEPPERIDGE FARM	85	310	42.8	14.8					295	35	123	.11	1.1	85
1 piece		3.6	0.3						88	31	1.2	.07		
CHOC MALT W/ UNCKD WHITE ICING, FROM MIX	50	173	33.3	4.4					159	31	83	.01	0.1	95
1 piece		1.7	tr						40		0.3	.03		
CHOC 'N CREAM, SARA LEE	50	164	18.6	9.7					106	27	68	.04	0.3	95
1 piece		1.6	0.2						39	95	0.3	.07	0	
CHOC ROLL, RICHS	28	101	14.2	4.9	15	91	58	26	61	20	26	.01	0.1	51
1 oz		1.2	0.1	3.1	57	63	68	46	34	2	0.3	.02		2

	WT g	CAL / PRO	CHO / FIB	FAT / PUFÁ	TRY / PHE	LEU / ISO	LYS / VAL	MET / THR	Na / K	Ca / Mg	P / Fe	THI / RIB	NIA / ASC	A / D
		g	g	g	mg	mg	mg	mg	mg	mg	mg	mg	mg	iu
CHOC (DEVILS FOOD) W/O ICING	50	183	26.0	8.6					147	37	68	.01	0.1	75
1 piece		2.4	0.1						70	12	0.4	.05		
CHOC (DEVILS FOOD), W/ UNCKD WHITE ICING 1 piece	50	184	29.6	7.3					117	29	53	.01	0.1	90
		1.9	0.1						55	12	0.3	.04		
CHOC (DEVILS FOOD), W/ CHOC ICING, FROM MIX 1 piece	50	184	27.9	8.2					117	35	65	.01	0.1	80
		2.2	0.1						77	12	0.5	.05		
CHOC (DEVILS FOOD), W/ CHOC ICING, FRZN 1 piece	50	185	26.2	7.9					173	18	60	.06	0.6	50
		1.8	0.1						35	10	0.7	.04	tr	
CHOC (DEVILS FOOD), FRZN, W/WHIP CRM FILL & CHOC ICING 1 piece	85	315	37.2	18.6					162	68	104	.02	0.2	230
		3.0	0.2						96		0.5	.07		
COCNT, FRZN, PEPPERIDGE FARM	85	317	45.2	14.9					247	25	90	.11	1.1	85
1 piece		2.7	0.1						56	9	1.2	.07		
COTTAGE PUDDING W/O SCE	50	172	27.2	5.6					150	45	58	.07	0.6	70
1 serving		3.2	0.1						44		0.7	.08	tr	
COTTAGE PUDDING W/ CHOC SCE	70	223	39.6	6.2					163	50	76	.08	0.7	70
1 serving		3.7	0.2						98		1.0	.10	tr	
COTTAGE PUDDING W/ FRUIT SCE	70	204	33.8	6.2					163	51	65	.08	0.8	84
1 serving		3.6	0.2						6.5		0.8	.10	8	
CRUMB BLUEBERRY, STOUFFERS	50	185	28.4	6.8					172	31		.04	0.6	40
1 piece		2.5									0.5	.08		
CRUMB CHEESE, STOUFFERS	50	173	22.5	8.2					163	24		.01	0.1	109
1 piece		3.8									0.5	.07		
CRUMB CHERRY, STOUFFERS	50	170	25.6	6.4					148	33		.04	0.5	140
1 piece		2.7									0.6	.09		
CRUMB FRENCH, FRZN	50	177	28.6	6.2					207	17	64	.05	0.4	150
1 piece		2.2	0.1						28	5	0.8	.04	tr	
FRUITCAKE, DARK	40	152	23.9	6.1					63	29	45	.05	0.4	48
1 piece		1.9	0.2						198	6	1.0	.06	tr	
FRUITCAKE, LIGHT	40	156	23.0	6.6					77	27	46	.04	0.3	28
1 piece		2.4	0.3						93	6	0.6	.05	tr	
GERMAN CHOC, FRZN	50	195	23.5	10.2					183	29	68	.04	0.3	77
1 piece		2.7	0.4						71	7	0.7	.06	tr	
GINGERBREAD, HOMEMADE	50	158	26	5.3					118	34	32	.06	0.4	45
1 piece		1.9	tr						227		1.1	.05	0	
GINGERBREAD, FROM MIX	50	138	25.5	3.4					152	45	50	.01	0.4	
1 piece		1.6							137		0.8	.04		
GOLDEN, FRZN, PEPPERIDGE FARM	85	320	43.6	15.6					252	27	105	.11	1.1	85
1 piece		3.1							66	19	1.2	.07		
HONEY SPICE W/ CARAMEL ICING FROM MIX 1 piece	50	176	30.4	5.4					122	35	96	.01	0.1	80
		2.0	0.1						41		0.4	.04		
MARBLE LAYER, FRZN, BAVARIAN FOODS 1 piece	80	276	42.9	18.4						67				
		1.8									1.6			
MARBLE W/ BOILED WHITE ICING, FROM MIX 1 piece	50	165	31	4.3					129	39	85	.01	0.1	45
		2.2	tr						61		0.4	.04		
ORANGE, FRZN, SARA LEE	60	221	35.9	7.8					191	24	77	.05	0.4	124
1 piece		2.2	0.1						45	3	0.6	.05	4	
ORANGE W/ ICING	50	183	29.9	6.6					141	37	56	.01	0.4	286
1 piece		1.5	tr						79		0.1	.02	tr	
PEACH COBBLER, FRZN, CAMPBELLS	100	210	34.0	7.0					140	15	83	.09	1.1	292
1 serving		3.0	0.2						94		2.1	.07	12	
PINEAPPLE UPSIDE-DOWN, FRZN	100	300	42.0	13.0					170	31	67	.07	1.3	245
1 serving		3.0	0.2						101		1.5	.06	2	
PLAIN W/O ICING	50	182	27.9	6.9					150	32	51	.01	0.1	85
1 piece		2.2	0.1						39		0.2	.04		
PLAIN W/ UNCKD WHITE ICING	50	183	31.6	5.9					113	25	37	.01	tr	100
1 piece		1.7							30		0.1	.03		
PLAIN W/ BOILED WHITE ICING	50	176	30.9	5.2					131	24	38	.01	0.1	65
1 piece		1.9							32		0.1	.03		
PLAIN W/ CHOC ICING	50	184	29.7	6.9					114	31	52	.01	0.1	90
1 piece		2.1	0.1						57		0.3	.05		
POUND (OLD-FASHIONED)	30	142	14.1	8.8					33	6	24	.01	0.1	84
1 piece		1.7	tr						18		0.2	.03	0	
POUND (MODIFIED)	30	123	16.4	5.6					53	12	31	.01	0.1	87
1 piece		1.9	tr						23		0.2	.03	tr	

	WT	Proximate			Amino Acids				Minerals			Vitamins		
		CAL	CHO	FAT	TRY	LEU	LYS	MET	Na	Ca	P	THI	NIA	A
		PRO	FIB	PUFA	PHE	ISO	VAL	THR	K	Mg	Fe	RIB	ASC	D
	g	g	g	g	mg	mg	mg	mg	mg	mg	mg	mg	mg	iu
POUND, FRZN, MORTON	30	126	15.8	6.3					146	17	37	.01	0.1	
1 piece		1.7							99		tr	.03	tr	
SHORTCAKE	25	86	12.2	2.0						10	13	.03	0.1	85
1 piece		1.0									0.1	.03	tr	
SHORTCAKE W/ BLACKBERRIES	149	347	57.7	8.3						105	101	.04	0.4	490
1 serving		4.6									0.9	.08	30	
SHORTCAKE W/ PEACHES	150	266	42.4	6.3						31	66	.12	1.6	1109
1 serving		3.2									1.2	.13	8	
SHORTCAKE W/ RASPBERRIES	160	290	47.4	6.7						72	90	.12	1.1	389
1 serving		4.2									1.7	.16	29	
SHORTCAKE W/ STRAWBERRIES	175	399	61.2	8.9						73	84	.17	1.3	429
1 serving		4.8									2.0	.21	89	
SPICE W/O ICING	50	175		5.9						31	115	.07	0.6	60
1 piece		2.2		28.3						6	0.5	.09	0	
SPICE W/ ICING	50	176	30.4	5.4					193	35	96	.06	0.5	80
1 piece		2.0	tr						102	9	0.6	.08	0	
SPONGE	50	149	27.0	2.8					84	15	56	.03	0.1	225
1 piece		3.8	0.0						44	6	0.6	.07	tr	
SPONGE, FRZN, SARA LEE	50	178	27.3	6.7					217	25	113	.06	0.5	105
1 piece		2.7	tr						40	2	0.6	.05	tr	
SPONGE W/ STRAWBERRIES & WHIP CRM 1 serving	153	380	60.4	7.8						45	77	.04	0.3	485
		4.1									1.2	.13	45	
VANILLA, FRZN, PEPPERIDGE FARM	85	325	46.3	14.3					252	24	84	.11	1.1	85
1 piece		2.5	tr						39	6	1.2	.07		
WALNUT LAYER, FRZN, SARA LEE	85	280	30.3	16.7					191	41	133	.06	0.5	174
1 piece		3.2	0.1						52	15	0.9	.09	tr	
WHITE MOUNTAIN DEVILS FOOD, FRZN, SARA LEE 1 piece	60	249	32.9	12.4					195	32	75	.05	0.4	77
		2.3	0.1						62	8	0.6	.06	tr	
WHITE W/O ICING	50	187	27.0	8.0					161	31	45	tr	0.1	15
1 piece		2.3	tr						38	4	0.1	.04		
WHITE W/ CHOC ICING, FROM MIX	50	175	31.4	5.3					113	49	89	.01	0.1	30
1 piece		1.9	0.1						58		0.2	.04		
WHITE W/ COCNT ICING, FROM MIX	50	185	30.3	6.1					128	22	36	tr	0.1	10
1 piece		1.8	0.1						53	4	0.1	.03		
WHITE W/ UNCKD WHITE ICING	50	187	31.4	6.4					117	24	32	tr	tr	55
1 piece		1.6	0.0						49	4	tr	.03		
YELLOW W/O ICING	50	182	29.1	6.4					129	36	56	.01	0.1	75
1 piece		2.3	0.1						39	4	0.2	.04	tr	
YELLOW W/ CARAMEL ICING	50	181	30.6	5.8					113	38	51	.01	0.1	85
1 piece		2.0	tr						36	4	0.3	.04		
YELLOW W/ CHOC ICING, FROM MIX	50	163	28.8	5.6					113	45	91	.01	0.1	70
1 piece		2.0	0.1						54		0.3	.04		
YELLOW W/ CHOC ICING, HOMEMADE 1 piece	50	182	30.2	6.5					104	34	56	.01	0.1	80
		2.1	0.1						54	4	0.3	.04		
CHERRY CRISP	140	226	55.0	0.1						26		.06	0.4	
1 serving		1.2									1.5	.05		
COBBLER, PEACH	100	160	25.4	6.4					158	7	15	.03	0.9	479
1 serving		1.0	0.2						88		0.5	.07	26	
COOKIES														
ASSORTED, PKGD, COMMERCIAL	20	96	14.2	4.0					73	7	33	.01	0.1	16
1 avg		1.0	tr						13		0.1	.01		
ALMOND CHEWS, PILLSBURY	10	47	6.3	2.3					38					
1 avg		0.3												
ANIMAL CRACKERS	10	43	8.0	0.9					30	5	11	tr	tr	13
5 pieces		0.7	tr						10		tr	.01		
ANIMALS, ICED, SUNSHINE	10	49	7.1	2.2						1	6			
2 pieces		0.3									0.1			
APPLE CINN, PILLSBURY	13	55	7.5	2.7					72					
1 avg		0.3												
APPLESAUCE, SUNSHINE	18	88	12.2	3.9						3	tr			
1 cookie		0.9									0.4			
APPLESAUCE, ICED, SUNSHINE	17	84	11.6	3.7						6	12			
1 cookie		0.8									0.3			

Values are stacked two per cell: the upper label (CAL, CHO, FAT / TRY, LEU, LYS, MET / Na, Ca, P / THI, NIA, A) on the first row; the lower label (PRO, FIB, PUFA / PHE, ISO, VAL, THR / K, Mg, Fe / RIB, ASC, D) on the second row.

Item	WT g	CAL / PRO	CHO / FIB	FAT / PUFA	TRY / PHE mg	LEU / ISO mg	LYS / VAL mg	MET / THR mg	Na / K mg	Ca / Mg mg	P / Fe mg	THI / RIB mg	NIA / ASC mg	A / D iu
ARROWROOT, NABISCO	10	47	7.3	1.7					24	6	13	.04	0.2	
2 cookies		0.6							11		0.3	.03		
AUNT SALLY, ICED, SUNSHINE	23	94	19.4	1.5						7	10			
1 cookie		0.9									0.4			
BIG TREAT, SUNSHINE	38	153	26.5	4.9						11	28			
1 cookie		1.2									0.4			
BORDEAUX, PEPPERIDGE FARM	16	48	11.0	3.5					50	11	13	.03	0.2	34
2 cookies		0.8	tr						26		0.3	.02		
BUTTER FLAVORED, SUNSHINE	10	46	7.0	1.8						5	18			
2 cookies		0.7									0.2			
BUTTERCUP, KEEBLER	16	76	11.3	3.0					95	5	27			
2 cookies		0.9							32		0.1			
BUTTERSCOTCH NUT, PILLSBURY	12	55	2.7	6.9					60					
1 cookie		0.3												
BUTTER THIN, RICH	11	50	7.8	1.9					46	14	10	tr	tr	72
1 cookie		0.7	tr						7		0.1	tr	0	
CAPRI, PEPPERIDGE FARM	16	82	9.7	4.6					39	6	16	.02	0.2	
1 cookie		0.8	0.1						26	7	0.3	.01		
CHERRY COOLERS, SUNSHINE	12	58	8.7	2.1						3	11			
2 cookies		0.5									0.2			
CHOCOLATE	21	93	15.0	3.3					29	11	27	.01	0.1	34
1 avg		1.5	0.1						27	3	0.2	.02		
CHOC CHIP, HOMEMADE	11	57	6.6	3.3					38	4	11	.01	0.1	12
1 avg		0.6	tr						13	2	0.2	.01		
CHOC CHIP, COMMERCIAL	11	52	7.7	2.3					44	4	13	tr	tr	13
1 avg		0.6	tr						15	2	0.2	.01	0	
CHOC CHIP COCNT, SUNSHINE	16	82	10.0	4.4						8	22			
1 cookie		0.9									0.3			
CHOC CUSTARD CUP, SUNSHINE	14	69	9.2	3.3						10	20			
1 cookie		0.8									0.4			
CHOC FUDGE SANDWICH	14	70	9.2	3.3					69	6	15			
1 cookie		0.8							23		0.2			
CHOC SNAPS, SUNSHINE	12	53	8.8	1.8						5	15			
4 cookies		0.7									0.3			
CINN SUGAR, PILLSBURY	12	50	7.3	2.3					60					
1 cookie		0.3												
COCONUT BARS	22	109	14.1	5.4					33	16	26	.01	0.1	35
1 avg		1.4	0.1						50	3	0.3	.01	0	
COCNT MACAROONS	14	67	9.3	3.2					5	4	12	.01	0.1	0
1 cookie		0.7	0.3						65		0.1	.02	0	
CREAM LUNCH, SUNSHINE	10	45	7.4	1.5						9	19			
1 cookie		0.7									0.3			
FIG BARS	14	50	10.6	0.8					35	11	8	.01	tr	15
1 bar		0.5							28	2	0.1	.01		
GINGERSNAPS	12	50	9.6	1.1					69	9	6	tr	tr	8
3 small		0.7	tr						55	2	0.3	tr		
GOLDEN FRUIT, SUNSHINE	19	63	14.8	0.6						13	21			
1 cookie		0.8									0.6			
GRAHAM CRACKERS, CHOC-COVERED 1 cookie	13	62	8.8	3.1					53	15	27	.01	0.2	8
		0.7	0.1						42		0.3	.04	0	
HYDROX, SUNSHINE	10	48	7.0	2.2						1	7			
1 cookie		0.4									0.2			
HYDROX MINT, SUNSHINE	10	48	7.0	2.2						1	7			
1 cookie		0.4									0.2			
HYDROX VANILLA, SUNSHINE	10	50	7.0	2.3						2	8			
1 cookie		0.4												
KRISPY, SUNSHINE	12	49	8.4	1.1						4	13			
4 cookies		1.1									0.4			
KRISPY KREEM, KEEBLER	10	54	6.5	3.1					24		7			
		0.2							12		tr			
LADY FINGERS	14	50	9.0	1.1					10	6	23	.01	tr	91
1 large		1.1	tr						10	2	0.2	.02	0	
LADY JOAN, SUNSHINE	10	49	6.4	2.5						2	5			
1 cookie		0.7									tr			

	WT	CAL	CHO	FAT	TRY	LEU	LYS	MET	Na	Ca	P	THI	NIA	A
		PRO	FIB	PUFA	PHE	ISO	VAL	THR	K	Mg	Fe	RIB	ASC	D
	g	g	g	g	mg	mg	mg	mg	mg	mg	mg	mg	mg	iu
LADY JOAN, ICED, SUNSHINE	10	49	6.7	2.3						2	5			
1 cookie		0.7									tr			
LALANNE SOYA, SUNSHINE	12	60	7.2	3.4						3	13			
4 cookies		0.9									0.4			
LEMON, SUNSHINE	15	76	10.0	3.6						6	13			
1 cookie		0.9									0.3			
LEMON COOLERS, SUNSHINE	12	57	8.7	2.2						4	8			
2 cookies		0.5									0.2			
LEMON SUGAR WAFERS	10	49	7.3	2.2						2	4			
1 cookie		0.3									0.1			
LIDO, PEPPERIDGE FARM	17	90	9.9	5.3					32	tr	17	.02	0.3	
1 cookie		0.9	tr						23	5	0.3	.02		
MALLO PUFFS, SUNSHINE	17	65	12.5	1.6						2	8			
1 cookie		0.5									0.2			
MARSHMALLOW	28	114	20.2	3.7					59	6	16	.01	0.1	73
1 cookie		1.1	0.1						25	4	0.1	.02	tr	
MCDONALDLAND	63	292	45.1	10.5					328	10	51	.28	0.8	47
1 box		4.3	0.2						57	10	1.4	.23	1	
MILANO, PEPPERIDGE FARM	12	63	7.3	3.6					22	5	12	.02	0.2	
1 cookie		0.7	tr						18	4	0.2	.01		
MOLASSES	15	71	10.2	2.9					58	15	11	.03	0.2	14
1 cookie		0.8	tr						21		0.5	.02	0	
OATMEAL, NABISCO	18	80	12.2	3.2					69	7	22	.05	0.3	
1 cookie		1.1							20		0.4	.03		
OATMEAL CHOC CHIP, PILLSBURY	13	57	7.3	2.7					23					
1 cookie		0.7												
OATMEAL W/ RAISINS	14	63	10.3	2.2					23	3	14	.02	0.1	7
1 cookie		0.9	0.1						52	2	0.4	.01	tr	
OATMEAL, ICED, SUNSHINE	15	68	11.4	2.2						3	16			
1 cookie		0.8									0.3			
OATMEAL PNT BTR, SUNSHINE	16	78	10.4	3.6						4	30			
1 cookie		1.2									0.4			
ORBIT CREME SANDWICH, SUNSHINE	10	50	6.8	2.3						4	10			
1 cookie		0.4									0.2			
ORLEANS, PEPPERIDGE FARM	12	61	7.3	3.6					15	4	13	.02	0.2	
2 cookies		0.6	0.1						23	6	0.2	.01		
PEANUT	12	57	8.0	2.3					21	5	14	.01	0.3	24
1 cookie		1.2	0.1						21		0.1	.01	tr	
PEANUT BUTTER BAR	50	198	26.1	9.7					116	28	87	.06	1.8	97
1 bar		4.4	0.2						119		0.6	.06	0	
PNT BTR & CHOC CHIP, PILLSBURY	10	45	5.2	2.4					53					
1 cookie		0.6												
PNT BTR PATTIES, SUNSHINE	14	69	8.8	2.9						4	22			
2 cookies		1.6												
PECAN SANDIES, KEEBLER	10	54	5.8	3.2					33		8			
		0.5							16		tr			
PLAIN, FROM MIX/CHILLED DOUGH	12	60	7.8	3.0					66	4	9	tr	tr	8
1 avg		0.5	tr						6	2	tr	tr	0	
RAISIN	15	57	12.1	0.8					8	11	24	.01	0.1	31
1 cookie		0.7	0.1						41	2	0.3	.01	tr	
ROCHELLE, PEPPERIDGE FARM	16	82	9.7	4.6					39	7	18	.03	0.3	
1 cookie		0.8	0.1						29	8	0.3	.02		
SCOTTIES, SUNSHINE	16	80	10.2	3.8						9	22			
2 cookies		1.2									0.3			
SHORTBREAD	14	70	9.2	3.2					8	10	22	.01	tr	12
2 cookies		1.0	tr						10	2	tr	.01	0	
SOCIAL TEA, NABISCO	10	43	7.5	1.1					33	2	8	.03	0.2	
2 cookies		0.7							6		0.2	.03		
SPRINKLES, SUNSHINE	16	57	11.4	1.5						3	11			
1 cookie		0.6									0.2			
SUGAR, HOMEMADE	20	89	13.6	3.4					64	16	21	.03	0.3	22
1 cookie		1.2	tr						15		0.3	.03	tr	
SUGAR WAFERS	11	53	8.0	2.1					21	4	9	tr	0.1	15
2 wafers		0.5	tr						7	2	tr	tr	0	

	WT		Proximate			Amino Acids				Minerals			Vitamins	
		CAL	CHO	FAT	TRY	LEU	LYS	MET	Na	Ca	P	THI	NIA	A
		PRO	FIB	PUFA	PHE	ISO	VAL	THR	K	Mg	Fe	RIB	ASC	D
	g	g	g	g	mg	mg	mg	mg	mg	mg	mg	mg	mg	iu
TOY, SUNSHINE	10	41	7.4	1.5					160	1	9			
3 cookies		0.6	tr						7	4	tr		11	
VANILLA CREME SANDWICH	14	69	9.7	3.1					68	4	34	.01	0.1	0
1 cookie		0.7	tr						5	2	0.1	.01	0	
VANILLA CUSTARD CUP, SUNSHINE	14	70	9.2	3.3						12	18			
1 cookie		0.8									0.2			
VANILLA WAFERS	11	51	8.2	1.8					28	4	7	tr	tr	14
3 wafers		0.6	tr						8	2	tr	.01	0	
VANILLA DIXIE, SUNSHINE	13	58	9.9	1.7						2	9			
1 cookie		0.9									0.3			
VIENNA FINGER SANDWICH, SUNSHINE 1 cookie	15	72	10.7	3.0						6	14			
		0.7									0.3			
YUM YUM, SUNSHINE	15		10.2	3.2						9	15			
1 cookie		0.5									0.2			
CREAM PUFF W/ CUSTARD FILL	105	245	21.5	14.6					87	85	120	.04	0.1	368
1 avg		6.8	0.0						127	14	0.7	.18	0	
CUPCAKES														
BIG WHEELS, HOSTESS	39	187	22.2	10.5					121	48	23	.06	0.8	0
1 cupcake		1.2									1.1	.11	0	
CHOC W/ ICING, FROM MIX	36	129	21.3	4.5					121	47	71	.01	0.1	61
1 avg		1.6	0.1						42		0.3	.04		
CHOC (DEVILS FOOD) W/ CHOC ICING, HOSTESS 1 cupcake	50	185	30.5	6.0					282	23	85	.10	0.9	0
		1.5									1.7	.08	0	
CHOC (DEVILS FOOD) W/ CHOC ICING, STOUFFERS 1 cupcake	40	163	24.2	6.5					170	14		.01	0.1	30
		1.6									0.8	.07		
DING DONGS, HOSTESS	39	187	22.2	10.5					121	8	58	.06	0.8	0
1 cupcake		1.2									1.1	.11	0	
HO HO'S, HOSTESS	25	118	15.0	6.0					63	15	25	.03	0.5	21
1 cupcake		0.8									0.5	.05	0	
ORANGE, HOSTESS	42	164	26.5	5.9					168	20	52	.09	0.8	0
1 cupcake		1.3									1.1	.10	0	
PLAIN W/O ICING, FROM MIX	25	88	14.0	3.0					113	40	59	.03		38
1 avg		1.2	0.1						21		0.1	.05		
SUZY Q'S, HOSTESS	64	256	37.1	10.9					301	20	116	.06	2.0	0
1 cupcake		1.9									1.1	.10	0	
YELLOW W/ CHOC ICING, STOUFFERS	40	155	25.2	5.3								.04	0.2	
1 cupcake		1.6										.07		
YELLOW W/ VAN ICING, STOUFFERS	40	160	24.8	6.2					184	16		tr	0.1	96
1 cupcake		1.3									0.2	.08		
CUSTARD, BAKED	157	205	22.8	9.1					124	163	189	.08	0.1	607
3/5 cup		8.8	0.0						229	22	1.1	.32	0	
CUSTARD, BOILED	130	164	18.2	7.3					103	131	151	.07	0.1	486
1/2 cup		7.1	0.0						190	18	0.8	.25	0	
CUSTARD, BANANA	112	143	19.9	4.6					125	142				
1/2 cup		5.6							180					
CUSTARD, CHOC	112	142	19.6	4.4					153	144				
1/2 cup		5.9							186					
CUSTARD, COCNT	112	144	20.3	4.4					124	141				
1/2 cup		5.7							174					
CUSTARD, LEMON	112	143	19.8	4.6					92	142				
1/2 cup		5.6							179					
CUSTARD, VANILLA	112	143	19.7	4.6					128	142				
1/2 cup		5.6							174					
DOUGHNUTS														
CAKE TYPE, PLAIN	32	125	16.4	6.0	19	122	53	26	160	13	61	.05	0.4	26
1 avg		1.5	tr	0.3	82	77	76	51	29	7	0.4	.05	tr	
CAKE TYPE W/ SUGAR ICING	37	151	21.7	6.5	27	171	75	36		12	26	.07	0.5	41
1 avg		2.1	tr	0.3	115	108	107	72		8	0.6	.06	tr	
CINN, HOSTESS	27	105	13.8	5.4					105	9		.04	0.4	3
1 doughnut		0.8									0.3	.03		
CREAM FILLED	35	122	16.2	4.7	24	149	65	31		15	23	.06	0.5	65
1 avg		1.8	0.0	0.2	100	94	93	62		9	0.4	.05	0	

	WT	CAL	CHO	FAT	TRY	LEU	LYS	MET	Na	Ca	P	THI	NIA	A
		PRO	FIB	PUFA	PHE	ISO	VAL	THR	K	Mg	Fe	RIB	ASC	D
	g	g	g	g	mg	mg	mg	mg	mg	mg	mg	mg	mg	iu
HONEY BUNS, FRZN, MORTON	64	170	24.8	6.5					56	3	16	.03	0.6	0
1 doughnut		3.5									0.2	.06	tr	
PLAIN, HOSTESS	27	111	12.2	6.8					128	12		.05	0.5	3
1 doughnut		1.1									0.4	.04	0	
POWDERED	30	154	17.3	8.9					71	11	23	.02	0.1	34
1 avg		1.6									0.2	.02	tr	
RAISED/YEAST	30	124	11.3	8.0	24	154	67	32	70	11	23	.05	0.4	18
1 avg		1.9	0.1	0.3	104	97	96	65	24	6	0.5	.05	0	
RAISED, JELLY CENTER	65	226	30.0	8.8	44	276	120	58		28	42	.12	0.9	121
1 avg		3.4	tr	0.3	186	174	172	116		16	0.8	.10	tr	
SUGAR & SPICE	30	153	16.2	9.5					64	23	21	.02	0.1	31
1 avg		1.4									0.2	.02	tr	
ECLAIR, CHOC, FRZN, RICHS	78	246	31.4	11.9					195	10	46	.02	0.1	
1 eclair		3.1	0.0						33	11	0.5	.06		
ECLAIR W/ CUSTARD FILL, CHOC	110	316	39.1	15.4						90	150	.12	1.0	730
ICING 1 avg		7.6	0.0								1.3	.24	0	
ECLAIR W/ WHIP CREAM FILL	105	296	15.0	25.7						48	70	.07	0.3	1120
1 avg		3.9	0.0								0.7	.12	0	
GELATIN DESSERT, ALL FLAVORS	120	97	23.5	tr					55	2	5	.01	0.2	
½ cup		1.7	0.0						91		0.1	.01	1	
GELATIN DESSERT W/ FRUIT	125	93	14.4	3.8					43	15	16	.03	0.2	260
½ cup		1.4	0.4								0.4	.03	11	
GELATIN DESSERT, LOW-CAL, D-ZERTA	120	8	0.0	0.0					8	0	0	.00	0.0	0
½ cup		2.0	0.0						50		0.0	.00	0	
GELATIN DESSERT, 1,2,3 SELF-LAYERING,	114	102	19.0	2.5					33	0	0	.00	0.0	0
JELLO ½ cup		1.1							1		0.0	.00	0	
ICE, LIME	185	144	60.3	0.0						2	4	.01	tr	0
1 cup		0.7							6				2	0
ICE, ORANGE	123	96	40.0	tr					tr	tr	tr	tr	tr	0
⅙ quart		0.5	tr						4		tr	tr	1	
ICE CREAM														
CHOCOLATE	133	295	32.8	16.0					75	186	168	.07	0.2	569
1 cup		5.0								18		.27		
FRENCH FRZN CUSTARD	133	257	27.7	14.4	83	585	462	139	84	194	153	.05	0.1	585
1 cup		6.0	0.0		283	378	408	271	241	10	0.1	.28	1	
FRENCH VANILLA, SOFT-SERVE	173	377	38.2	22.5	100	680	548	177	154	235	199	.08	0.2	794
1 cup		7.0	0.0	1.0	337	422	466	325	337	24	0.4	.45	1	
REGULAR (10% FAT)	133	257	31.7	14.1	68	471	380	120	116	176	134	.05	0.1	543
1 cup		4.8	0.0	0.5	231	290	322	217	257	19	0.1	.28	1	
REGULAR (12% FAT)	133	278	27.4	16.4	75	527	415	127	53	164	132	.05	0.1	692
1 cup		5.3	0.0	0.6	255	340	367	245	149	19	0.1	.25	1	
RICH (16% FAT)	148	349	32.0	23.8	53	374	327	104	108	151	115	.04	0.1	897
1 cup		3.8	0.0	0.9	200	241	262	175	221	16	0.1	.16	1	
STRAWBERRY	133	250	31.3	12.0					59	146	124	.05	0.1	503
1 cup		4.3								19		.20		
VANILLA	133	290	30.2	16.2					82	208	156	.08	0.2	681
1 cup		5.7	0.0	0.3						19		.30	1	
VANILLA, SOFT-SERVE	227	328	67.4	9.1						342				
1 cup		10.5												
ICE CREAM BARS														
CHOC COATED VANILLA	60	162	14.5	10.6					28	70	52	.02	0.1	209
1 bar		2.1							107	8		.10		
CREAMSICLE, SEALTEST	66	103	17.6	3.1					27	46	37	.02	0.3	125
1 bar		1.2							82	5	0.0	.08		0
DRUMSTICK	60	186	21.5	9.9					57	67	59	.02	0.5	185
1 avg		2.6							99	7	0.1	.09		0
FUDGESICLE	73	91	18.6	0.2					55	129	99	.03	0.7	
1 avg		3.8							173	14	0.1	.18		0
SANDWICH	62	167	26.1	6.2					92	73	72	.02	0.5	193
1 avg		3.1		0.3					102	7	0.1	.10		2
ICE CREAM CONE	12	4.5	9.3	0.3					28	19	24	.01	0.1	tr
1 cone		1.2	tr						29		0.1	.02	tr	

	WT	Proximate			Amino Acids				Minerals			Vitamins		
		CAL	CHO	FAT	TRY	LEU	LYS	MET	Na	Ca	P	THI	NIA	A
		PRO	FIB	PUFA	PHE	ISO	VAL	THR	K	Mg	Fe	RIB	ASC	D
	g	g	g	g	mg	mg	mg	mg	mg	mg	mg	mg	mg	iu
ICE CREAM SODA	345	262	48.7	7.2						69	55	.03	0.1	297
1 cup		2.3									0.0	.10	1	0
ICE MILK, CHOC	90	137	20.2	4.6	50	347	281	89	61	140	111	.04	0.1	190
⅙ quart		4.3	0.0	0.1	171	214	238	160	175	12	0.1	.20	1	
ICE MILK, CHOC-COVERED BAR	60	144	16.0	7.6					38	98	73	.04	0.1	83
1 bar		2.8							126	8		.14		
ICE MILK, STRAWBERRY	90	133	22.1	3.1					64	161	121	.06	0.1	129
⅙ quart		4.3								12		.23		
ICE MILK, VANILLA	90	136	20.8	7.6					75	189	142	.07	0.1	151
⅙ quart		5.1		0.2						12		.27		
ICE MILK, VAN, SOFT SERVE	87	111	19.2	2.3	113	787	637	201	163	274	202	.12	0.2	175
½ cup		4.0	0.0	0.2	388	486	538	363	412	29	0.3	.54	1	
JUNKET, CHOC (W/ SKIM MILK)	135	95	17.8	0.5					63	163	128	.08	0.7	
½ cup		4.5	0.1						186	19	0.3	.23		1
JUNKET, CHOC (W/ WHOLE MILK)	135	127	17.4	4.5					59	159	124	.07	0.7	169
½ cup		4.3	0.1	0.0					185	19	0.3	.22		1
JUNKET, FRUIT & VAN FLAVORS (W/ SKIM MILK) ½ cup	116	88	17.1	0.1					58	143	106	.05	0.3	1
		4.9	0.1						159	16	0.2	.19	1	1
JUNKET, FRUIT & VAN FLAVORS (W/ WHOLE MILK) ½ cup	135	140	19.4	4.5					65	161	119	.04	0.3	169
		5.5	0.1	0.0					182	19	0.3	.22		1
LADY FINGER W/ WHIP CREAM FILL	114	326	38.2	16.6					49	52	104	.03	0.1	844
1 avg		6.7									0.8	.11		
PEACH CRISP	100	172	25.3	6.4					162	7	15	.03	0.9	479
		4.0	0.2						88	1	0.5	.04	26	

PIE

	WT	CAL	CHO	FAT	TRY	LEU	LYS	MET	Na	Ca	P	THI	NIA	A
		PRO	FIB	PUFA	PHE	ISO	VAL	THR	K	Mg	Fe	RIB	ASC	D
APPLE, HOMEMADE	160	410	61.0	17.8					482	1	35	.03	0.6	48
1 piece		3.4	0.6						128	10	0.5	.03	2	
APPLE, FRZN	160	432	60.8	19.2					352	13	40	.14	1.1	16
1 piece		3.2	0.6						76		1.4	.10	5	
APPLE, HOSTESS	127	419	52.1	22.9					406	20	48	.23	1.7	0
1 piece		2.5									0.8	.10	2	
APPLE, MCDONALDS	90	295	30.5	18.3					408	12	23	.02	1.3	68
1 pie		2.2	0.5						38	7	0.6	.03	3	5
APPLE DUTCH BAVARIAN CREAM, FRZN, SARA LEE 1 piece	100	240	36.7	9.8					280	46	47	.05	0.3	170
		1.9	0.2						107	13	0.6	.07	16	
BANANA CREAM, FRZN, MORTON	100	260	34.0	14.0					150	25	42	.02	0.2	20
1 piece		2.0									0.3	.08	0	
BANANA CUSTARD, HOMEMADE	160	353	49.2	14.8					310	106	131	.06	0.5	400
1 piece		7.2	0.3						325	21	0.8	.20	2	
BLACKBERRY, HOMEMADE	160	389	55.0	17.6					429	30	42	.03	0.5	144
1 piece		4.2	3.0						160	10	0.8	.03	6	
BLUEBERRY, HOMEMADE	160	387	56.0	17.3					429	18	37	.03	0.5	48
1 piece		3.8	1.1						104	10	1.0	.03	5	
BLUEBERRY, FRZN, MORTON	100	270	37.0	12.0					220	10	26	.09	0.8	20
1 piece		2.0	0.7						53		1.0	.07	2	
BUTTERSCOTCH, HOMEMADE	160	427	61.3	17.6					342	120	130	.05	0.3	416
1 piece		7.0	tr						152	21	1.4	.16	tr	
BUTTERSCOTCH CREAM	100	264	29.0	15.3					139	69	69	.06	0.9	319
1 piece		3.1	tr						124		1.0	.11	tr	
CHERRY, HOMEMADE	160	418	61.5	18.1					486	22	40	.03	0.8	705
1 piece		4.2	0.2						168	10	0.5	.03	tr	
CHERRY, FRZN	100	29	44.4	12.0					229	12	23	.02	0.2	290
1 piece		2.2	0.1						82		0.3	.02	2	
CHERRY, HOSTESS	127	445	57.2	21.6					406	20	42	.23	1.7	
1 piece		3.8									0.8	.10	2	
CHERRY, MCDONALDS	92	296	32.4	17.6					454	12	23	.02	0.4	212
1 pie		2.2	0.1						56.5	7	0.4	.03	1	5
CHOC BAVARIAN CREAM, FRZN SARA LEE 1 piece	85	299	25.6	21.3					68	53	67	.03	0.3	701
		2.5							119	16	0.6	.09	1	
CHOC CHIFFON, HOMEMADE	80	262	35.0	12.2					201	19	77	.03	0.1	248
1 piece		5.4	0.1						88		0.9	.08	0	
CHOC CHIP BAVARIAN CREAM, FRZN, SARA LEE 1 piece	80	246	21.6	17.0					68	58	56	.03	0.2	574
		2.4	0.2						83	18	0.4	.10	6	

	WT g	CAL / PRO	CHO / FIB	FAT / PUFA	TRY / PHE	LEU / ISO	LYS / VAL	MET / THR	Na / K	Ca / Mg	P / Fe	THI / RIB	NIA / ASC	A / D
				g	mg	mg	mg	mg	mg	mg	mg	mg	mg	iu
CHOC MERINGUE, HOMEMADE	50	378	50.0	18.0					385	103	147	.05	0.3	286
1 piece		7.2	0.3						209		1.0	.18	tr	
COCONUT CREAM, FRZN, MORTON	100	290	37.0	15.0					150	23	44	.02	0.1	5
1 piece		2.0	0.2						90		0.3	.07	0	
COCNT CUSTARD, HOMEMADE	155	365	38.5	19.4					284	145	180	.09	47	357
1 piece		9.4	0.2						253	20	1.1	.29	0	
COCNT CUSTARD, FROM MIX	100	203	29.1	7.9					235	93	103	.03	0.2	210
1 piece		4.3	0.3						154		0.4	.14		
COCNT CUSTARD, FRZN	100	200	26.0	9.0					250	55	59	.06	0.4	165
1 piece		4.0	0.2						172		0.6	.12	0	
CUSTARD, HOMEMADE	150	327	35.1	16.6					430	144	170	.08	0.4	345
1 piece		9.2	tr						205		0.9	.24	0	
DUTCH APPLE BAVARIAN CREAM, FRZN, SARA LEE 1 piece	100	240	36.7	9.8					280	46	47	.05	0.3	170
		1.9	0.2						107	13	0.6	.07	16	
LEMON, ARTHUR TREACHERS	85	275	35.0	13.9					314	10	60	.18	1.5	
1 serving		2.6		1.9					43	8	0.9	.09	1	4
LEMON, HOSTESS	127	432	57.2	21.6					406	20	48	.15	1.1	0
1 piece		2.5									0.4	.10	1	
LEMON CHIFFON, HOMEMADE	107	335	46.9	13.5					279	25	89	.03	0.2	182
1 piece		7.5	tr						87		1.0	.08	3	
LEMON CHIFFON, FROM PUDDING MIX 1 piece	100	288	45.0	9.6					261	23	83	.03	0.2	170
		2.5							81		0.9	.08	3	
LEMON CREAM, FRZN, MORTON	100	270	37.0	14.0					160	22	40	.02	0.1	5
1 piece		2.0									0.3	.07	0	
LEMON MERINGUE, HOMEMADE	140	357	52.8	14.3					395	20	69	.05	0.3	238
1 piece		5.2	tr						70		0.7	.11	4	
LEMON MERINGUE, FROM PUDDING MIX 1 piece	100	227	36.4	7.5					282	14	49	.03	.02	170
		2.7							50		0.5	.08	3	
LIME BAVARIAN CREAM, SARA LEE	100	268	27.9	17.1					44	30	33	.03	0.2	590
1 piece		1.5	0.2						45	11	0.4	.05	tr	
MINCE, HOMEMADE	160	434	66.0	18.4					716	45	61	.11	0.6	tr
1 piece		4.0	0.6						285	29	1.6	.06	2	
MINCE, FRZN, MORTON	100	270	40.0	12.0					310	11	27	.09	0.7	5
1 piece		2.0									1.0	.06	1	
NEOPOLITAN, FRZN, MORTON	100	280	38.0	15.0					160	24	47	.02	0.1	5
1 piece		2.0									0.4	.07	0	
PEACH, HOMEMADE	165	421	63.0	17.7					246	16	48	.03	1.2	1200
1 piece		4.0	0.7							10	0.8	.07	5	
PEACH, FRZN, MORTON	100	260	35.0	12.0					230	8	27	.09	0.9	195
1 piece		2.0									0.9	.06	1	
PECAN, HOMEMADE	80	334	41.0	18.3					177	37	82	.12	0.2	128
1 piece		4.1	0.4						98		2.2	.05	tr	
PECAN, FRZN, MORTON	100	365	52.6	16.4					339	38	74	.12	0.3	222
1 piece		4.0									1.9	.06	tr	
PINEAPPLE, HOMEMADE	160	404	61.0	17.1					434	21	34	.06	0.6	32
1 piece		3.5	0.3						115		0.8	.03	2	
PINEAPPLE CHEESE, HOMEMADE	160	270	39.9	9.5						81	107	.08	0.2	286
1 piece		7.1	0.1								0.9	.16	1	
PINEAPPLE CHIFFON, HOMEMADE	107	308	41.8	12.9					274	26	81	.04	0.4	374
1 piece		7.0	0.1						105		1.0	.09	1	
PINEAPPLE CUSTARD, HOMEMADE	150	330	48.1	13.0					279	75	98	.06	0.6	270
1 piece		6.0	0.2						145	20	0.6	.14	2	
PUMPKIN, HOMEMADE	150	317	36.7	16.8					321	76	104	.04	0.8	3700
1 piece		6.0	0.8						240	20	0.8	.15	tr	
PUMPKIN, FRZN, MORTON	100	210	30.0	8.0					270	69	59	.08	0.6	2150
1 piece		4.0	0.5						133		0.7	.15	2	
PUMPKIN BAVARIAN CREAM, FRZN, SARA LEE 1 piece	100	260	24.1	17.8					118	47	42	.04	0.3	2392
		1.9	0.2						138	14	0.7	.08	2	
RAISIN, HOMEMADE	120	325	32.5	12.9					342	22	48	.04	0.4	tr
1 piece		3.1	0.4						231		1.1	.04	1	
RHUBARB, HOMEMADE	160	405	61.2	17.1					432	102	42	.03	0.5	80
1 piece		4.0	1.0						254		1.1	.06	5	
SHOO-FLY, PENNSYLVANIA DUTCH HOMEMADE 1 piece	110	441	70.6	16.4						129	65	.11	0.9	411
		4.0	0.1								3.2	.07	0	

	WT	Proximate			Amino Acids				Minerals			Vitamins		
		CAL	CHO	FAT	TRY	LEU	LYS	MET	Na	Ca	P	THI	NIA	A
		PRO	FIB	PUFA	PHE	ISO	VAL	THR	K	Mg	Fe	RIB	ASC	D
	g	g	g	g	mg	mg	mg	mg	mg	mg	mg	mg	mg	iu
STRAWBERRY, HOMEMADE	115	228	35.6	9.1					227	18	29	.02	0.4	46
1 piece		2.2	0.9						138		0.8	.05	2.9	
STRAWBERRY, FRZN, MORTON	100	269	39.1	11.6					246	13	30	.03	0.5	13
1 piece		2.4									0.6	.04	23	
STRAWBERRY W/ WHIP CREAM	145	378	56.2	16.8						47	52	.03	0.3	485
HOMEMADE 1 piece		2.9	0.9								0.7	.05	22	
STRAWBERRY BAVARIAN CREAM,	100	276	32.4	16.0					68	47	44	.04	0.3	575
FRZN, SARA LEE 1 piece		2.0	0.3						86	15	0.5	.09	13	
STRAWBERRY CREAM, FRZN,	100	270	37.0	13.0					160	24	39	.02	0.1	5
MORTON 1 piece		2.0									0.4	.07	6	
SWEET POTATO, HOMEMADE	160	342	37.8	18.2					349	110	134	.08	0.5	3840
1 piece		7.2	0.3						261		0.8	.19	6	
PIECRUST														
CRUMB	165	866	64.1	64.4						113	126	.24	2.8	2460
1 crust		10.9	0.3								2.3	.20	0	
PASTRY, ENR	135	675	59.1	45.2					825	19	67	.27	2.4	0
1 crust		8.2	0.3						67	19	2.3	.19	0	
PASTRY, UNENR	135	675	59.1	45.2					825	19	67	.04	0.7	
1 crust		8.2	0.3						67	19	0.7	.04	0	
PASTRY, UNENR	270	1350	118.2	90.4					1650	38	134	.08	0.7	0
1 double crust		16.4	0.6						134		1.4	.08	0	
PASTRY, FROM MIX	135	626	59.4	39.3					1098	55		.04	0.7	0
1 crust		8.6	0.3						76		0.5	.04	0	
GRAHAM CRACKER, HOMEMADE	32	159	17.5	10.0	1	9	7	2	184	8	24	.01	0.2	326
1 serving		1.2	0.2		4	5	6	4	57	3	0.2	.03	0	0
STICKS, PILLSBURY	28	153	14.3	9.5					232					
1 oz		2.1												
POPSICLES														
POPSICLE	88	65	16.7	0.0						0		.00	0.0	0
1 avg		0.0										.00	0	
POPSICLE, KOOL-POPS, ORANGE	33	24	6.3	0.0					9					
1 bar		0.0												
POPSICLE, TWIN POP	128	95	23.7	tr										
1 bar		tr												
POPTARTS														
BLUEBERRY, FROSTED	50	197	34.1	5.6					70		34	.14	2.0	482
1 poptart		2.5	1.1						26	7	1.8	.16		
CHOC FUDGE	50	200	33.5	6.0					249	97	116	.15	1.9	482
1 poptart		3.5	0.4						107	16	1.7	.17		
DUTCH APPLE	50	200	35.0	5.5					232	97	97	.15	1.9	482
1 poptart		0.3	0.2						83	8	1.7	.17		
STRAWBERRY	50	205	35.0	6.0					218	97	97	.15	1.9	482
1 poptart		3.0	0.1						72	8	1.7	.17		
PRUNE WHIP	80	125	29.5	0.2					131	18	26	.02	0.4	368
1 serving		3.5	0.5						232		1.0	.11	2	
PUDDING														
BANANA, CND, DEL MONTE	100	125	21.2	3.7					195	84	70	.02	1.6	
3½ oz		2.4							110	8	0.1	.17		
BANANA, RICHS	84	142	19.2	6.4					117	56				
3 oz		1.7							76					
BANANA CREAM, FROM MIX	145	180	27.8	6.1					280					
½ cup		4.4							240					
BREAD W/ RAISINS, HOMEMADE	165	314	47.8	10.0						191	194	.15	0.9	457
¾ cup		8.9	0.2								1.7	.31		
BUTTERSCOTCH, CORNSTARCH BASE	140	207	37.4	4.7						165	124	.04	0.1	194
½ cup		4.3	0.0								0.8	.21	0	
BUTTERSCOTCH (W/ SKIM MILK)	130	117	25.0	0.2					644	153	116	.05	0.1	226
½ cup		3.6	tr						239	25	0.6	.16	1	45
BUTTERSCOTCH, CND, DEL MONTE	100	124	23.0	3.6					171	89	68	.02	0.1	2
3½ oz		2.5	0.0						118	9	0.1	.12	1	

	WT	CAL	CHO	FAT	TRY	LEU	LYS	MET	Na	Ca	P	THI	NIA	A
	g	PRO	FIB	PUFA	PHE	ISO	VAL	THR	K	Mg	Fe	RIB	ASC	D (iu)
BUTTERSCOTCH, COOL 'N CREAMY	120	170	30.0	5.0					185					
½ cup		2.0							188					
BUTTERSCOTCH, RICH'S	128	200	27.4	8.8	26	192	179	51	192	38	70	.03	0.0	129
4.5 oz		2.6	0.0	8.5	90	128	154	90	128	7	0.3	.13		
BUTTERSCOTCH, SWISS MISS	84	127	16.0	6.1					101					
3 oz		2.1							34					
CHOCOLATE, CORNSTARCH BASE	144	219	37.1	6.6					81	147	129	.04	0.2	196
½ cup		4.5	0.1						246		0.2	.22	0	
CHOC, FROM MIX	130	163	31.7	3.9					161	187	118	.04	0.1	169
½ cup		3.9	0.1						168	18	0.7	.20		
CHOC, CND, DEL MONTE	100	131	23.3	4.0					160	95	74	.02	0.1	2
3½ oz		3.0	0.0						130	23	0.1	.13	1	
CHOC, LIGHT, COOL 'N CREAMY	120	202	28.0	8.8					148					
½ cup		3.6							212					
CHOC, LIGHT, RICH'S	126	213	26.7	10.5					202	33	78	.03	0.3	
4.5 oz		2.5	0.0						164	19	1.0	.11		
CHOC, LIGHT, SWISS MISS	84	129	16.0	6.3					76					
3 oz		2.3							34					
CHOC, DARK, COOL 'N CREAMY	120	190	31.0	5.0					222					
½ cup		3.0							258					
CHOC, DARK, RICH'S	128	213	27.1	10.6	26	179	166	51	177	76	79	.03	0.7	
4.5 oz		2.7	0.0	2.0	90	128	141	90	176	19	1.0	.12		
CHOC, DARK, SWISS MISS	84	137	17.6	6.3					76					
3 oz		2.6							34					
CHOC, DIET, SEGO	240	250	39.0	6.0										
1 serving		11.0												
CHOCOLATE FUDGE	100	135	22.3	4.0					135	100	84	.03	0.7	
3½ oz		2.4	0.1						106		0.6	.19		
CHOC FUDGE, CND, DEL MONTE	100	123	22.7	4.0					168	79	78	.02	1.0	3
3½ oz		2.8	0.0						119	26	0.4	.12	0	
CHOC NUT, CND, DEL MONTE	100	128	21.3	3.9					170	85	78	.02	0.4	
3½ oz		2.7							174	19	0.5	.14		
CHOC W/ WHIP TOP, RICH'S	100	194	22.5	10.6					129	49		.08	0.3	
3½ oz		2.0							105			.18		
D-ZERTA LOW-CAL (W/ SKIM MILK)	130	76	14.1	0.0					146	159	126	.07	0.1	1
½ cup		5.5							240		0.0	.26	1	
INDIAN, BAKED	158	161	22.6	5.6						221	151	.08	0.4	395
⅔ cup		5.4									1.4	.26	0	
LEMON, FROM MIX	100	153	31.4	2.9					100	7	12	.03	0.4	
3½ oz		0.3							8		0.7	.10		
LEMON, CND, JELLO	113	153	31.6	2.9					141	7	22	.02	0.1	20
½ cup		0.0							10		0.6	.07	1	
LEMON, RICH'S	84	134	20.9	5.4					113	8				
3 oz		0.3							13					
LEMON, SPONGE OR SNOW, HOMEMADE	130	114	26.5	tr						4	5	.01	tr	0
1 serving		3.1	0.0								0.1	.02	10	
LEMON SPONGE OR SNOW W/ CUSTARD SCE, HMDE	225	249	41.3	6.0						129	136	.06	0.1	370
1 serving		8.9	0.0								0.7	.24	10	
RICE, CND, JELLO	130	185	36.8	3.1					306	59	51	.04	0.1	21
½ cup		2.3	0.0						79		0.5	.13	1	
RICE, FRZN, CAMPBELLS	100	160	28.0	4.0					120	82	76	.05	0.3	10
1 serving		3.0	0.0						116		0.4	.07	0	
RICE W/ RAISINS, HOMEMADE	145	212	38.7	4.5					103	142	136	.04	0.3	160
¾ cup		5.2	0.1						257		0.6	.20	tr	
RICE PINEAPPLE	130	217	430	5.1					15	10	12	.08	1.2	24
½ cup		0.9	0.2						55		0.4	.01	3	
STRAWBERRY, CND, DEL MONTE	100	115	2.2	2.9					176	30	52		0.3	
3½ oz		0.7							54	3	0.2	.04	23	
SUSTACAL, ALL FLAVORS, MEAD JOHNSON	140	239	31.8	9.5					119	218	218	.23	3.0	746
1 serving		6.8							288	60	2.7	.26	9	60
TAPIOCA, HOMEMADE	105	133	17.3	5.0						105	109	.05	0.1	313
½ cup		4.9									0.5	.19	1	
TAPIOCA, FROM MIX	127	180	29.6	5.3					180	150	138	.05	0.1	332
½ cup		4.2							180		0.3	.25	1	

| | WT | CAL | CHO | FAT | TRY | LEU | LYS | MET | Na | Ca | P | THI | NIA | A |
| | | PRO | FIB | PUFA | PHE | ISO | VAL | THR | K | Mg | Fe | RIB | ASC | D |
	g	g	g	g	mg	mg	mg	mg	mg	mg	mg	mg	mg	iu
TAPIOCA, CND, GENERAL FOODS	100	129	23.6	3.1					185	66	75	.02	0.1	16
3½ oz		1.5	0.0						74	8	1.0	.12	1	
TAPIOCA, APPLE	122	143	35.8	0.1					62	4	5	tr	tr	12
½ cup		0.2	0.1						32		0.2	tr	tr	
TAPIOCA, CHOCOLATE	132	181	28.0	5.3						151	136	.04	0.6	195
½ cup		5.2									0.6	.21	0	
TAPIOCA, CREAM	130	174	22.2	0.7					203	137	142	.05	0.1	376
½ cup		6.5	0.0						176		0.5	.23	1	
VANILLA (BLANC MANGE)	125	139	19.9	4.9					81	146	114	.04	0.1	200
½ cup		4.4	tr						173		tr	.20	1	
VANILLA, CORNSTARCH BASE	125	152	23.8	4.7						144	113	.04	0.1	195
½ cup		4.2	tr								0.1	.21	0	
VAN, FROM MIX (W/ WHOLE MIX)	148	175	30.5	4.1	58	399	326	105	251	149	108	.05	0.6	153
½ cup		4.1	0.0	0.2	193	242	267	225	186	tr	0.2	.22	2	50
VAN, FROM MIX (W/ SKIM MIX)	148	147	31.2	0.3	61	427	338	107	258	157	127	.06	0.6	254
½ cup		4.1	0.0	tr	211	256	283	198	207	14	0.2	.19	2	51
VAN, CND, DEL MONTE	100	124	22.8	3.8					171	89	68	.02	0.1	2
3½ oz		2.5	0.0						118	8	0.1	.12	1	
VANILLA, COOL 'N CREAMY	120	170	28.0	5.0					186					
½ cup		3.0							187					
VANILLA, RICH'S	128	195	27.5	8.8	26	192	179	51	127	38	73	.03	tr	129
4½ oz		2.4	0.0	0.4	90	128	154	90	38	8	0.3	.12		
VANILLA, SWISS MISS	84	119	14.3	6.1					101					
3 oz		2.1							34					
VANILLA & BUTTERSCOTCH, FROM	112	143	22.8	4.1					208	124				
MIX ½ cup		tr							160					
VANILLA TAPIOCA, FROM MIX	145	180	30.0	5.1					130					
½ cup		4.5							230					
RENNIN DESSERT														
HOMEMADE	127	113	14.7	4.4					104	141	105	.04	0.1	178
½ cup		3.9	0.0						160			.19	1	
CHOCOLATE, FROM MIX	127	130	17.9	4.8					66	155	122	.04	0.1	178
½ cup		4.3	0.1						159			.19	1	
RASPBERRY, STRAWBERRY, FROM	135	128	17.3	4.9					62	163	113	.04	0.1	203
MIX ½ cup		4.3	0.0						173			.22	1	
VAN, CARAMEL OR FRUIT FLAVOR,	125	119	16.0	4.5					58	146	115	.04	0.1	188
FROM MIX ½ cup		4.0	0.0						160			.20	1	
SHERBET														
SHERBET, LEMON, HOMEMADE	135	241	44.9	5.4						168	135	.04	0.1	274
½ cup		5.0								19	0.1	.25	5	
SHERBET, ORANGE	100	134	30.8	2.0	16	110	89	28	46	54	38	.01	0.1	96
3½ oz		0.9	0.0	tr	54	68	75	51	103	7	0.1	.05	2	
SHERBET, VARIOUS FLAVORS	96	118	28.8	0.0						48	38	.02	0.0	0
½ cup		1.4									0.0	.07	0	
SNOBALLS, HOSTESS	42	160	26.9	5.0					155	20	47	.06	0.8	0
1 snoball		1.3									1.1	.07	0	
TURNOVERS														
APPLE, PEPPERIDGE FARM	79	296	29.2	19.0					217	2	18	.16	0.6	8
1 turnover		3.2	0.2						55		1.1	.06	2	
APPLE & CHERRY, PILLSBURY	50	150	23.0	6.0					300					
1 turnover		1.0												
BLUEBERRY, PEPPERIDGE FARM	78	316	32.0	19.5					261	3	20	.16	0.6	8
1 turnover		3.9	0.3						34		0.9	.09	4	
BLUEBERRY, PILLSBURY	50	150	23.0	6.0					305					
1 turnover		2.0												
CHERRY, PEPPERIDGE FARM	78	293	28.1	18.7					211	4	20	.16	0.6	240
1 turnover		3.9	0.1						47		1.2	.06	0	
LEMON, PEPPERIDGE FARM	78	324	32.8	18.3					277	1	19	.17	0.6	8
1 turnover		3.9	0.1						31	6	1.1	.11	3	
PEACH, PEPPERIDGE FARM	78	296	31.2	16.6					250	3	21	.21	0.8	362
1 turnover		3.9	0.2						63	7	1.1	.05	11	

	WT	CAL	CHO	FAT	TRY	LEU	LYS	MET	Na	Ca	P	THI	NIA	A
		PRO	FIB	PUFA	PHE	ISO	VAL	THR	K	Mg	Fe	RIB	ASC	D
	g	g	g	g	mg	mg	mg	mg	mg	mg	mg	mg	mg	iu
RASPBERRY, PEPPERIDGE FARM	79	320	33.2	20.0					218	1	21	.19	0.6	8
1 turnover		4.0	0.5						48	9	1.4	.05	2	
STRAWBERRY, PEPPERIDGE FARM	78	308	34.3	17.9					246	4	21	.18	0.6	8
1 turnover		3.9	0.2						45		1.2	.09	10	
TWINKIES, HOSTESS	39	152	22.6	6.2					203	18	108	.06	0.7	0
1 twinkie		1.2									1.0	.09	0	
WHIP 'N CHILL, CHOC	85	71	11.2	2.0					61					
½ cup		2.0							87					
WHIP 'N CHILL, VAN	85	71	11.2	2.0					20	52	42	.02	0.0	47
½ cup		2.0							26		0.0	.11	0	
YOGURT, FRZN														
FRUITED VARIETIES	113	108	20.9	1.0						100	100	.01	0.0	0
½ cup		3.5								12	0.0	.13	tr	
LITE-LINE, BORDEN	100	97	19.5	0.9										
3½ oz		3.5												
RASPBERRY BAR, CHOC-COATED	70	127	14.7	6.9						98	59	.00	0.0	0
1 bar		2.0									0.0	.07	0	
SOFT-SERVE, BISON	100	133	27.3	1.5						130	110	.00	0.0	0
3½ oz		3.9									0.0	.15	0	
STRAWBERRY BAR	70	69	12.7	1.0						78	59	.00	0.0	0
1 bar		2.0									0.0	.07		

DESSERT SAUCES, SYRUPS, AND TOPPINGS

	WT	CAL	CHO	FAT	TRY	LEU	LYS	MET	Na	Ca	P	THI	NIA	A
		PRO	FIB	PUFA	PHE	ISO	VAL	THR	K	Mg	Fe	RIB	ASC	D
BUTTERSCOTCH SCE	44	203	40.5	7.2						41	23	tr	tr	296
2 T		0.5									1.4	.02	tr	
BUTTERSCOTCH TOP, SMUCKERS	50	156	39.5	0.1					111	24	23	.00	0.0	1
		0.7							34	3	0.1	.04	0	
CARAMEL TOP, SMUCKERS	50	155	39.0	tr					152	28	23	.00	0.0	1
		0.8							33	3	0.1	.05	0	
CHERRY TOP, SMUCKERS	50	147	37.5	0.1					17	14	8	tr	tr	0
		0.2							380	2	0.1	.01	tr	
CHOCOLATE SAUCE	39	87	4.1	4.5						40	45	.01	0.1	109
2 T		1.4									0.2	.06	tr	
CHOCOLATE SAUCE	78	173	8.2	8.9						80	90	.02	0.2	217
¼ cup		2.7									0.4	.11	tr	
CHOC SYRUP, FUDGE TYPE	40	132	21.6	5.5					36	51	64	.02	0.2	60
2 T		2.0	0.2						114		0.5	.09		
CHOC SYRUP, THIN TYPE	40	98	25.1	0.8					21	7	37	.01	0.2	
2 T		0.9	0.2						113	25	0.6	.03	0	
CHOC SYRUP, HERSHEY	40	106	23.0	0.5					28	6	52	.01	0.2	8
2 T		1.3	0.2						68		0.7	.02	tr	
CHOC SYRUP, SMUCKERS	40	117	25.6	1.5					29	12	27	.00	0.1	1
2 T		1.1							56	15	0.4	.02		
CHOC FUDGE TOP, SMUCKERS	50	151	36.0	0.6					69	29	36	.00	tr	2
		1.4							67	13	0.4	.05	0	
CHOC MINT TOP, SMUCKERS	50	151	36.0	0.7					75	30	.26	.00	0.0	tr
		1.4							44	6	0.2	.05	0	
CUSTARD SAUCE	72	85	9.3	3.8						78	82	.04	0.1	235
¼ cup		3.7									0.4	.24	tr	
CUSTARD SAUCE	144	169	18.6	7.5						157	164	.07	0.2	470
½ cup		7.3									0.8	.28	1	
HARD SAUCE	21	97	12.0	5.7						2	1	tr	tr	231
2 T		0.1									0.0	tr	0	
HARD SAUCE	42	193	23.9	11.3						3	2	tr	tr	462
¼ cup		0.1									0.0	tr	0	
LEMON SAUCE	54	133	27.8	2.8						3	2	tr	tr	124
¼ cup		0.1									tr	tr	3	
MARSHMALLOW CREME TOP, KRAFT	50	158	40.3	0.0					29	16	6			
		0.5	0.0						16				5	

	WT	Proximate			Amino Acids				Minerals			Vitamins		
		CAL	CHO	FAT	TRY	LEU	LYS	MET	Na	Ca	P	THI	NIA	A
		PRO	FIB	PUFA	PHE	ISO	VAL	THR	K	Mg	Fe	RIB	ASC	D
	g	g	g	g ·	mg	mg	mg	mg	mg	mg	mg	mg	mg	iu
MILK CHOC FUDGE TOP, HERSHEY	30	104	15.3	4.0					35	33	51	.01	0.1	27
2 T		1.3	0.7						65		0.4	.07	tr	
MILK CHOC FUDGE TOP, SMUCKERS	50	161	36.0	0.5					90	58	48	.02	tr	6
		4.0							82	9	0.2	.14	0	
PNT BTR CARAMEL TOP, SMUCKERS	50	168	34.5	2.1					175	20	63	.02	1.6	tr
		3.8							101	2	0.4	.06	0	
PECANS IN SYRUP TOP, SMUCKERS	50	168	39.0	0.8					0	15	60	.13	0.2	27
		2.3							125	29	0.5	.06	0	
PINEAPPLE TOP, SMUCKERS	50	146	37.0	0.2					17	14	5	.01	tr	11
		0.1							28	1	0.2	tr	1	
RAISIN SAUCE	48	126	25.9	3.0						15	23.5	.03	0.6	161
¼ cup		0.6										.02	5	
STRAWBERRY TOP, SMUCKERS	50	139	35.5	0.1					17	12	4	.00	0.0	1
		0.2							4	0	0.2	.00	tr	
WALNUTS IN SYRUP TOP, SMUCKERS	50	169	37.5	1.3					tr	19	74	.04	0.1	6
		2.9							88	25	0.5	.04	0	
WHIPPED CREAM, HEAVY	4	14	0.1	1.5						3	2	tr	0.0	59
1 T		0.1								tr	0.0	tr	tr	
WHIPPED CREAM, PRESSURIZED	3	8	0.4	0.7	1	9	8	2	4	3	3	tr	tr	27
1 T		0.1	0.0		5	6	6	4	4	tr	tr	tr	0	
WHIP TOP, RICH'S	7	19	1.2	1.6					4	tr	0	.00	0.0	11
1 serving		0.0	0.0	0.4					tr	0	tr	.00	0	
WHIP TOP, FROM MIX, DREAM WHIP	5	26	3.1	1.7					3	4	4	tr	tr	tr
1 T		0.1	0.0						1		tr	.02	tr	
WHIPPED TOP, FRZN	15	13	0.9	1.0	1	5	4	2	1	tr	tr	.00	0.0	34
1 T		0.1	0.0	tr	3	3	4	2	1	tr	tr	.00	0	
WHIP TOP, FRZN, COOL WHIP	5	14	1.0	1.0					1	0	0			0
1 T		0.0							1	tr	0.0			
WHIP TOP, FRZN, PET	5	14	1.0	1.0										
1 T		0.0												
WHIPPED TOP, POWDERED	1	6	0.5	0.4	1	5	4	1	1	tr	1	.00	0.0	11
1 T		tr	0.0	tr	3	3	4	2	2			.00	0	
WHIP TOP, POWDERED	4	8	0.7	0.5	2	14	11	4	3	4	3	tr	tr	14
(W/ WHOLE MILK) 1 T		0.1	0.0		7	9	10	6	6	tr	tr	.01	tr	
WHIPPED TOP, PRESSURIZED	4	11	0.6	0.9	1	4	3	1	2	tr	1	.00	0.0	19
1 T		tr	0.0	tr	2	2	3	2	1	tr	tr	.00	0	
WHIP TOP, PRESSURIZED, LUCKY WHIP	5	14	0.6	1.3					4	1	2	.00	0.0	0
1 T		tr							0			.00	0	

	WT	Proximate			Amino Acids				Minerals			Vitamins		
		CAL	CHO	FAT	TRY	LEU	LYS	MET	Na	Ca	P	THI	NIA	A
		PRO	FIB	PUFA	PHE	ISO	VAL	THR	K	Mg	Fe	RIB	ASC	D
	g	g	g	g	mg	mg	mg	mg	mg	mg	mg	mg	mg	iu
EGGS, CHICKEN														
BOILED (HARD/SOFT)	48	78	0.4	5.5	99	546	397	192	59	26	98	.04	tr	570
1 med		6.2	0.0	0.5	360	409	459	310	62	5	1.1	.13	0	23
FRIED W/ 1 t MARGARINE	50	108	0.4	8.6	99	546	397	192	169	30	111	.05	0.1	710
1 med		6.2	0.0	0.5	360	409	459	310	70	5	1.2	.15	0	27
OMELET, FRZN, HEAT & SERVE	100	218	2.7	18.2					338	71	176	.09	2.8	913
1 serving		10.5							137		1.7	.26	tr	
OMELET, PLAIN	62	107	1.5	8.0	115	634	461	223	160	50	117	.05	0.1	670
1 med		7.2	0.0	0.6	418	475	533	360	91	7	1.0	.17	0	31
OMELET, SPANISH	152	329	7.6	26.9	229	1257	921	438	330	103	280	.11	1.0	2000
2 eggs, 4 T sauce		14.6	0.5	1.5	829	941	1053	720	128	17	2.2	.26	13	31
POACHED	48	78	0.4	5.6	98	537	390	189	130	26	98	.04	tr	560
1 med		6.1	0.0	0.5	354	403	451	305	62	5	1.1	.12	0	27
SCRAMBLED, FRZN, LAND O' LAKES	70	86	1.7	5.7					64	48	97	.06	tr	539
1 serving		7.1							96		1.2	.18		
SCRAMBLED, MCDONALDS	77	161	1.9	11.9					206	49	166	.07	0.4	512
1 serving		11.6	0.2						143	11	2.2	.60	1	59
SCRAMBLED, MILK & FAT ADDED	65	112	1.6	8.4	114	654	483	218	167	52	123	.05	0.1	700
1 med		7.3	0.0	0.6	413	479	535	361	95	7	1.1	.18	0	31
WHITE, FRESH/FRZN	31	16	0.3	tr	51	296	204	133	47	3	5	tr	tr	0
1 med		3.4	0.0	0.0	214	218	262	150	43	3	tr	.08	0	0
WHOLE, DRIED, STABILIZED	7	41	0.3	2.9	53	284	208	102	30	13	56	.02	tr	300
(GLUCOSE-REDUCED) 1 T		3.3	0.0	0.2	188	214	241	162	32	3	0.6	.08	0	
WHOLE, FRESH/FRZN	48	78	0.4	5.5	99	546	397	192	59	26	50	.05	tr	570
1 med		6.2	0.0	0.5	360	409	459	310	62	5	1.1	.14	0	27
WHOLE, FRESH/FRZN	54	88	0.5	6.2	112	616	448	217	66	29	110	.06	0.1	640
1 large		7.0	0.0	0.5	406	462	518	350	70	6	1.2	.16	0	27
YOLK, FRESH/FRZN	17	59	0.1	5.2	39	235	185	70	12	24	97	.03	tr	580
1 med		2.8	0.0	0.3	123	171	190	140	19	3	0.9	.07	0	27
EGGS, OTHER														
DUCK, WHOLE	74	142	0.5	10.7	192	812	704	426	90	41	144	.13	0.1	910
1 med		9.9	0.0	0.9	622	443	655	545	96	12	2.1	.22	0	
GOOSE, WHOLE	90	166	1.2	12.0										
1 med		12.5	0.0	1.5									0	
QUAIL, WHOLE	10	16	tr	1.1						6	23	.01	tr	30
1 med		1.3	0.0								0.4	.08	0	
TURKEY, WHOLE	80	136	1.4	9.4						79	136	.09	0.4	
1 med		10.5	0.0	1.1							3.3	.38	0	
TURTLE, WHOLE	100	115	0.9	6.3	139		882	315		62	180	.28	0.1	65
3-5 eggs		12.6	0.0		580		554				1.6	.31		
EGG SUBSTITUTES														
BUD ANHEUSER-BUSCH	60	38	2.7	0.1					86	23	25	.01		708
¼ cup		6.5							126		0.1	.18	tr	
CHONO, POWDER, GENERAL MILLS	7	27	2.5	0.7					44	47				
1 T		2.7	0.0	0.5					67		0.1			
CHONO, RECON, GENERAL MILLS	50	39	3.6	1.0					95	67				
1 serving		3.8	0.0											
EGG BEATERS, FLEISHMANNS	60	100	1.8	7.5					108	48	42	.08		810
¼ cup		6.6		4.0					128		1.1	.26		26
EGGSTRA, POWDER, TILLIE LEWIS	7	30	1.5	0.8					56	23	34	.01	tr	86
1 T		3.9	0.0						52		0.2	.13	tr	
EGGSTRA, RECON, TILLIE LEWIS	50	43	2.2	1.0					80	32	47	.02	tr	123
1 serving		5.5	0.0						74		0.3	.15	tr	
EGGTIME, POWDER, NESTLÉ	12	40	3.0	1.0					120					
1 serving		7.0												
EGGTIME, RECON, NESTLÉ	55	90	3.0	5.0					120					
1 serving		7.0												
SECOND NATURE, AVOSET FOODS	50	44	0.6	1.6	117	542	425	234	106	26	60	.05	tr	1079
1 serving		6.0		0.7	382	436	521	319	143		1.0	.14		25
SUPER SCRAMBLE, TRANS AMERICAN	60	38	1.8	0.1	1	7	5	2	134	41	22	.07	0.1	1104
¼ cup		7.4	0.1		4	4	5	3	132		1.4	.35		24

	WT	Proximate			Amino Acids				Minerals			Vitamins		
		CAL	CHO	FAT	TRY	LEU	LYS	MET	Na	Ca	P	THI	NIA	A
		PRO	FIB	PUFA	PHE	ISO	VAL	THR	K	Mg	Fe	RIB	ASC	D
	g	g	g	g	mg	mg	mg	mg	mg	mg	mg	mg	mg	iu
FRZN*	60	96	1.9	6.7					120	44	43	.07		810
¼ cup		6.8	0.0	3.7					128		1.2	.23		
LIQUID†	47	40	0.3	1.6					83	25	57	.05	0.1	1015
1½ fl oz		5.6	0.0	0.8					155		1.0	.14	0	
POWDER‡	10	44	2.2	1.3					79	32	47	.02	0.1	122
.35 oz		5.5	0.0	0.2					74		0.3	.17	tr	

*contains egg white, corn oil & nonfat dry milk.
†contains egg white, hydrogenated soybean oil & soy protein.

‡contains egg white solids, whole egg solids, sweet whey solids, nonfat dry milk solids & soy protein.

		Proximate			Amino Acids				Minerals			Vitamins		
	WT	CAL	CHO	FAT	TRY	LEU	LYS	MET	Na	Ca	P	THI	NIA	A
		PRO	FIB	PUFA	PHE	ISO	VAL	THR	K	Mg	Fe	RIB	ASC	D
	g	g	g	g	mg	mg	mg	mg	mg	mg	mg	mg	mg	iu
Arthur Treachers														
CHICKEN, FRIED FILET	136	369	16.5	21.6					326	11	192	.07	11.0	102
		27.1		5.6					326	27	0.8	.15	1	4
CHICKEN SANDWICH	156	413	44.0	19.2					708	59	147	.17	8.1	123
		16.2		6.7					279	27	1.7	.24	19	3
CHIPS (FRENCH FRIED POTATOES)	113	275	34.8	13.1					392	12	85	.17	2.4	85
		4.0		2.7					597	29	0.5	.03	6	6
COLE SLAW	82	118	10.7	8.0					256	22	21	.02	0.4	164
		1.0		4.3					157	7	0.2	.02	57	4
FISH CHOWDER	170	112	11.2	5.4					835	61	87	.07	0.4	340
		4.6		0.9					228	12	0.1	.14	2	19
FISH, FRIED	147	354	25.3	19.7					448	15	197	.10	2.1	110
		19.1		5.6					407	9	0.6	.02	2	15
FISH SANDWICH	156	440	44.0	19.2					836	89	170	.27	4.1	117
		16.4		6.9					248	22	1.3	.22	2	11
KRUNCH PUP (HOT DOG)	57	204	12.1	14.9					448	8	50	.05	1.2	43
		5.4		2.2					71	7	0.6	.05	3	9
SHRIMP, FRIED	115	381	27.1	24.4					537	57		.08	1.7	86
		13.1		6.2					99	29	0.6	.05	1	2
Burger Chef														
CHEESEBURGER	104	304	24.0	17.0						156		.22	3.2	266
		14.0									2.0	.23	1	
DOUBLE CHEESEBURGER	145	434	24.0	26.0						246		.25	4.8	430
		24.0									3.1	.34	1	
FRENCH FRIES	68	187	25.0	9.0						10		.09	2.1	tr
		3.0									0.9	.05	14	
HAMBURGER, BIG CHEF	186	542	35.0	34.0						189		.34	5.4	282
		23.0									3.4	.35	2	
HAMBURGER, REGULAR	91	258	24.0	13.0					496	69	114	.22	3.2	114
		11.0	0.3						207	20	1.9	.18	1	
HAMBURGER, SUPER CHEF	252	600	39.0	37.0						240		.37	6.7	763
		29.0									4.2	.43	9	
MARINER PLATTER	373	680	85.0	24.0						137		.37	7.3	448
		32.0									4.7	.40	24	
RANCHER PLATTER	316	640	44.0	38.0						57		.30	8.7	367
		30.0									5.1	.37	24	
SHAKE, VANILLA	305	326	47.0	11.0					307	411	356	.11	0.3	10
		11.0	tr						518	34	0.2	.57	2	
SKIPPERS TREAT	179	604	47.0	37.0						201		.29	3.7	303
		21.0									2.5	.30	1	
Burger King														
CHEESEBURGER	130	305	29.0	13.0					823	141	274	.01	2.2	195
		17.0	tr						303	24	2.0	.02	1	
FRENCH FRIES, LARGE	87	428	56.0	20.0					340	24	122	.02	4.8	0
		6.0	0.6						722		2.0	.02	32	
FRENCH FRIES, REGULAR	43	214	28.0	10.0					170	12	61	.01	2.4	0
		3.0	0.3						361		1.0	.01	16	
HAMBURGER	90	252	29.0	9.0					372	45	96	.01	2.2	21
		14.0	0.3						236	17	2.0	.01	1	
HAMBURGER, WHOPPER		606	51.0	32.0						37		.02	5.2	641
		29.0									6.0	.03	13	
HOT DOG		291	23.0	17.0						40		.04	2.0	0
		11.0									2.0	.02	0	
SHAKE, VANILLA	315	331	48.5	9.5					306	479	386	.01	0.3	9
1 avg		13.2	0.3						592	42	0.3	.05	tr	
WHALER		486	64.0	46.0						70		.01	1.0	141
		18.0									1.0	.01	1	
Dairy Queen														
BANANA SPLIT	383	540	91.0	15.0						350		.60	.8	750
		10.0									1.8	.60	18	

*Portions are those commonly served by the restaurant.

	WT	|	Proximate			Amino Acids				|	Minerals			|	Vitamins		
			CAL	CHO	FAT	TRY	LEU	LYS	MET		Na	Ca	P		THI	NIA	A
			PRO	FIB	PUFA	PHE	ISO	VAL	THR		K	Mg	Fe		RIB	ASC	D
	g		g	g	g	mg	mg	mg	mg		mg	mg	mg		mg	mg	iu
BUSTER BAR	149		390	37.0	22.0							200			.09	1.6	300
			10.0										0.7		.34	tr	
CHEESE DOG, BRAZIER	113		330	24.0	19.0							168				3.3	
			15.0										1.6		.18		
CHEESE DOG, SUPER BRAZIER	203		593	43.0	36.0							297			.43	8.1	
			26.0										4.4		.48	14	
CHILI DOG, BRAZIER	128		330	25.0	20.0							86			.15	3.9	
			13.0										2.0		.23	.11	
CHILI DOG, SUPER BRAZIER	210		555	42.0	33.0							158			.42	8.8	
			23.0										4.0		.48	18	
DILLY BAR	85		240	22.0	15.0							100			.06	tr	100
			4.0										0.4		.17	tr	
FISH SANDWICH	170		400	41.0	17.0							60			.15	3.0	tr
			20.0										1.1		.26	tr	
FISH SANDWICH W/ CHEESE	177		440	39.0	21.0							150			.15	.3	100
			24.0										0.4		.26	tr	
FLOAT	397		330	59.0	8.0							200			.12	tr	100
			6.0										tr		.17	tr	
FLOAT, MR. MISTY	404		440	85.0	8.0							200			.12	tr	120
			6.0										tr		.17	tr	
FREEZE	397		520	89.0	13.0							300			.15	tr	200
			11.0										tr		.34	tr	
FREEZE, MR. MISTY	411		500	87.0	12.0							300			.15	tr	200
			10.0										tr		.34	tr	
FRENCH FRIES, LARGE	113		320	40.0	16.0							tr			.09	1.2	tr
			3.0										0.4		.03	5	
FRENCH FRIES, REGULAR	71		200	25.0	10.0							tr			.06	0.8	tr
			2.0										0.4		tr	4	
HAMBURGER, BIG BRAZIER	184		457	37.0	23.0						909	113			.37	9.6	
			27.0										5.2		.39		
HAMBURGER, BIG BRAZIER DELUX	213		470	36.0	24.0							111			.34	9.6	
			28.0										5.2		.37		
HAMBURGER, BRAZIER	106		260	28.0	9.0						572	70			.28	5.0	
			13.0										3.5		.26		
HAMBURGER W/ CHEESE, BIG BRAZIER	213		553	38.0	30.0							268			.34	9.5	495
			32.0										5.2		.53		
HAMBURGER W/ CHEESE, BRAZIER	121		318	30.0	14.0						871	163			.29	5.7	
			18.0										3.5		.29		
HAMBURGER, SUPER BRAZIER	298		783	35.0	48.0						1624	282			.39	15.6	
			53.0										7.3		.69		
HOT DOG, BRAZIER	99		273	23.0	15.0						866	75			.12	2.6	
			11.0										1.5		.15	11	
HOT DOG, SUPER BRAZIER	182		518	41.0	30.0							158			.42	7.0	tr
			20.0										4.3		.44	14	
ICE CREAM IN CONE, SMALL	71		110	18.0	3.0							100			.03	tr	100
			3.0										tr		.14	tr	
ICE CREAM IN CONE, MED	142		230	35.0	7.0							200			.09	tr	300
			6.0										tr		.26	tr	
ICE CREAM IN CONE, LARGE	213		340	10.0	10.0							300			.15	tr	400
			52.0										tr		.43	tr	
ICE CREAM IN CONE, DIP IN CHOC small	78		150	20.0	7.0							100			.03	tr	100
			3.0										tr		.17	tr	
ICE CREAM IN CONE, DIP IN CHOC med	156		300	40.0	13.0							200			.09	tr	300
			7.0										0.4		.34	tr	
ICE CREAM IN CONE, DIP IN CHOC large	234		450	58.0	20.0							300			.12	tr	400
			10.0										0.4		.51	tr	
ICE CREAM PARFAIT	284		460	81.0	11.0							300			.12	0.4	400
			10.0										1.8		.43	tr	
ICE CREAM SANDWICH	60		140	24.0	4.0							60			.03	0.4	100
			3.0										.04		.14	tr	
ONION RINGS	85		300	33.0	17.0							20			.09	0.4	tr
			6.0										0.4		tr	2	

*Portions are those commonly served by the restaurant.

	WT	Proximate			Amino Acids				Minerals			Vitamins		
		CAL	CHO	FAT	TRY	LEU	LYS	MET	Na	Ca	P	THI	NIA	A
		PRO	FIB	PUFA	PHE	ISO	VAL	THR	K	Mg	Fe	RIB	ASC	D
	g	g	g	g	mg	mg	mg	mg	mg	mg	mg	mg	mg	iu
SHAKE, SMALL	241	340	51.0	11.0						300		.06	0.4	400
		10.0									1.8	.34	2	
SHAKE, MED	418	600	89.0	20.0						500		.12	0.8	750
		15.0									3.6	.60	4	
SHAKE, LARGE	588	840	125.0	28.0						600		.15	1.2	750
		22.0									5.4	.85	6	
SUNDAE, CHOC, SMALL	106	170	30.0	4.0						100		.03	tr	100
		4.0									0.7	.17	tr	
SUNDAE, CHOC, MED	184	300	53.0	7.0						200		.06	tr	300
		6.0									1.1	.26	tr	
SUNDAE, CHOC, LARGE	248	400	71.0	9.0						300		.09	0.4	400
		9.0									1.8	.43	tr	
SUNDAE, FIESTA	269	570	84.0	22.0						200		.23	tr	200
		9.0									tr	.26	tr	
SUNDAE, HOT FUDGE	266	570	83.0	22.0						300		.45	0.8	500
		11.0									1.1	.43	tr	
Kentucky Fried Chicken														
ORIGINAL RECIPE, DINNER†	425	830	56.0	46.0						150		.38	15.0	750
		52.0									4.5	.56	27	
EXTRA CRISPY DINNER†	437	950	63.0	54.0						150		.38	14.0	750
		52.0									3.6	.56	27	
ORIGINAL RECIPE, CHICKEN ONLY	100	290	9.0	17.8					535	47		.08	6.4	65
		23.4									1.3	.19	1	
EXTRA CRISPY, CHICKEN ONLY	100	323	12.0	20.8					440	34		.07	6.0	74
		22.1									1.2	.14	1	
INDIVIDUAL PIECES (ORIGINAL RECIPE)														
DRUMSTICK	54	136	2.0	8.0						20		.04	2.7	30
		14.0									0.9	.12	1	
KEEL	96	283	6.0	13.0								.07		50
		25.0									0.9	.13	1	
RIB	82	241	8.0	15.0						55		.06	5.8	58
		19.0									1.0	.14		
THIGH	97	276	12.0	19.0						39		.08	4.9	74
		20.0									1.4	.24		
WING	45	151	4.0	10.0						·		.03		
		11.0									0.6	.07		
9 PIECES	652	1892	59.0	116.0								.49		
		152.0									8.8	1.27		
COLESLAW	90	110	13.0	5.9					237	27		.04	0.2	527
		1.1									0.4	.04	20	
POTATO, MASHED W/ GRAVY	113	74	14.6	2.0					353	17		.09	1.1	63
		2.0									0.6	.12	4	
ROLL	17	52	9.2	1.1					83	17		.08	0.8	15
1 avg		1.4									0.4	.07	tr	
Long John Silver's														
BREADED OYSTERS		460	58.0	19.0										
6 pieces		14.0												
BREADED CLAMS		465	46.0	25.0										
5 oz		13.0												
CHICKEN PLANKS		458	35.0	23.0										
4 pieces		27.0												
COLESLAW		138	16.0	8.0										
4 oz		1.0												
CORN-ON-THE-COB		174	29.0	4.0										
1 piece		5.0												
FISH W/ BATTER		318	19.0	19.0										
2 pieces		19.0												

*Portions are those commonly served by the restaurant.
†Dinner consists of mashed potatoes w/ gravy, cole slaw, roll and 3 pieces of chicken, either 1) wing, rib and thigh; 2) wing, drumstick and thigh; or 3) wing, drumstick and keel.

	WT	Proximate			Amino Acids				Minerals			Vitamins		
		CAL	CHO	FAT	TRY	LEU	LYS	MET	Na	Ca	P	THI	NIA	A
		PRO	FIB	PUFA	PHE	ISO	VAL	THR	K	Mg	Fe	RIB	ASC	D
	g	g	g	g	mg	mg	mg	mg	mg	mg	mg	mg	mg	iu
FISH W/ BATTER		477	28.0	28.0										
3 pieces		28.0												
FRYES		275	32.0	15.0										
3 oz		4.0												
HUSH PUPPIES		153	20.0	7.0										
3 pieces		1.0												
OCEAN SCALLOPS		257	27.0	12.0										
6 pieces		10.0												
PEG LEG W/ BATTER		514	30.0	33.0										
5 pieces		25.0												
SHRIMP W/ BATTER		269	31.0	13.0										
6 pieces		9.0												
TREASURE CHEST		467	27.0	29.0										
2 pieces fish, 2 Peg Legs		25.0												
McDonalds														
BIG MAC	187	541	39.0	31.4					963	175	215	.35	8.2	327
		25.6	0.7						387	38	4.3	.37	2	37
CHEESEBURGER	114	306	30.4	13.3					724	158	133	.24	5.5	372
		15.6	0.2						244	24	2.9	.30	2	14
EGG MCMUFFIN	132	352	26.0	20.0					911	187	264	.36	4.3	361
		18.0	0.4						222	25	3.2	.60	2	40
EGGS, SCRAMBLED	77	161	1.9	11.9					206	49	166	.07	0.4	512
		11.6	0.2						143	11	2.2	.60	1	59
ENGLISH MUFFIN, BUTTERED	62	186	28.0	5.6					446	87	94	.22	6.4	106
		5.6	0.1						66	13	1.6	.14	1	8
FILET O' FISH	131	402	34.0	22.7					707	105	157	.28	3.9	152
		15.0	0.7						292	29	1.8	.28	4	37
FRENCH FRIES	69	211	25.0	10.6					112	10	49	.15	2.9	52
		3.1	0.5						567	22	0.5	.03	11	3
HAMBURGER	99	257	30.1	9.4					525	63	88	.23	5.1	231
		13.3	0.2						234	20	3.0	.23	2	10
HOT CAKES W/ BUTTER & SYRUP	206	472	89.0	9.0						54		.31	4.0	255
		8.0									2.4	.43		
MCDONALDLAND COOKIES	63	292	45.1	10.5					328	10	51	.28	0.8	47
1 box		4.3	0.2						57	10	1.4	.23	1	
PIE, APPLE	90	295	30.5	18.3					408	12	23	.02	1.3	68
		2.2	0.5						38	7	0.6	.03	3	5
PIE, CHERRY	92	296	32.4	17.6					454	12	23	.02	0.4	212
		2.2	0.1						57	7	0.4	.03	1	5
QUARTER POUNDER	164	418	33.0	20.5					278	79	179	.31	9.8	164
		25.6	0.8						443	38	5.1	.41	2	23
QUARTER POUNDER W/ CHEESE	193	518	33.0	28.6					1206	251	257	.35	15.1	683
		30.9	0.8						471	43	4.6	.59	3	36
SAUSAGE, PORK	50	191	tr	17.2					487	13	57	.23	6.1	37
		8.9	tr						130	8	0.9	.13	1	36
SHAKE, CHOCOLATE	289	324	51.7	8.4					329	338	292	.12	0.8	347
		10.7	0.3						656	51	1.0	.88	3	35
SHAKE, STRAWBERRY	293	346	57.5	8.5					257	340	299	.12	.5	322
		10.3							545	35	0.2	.66	3	31
SHAKE, VANILLA	289	324	51.7	7.8					250	361	266	.12	.6	347
		10.7	0.0						500	35	0.2	.66	3	35
Pizza Hut														
THIN 'N CRISPY														
BEEF		490	51.0	19.0						350		.30	7.0	750
3 slices†		29.0									6.3	.60		
CHEESE		450	54.0	15.0						450		.30	5.0	750
3 slices†		25.0									4.5	.51		
PEPPERONI		430	45.0	17.0						300		.30	6.0	1000
3 slices†		23.0									4.5	.51		
PORK		520	51.0	23.0						350		.38	7.0	1000
3 slices†		27.0									6.3	.68		

*Portions are those commonly served by the restaurant.   †3 slices = ½ of a 10 inch pizza.

	WT	Proximate			Amino Acids				Minerals			Vitamins		
		CAL	CHO	FAT	TRY	LEU	LYS	MET	Na	Ca	P	THI	NIA	A
		PRO	FIB	PUFA	PHE	ISO	VAL	THR	K	Mg	Fe	RIB	ASC	D
	g	g	g	g	mg	mg	mg	mg	mg	mg	mg	mg	mg	iu
SUPREME		510	51.0	21.0						350		.38	7.0	1250
3 slices†		27.0									7.2	.68	2	
THICK 'N CHEWY														
BEEF		620	73.0	20.0						400		.68	8.0	750
3 slices†		38.0									7.2	.60		
CHEESE		560	71.0	14.0						500		.68	7.0	1000
3 slices†		34.0									5.4	.68		
PEPPERONI		560	68.0	18.0						400		.68	8.0	1250
3 slices†		31.0									5.4	.75	4	
PORK		640	71.0	23.0						400		.90	9.0	750
3 slices†		36.0									7.2	.77	1	
SUPREME		640	74.0	22.0						400		.68	9.0	1000
3 slices†		36.0									7.2	.85	9	
Taco Bell														
BEAN BURRITO	166	343	48.0	12.0						98		.37	2.2	1657
		11.0									2.8	.22	15	
BEEF BURRITO	184	466	37.0	21.0						83		.30	7.0	1675
		30.0									4.6	.39	15	
BEEFY TOSTADA	184	291	21.0	15.0						208		.16	3.3	3450
		19.0									3.4	.27	13	
BELLBEEFER	123	221	23.0	7.0						40		.15	3.7	2961
		15.0									2.6	.20	10	
BELLBEEFER W/ CHEESE	137	278	23.0	12.0						147		.16	3.7	3146
		19.0									2.7	.27	10	
BURRITO SUPREME	225	457	43.0	22.0						121		.33	4.7	3462
		21.0									3.8	.35	16	
COMBINATION BURRITO	175	404	43.0	16.0						91		.34	4.6	1666
		21.0									3.7	.31	15	
ENCHIRITO	207	454	42.0	21.0						259		.31	4.7	1178
		25.0									3.8	.37	10	
PINTOS 'N CHEESE	158	168	21.0	5.0						150		.26	0.9	3123
		11.0									2.3	.16	9	
TACO	83	186	14.0	8.0						120		.09	2.9	120
		15.0									2.5	.16	0	
TOSTADA	38	179	25.0	6.0						191		.18	0.8	3152
		9.0									2.3	.15	10	
Wendys														
CHEESEBURGER, SINGLE	240	520	32.0	30.0						200		.37	6.0	200
		31.0									4.5	.42	0	
CHEESEBURGER, TRIPLE	400	940	28.0	59.0						350		.75	14.0	400
		73.0									9.6	.76	2	
CHILI CON CARNE	250	250	23.0	7.0						80		.23	3.0	1000
		22.0									3.6	.17	3	
FRENCH FRIES	120	340	42.0	17.0						0		.12	3.0	0
		4.0									1.1	.01	6	
HAMBURGER, DOUBLE	285	630	34.0	34.0						100		.38	10.0	100
		47.0									8.1	.51	1	
HAMBURGER, SINGLE	200	440	33.0	25.0						80		.22		
		25.0									4.5	.34	0	
HAMBURGER, TRIPLE	360	780	27.0	45.0						100		.45	14.0	200
		66.0									9.0	.68	1	
SHAKE, CHOCOLATE	250	390	55.0	15.0						250		.15	0.0	300
		8.0									0.7	.60	0	

*Portions are those commonly served by the restaurant.   †3 slices = ½ of a 10 inch pizza.

	WT	Proximate			Amino Acids				Minerals			Vitamins		
		CAL	CHO	FAT	TRY	LEU	LYS	MET	Na	Ca	P	THI	NIA	A
	g	PRO g	FIB g	PUFA g	PHE mg	ISO mg	VAL mg	THR mg	K mg	Mg mg	Fe mg	RIB mg	ASC mg	D iu
FATS														
BACON FAT	14	126		14.0										
1 T		0.0	tr	0.8										
BEEF, SEPARABLE FAT, RAW	28	216	0.0	23.3					18	1	10	.01	0.3	48
1 oz		1.3	0.0						99	tr	0.2	.01		
BEEF TALLOW	28	252		28.0										
1 oz				1.2										
CHICKEN FAT	14	126	0.0	14.0										0
1 T		0.0		2.8										
CRISCO SHORTENING	14	124	0.0	14.0	0	0	0	0	tr	0	0	.00	0.0	tr
1 T		0.0	0.0	3.9	0	0	0	0	0			.00	0	0
FLUFFO SHORTENING	14	124	0.0	14.0	0	0	0	0	0	0	0	.00	0.0	221
1 T		0.0	0.0	3.3	0	0	0	0	0		tr	.00	0	0
LAMB, SEPARABLE FAT, CKD	28	199	0.0	21.2					20	1	9	.02	0.5	0
1 oz		1.8	0.0						81		0.0	.02	0	0
LARD	14	126	0.0	14.0	0	0	0	0	0	0	0	.00	0.0	0
1 T		0.0	0.0	1.4	0	0	0	0	0	tr	0.0	.00	0	
LARD	220	1984	0.0	220.0	0	0	0	0	0	0	0	.00	0.0	0
1 cup		0.0	0.0	22.0	0	0	0	0	0	1	0	.00	0	
PORK BACKFAT, RAW	28	232	0.0	25.4					20	tr	0	.03	0.1	0
1 oz		0.6	0.0	2.8					80		0.1	.01	0	0
PORK, SEPARABLE FAT, RAW	28	192	0.0	20.5					20	1	12	.08	0.4	0
1 oz		1.6	0.0						80		0.3	.02	0	
PORK, SEPARABLE FAT, CKD	28	216	0.0	23.4					18					
1 oz		1.3	0.0						109				0	0
PRIMEX SHORTENING	15	133		15.0										
1 T				2.6										
SALT PORK, RAW	28	219	0.0	23.8	2	103	89	15	339			.05	0.3	0
1 oz		1.1	0.0	2.7	44	31	47	39	12		0.2	.01	0	0
SALT PORK, FRIED	28	191	0.0	19.6	6	317	274	47		2	34	.08	0.6	0
1 oz		3.4	0.0	2.4	136	95	145	178			0.5	.03	0	0
SPRY, LIGHT	14	124	0.0	14.0					0	0	0	.00	0.0	0
1 T		0.0							0		0.0	.00	0	
SUET (BEEF KIDNEY FAT)	14	120		13.1										
1 T		0.2		0.3										
VEGETABLE SHORTENING	14	123	0.0	13.8	0	0	0	0	0	0	0	.00	0.0	0
1 T		0.0	0.0	1.0	0	0	0	0	0	0	0.0	.00	0	
OILS														
CORN	14	126	0.0	14.0	0	0	0	0	0	0	0	.00	0.0	0
1 T		0.0	0.0	7.4	0	0	0	0	0	tr	0.0	.00	0	
COTTONSEED	14	126	0.0	14.0	0	0	0	0	0	0	0	.00	0.0	0
1 T		0.0	0.0	7.0	0	0	0	0	0	0	0	.00	0	
OLIVE	14	124	0.0	14.0	0	0	0	0	0	0	0	.00	0.0	0
1 T		0.0	0.0	1.0	0	0	0	0	0	0	0.0	.00	0	
PALM	14	124	0.0	14.0	0	0	0	0	0	0	0	.00	0.0	1831
1 T		0.0	0.0	1.3	0	0	0	0	0		0.0	.00	0	
PEANUT	14	124	0.0	14.0	0	0	0	0	0	0	0	.00	0.0	0
1 T		0.0	0.0	4.0	0	0	0	0	0	0	0.0	.00	0	
SAFFLOWER	14	124	0.0	14.0	0	0	0	0	0	0	0	.00	0.0	0
1 T		0.0	0.0	10.1	0	0	0	0	0		0.0	.00	0	
SESAME	14	126	0.0	14.0	0	0	0	0	0	0	0	.00	0.0	0
1 T		0.0	0.0	5.7	0	0	0	0	0		0.0	.00	0	
SOYBEAN	14	124	0.0	14.0	0	0	0	0	0	0	0	.00	0.0	0
1 T		0.0	0.0	7.3	0	0	0	0	0	0	0.0	.00	0	
SUNFLOWER	14	124	0.0	14.0	0	0	0	0	0	0	0	.00	0.0	0
1 T		0.0	0.0	8.9	0	0	0	0	0	0	0.0	.00	0	
VEGETABLE, CRISCO	14	124	0.0	14.0	0	0	0	0	tr	0	0	.00	0.0	0
1 T		0.0	0.0	5.6	0	0	0	0	0		tr	.00	0	
VEGETABLE, GOLD LABEL, KRAFT	14	124	0.0	14.0	0	0	0	0	0	0	0	.00	0.0	0
1 T		0.0	0.0	6.9	0	0	0	0	0		0.0	.00	0	
VEGETABLE, PURITAN	14	124	0.0	14.0	0	0	0	0	0	0	0	.00	0.0	0
1 T		0.0	0.0	9.5	0	0	0	0	0		0.0	.00	0	
WHEAT GERM	14	126	0.0	14.0	0	0	0	0	0	0	0	.00	0.0	0
1 T		0.0	0.0	8.6	0	0	0	0	0		0.0	.00	0	

	WT	Proximate			Amino Acids				Minerals			Vitamins		
		CAL	CHO	FAT	TRY	LEU	LYS	MET	Na	Ca	P	THI	NIA	A
		PRO	FIB	PUFA	PHE	ISO	VAL	THR	K	Mg	Fe	RIB	ASC	D
	g	g	g	g	mg	mg	mg	mg	mg	mg	mg	mg	mg	iu
ABALONE, RAW	100	98	3.4	0.5						37	191	.18		
3½ oz		18.7	0.0								2.4	.14		
ABALONE, CND	100	80	2.3	0.3						14	128	.12		
3½ oz		16.0	0.0											
ALBACORE, RAW*	100	177	0.0	7.6	253	1923	2226	734	40	26				
3½ oz		25.3	0.0	tr	936	1290	1341	1088	293					
ALEWIFE, RAW	100	127	0.0	4.9	194	1474	1707	563			218			
3½ oz		19.4	0.0		718	989	1028	834						
ALEWIFE, CND	100	141	0.0	8.0	162	1231	1426	470						
3½ oz		16.2	0.0		599	826	859	697						
ANCHOVY, CND	12	21	tr	1.2	23	175	202	67		20	25			
3 thin filets		2.3	0.0	0.7	85	117	122	99						
ANCHOVY PASTE	7	14	0.3	0.8	14	106	123	41						
1 t		1.4	0.0	0.5	52	71	74	60						
ANCHOVY, PICKLED	28	49	0.1	2.9	59	445	515	170		47	59			
1 oz		5.4	tr		216	299	311	252						
BARRACUDA, PACIFIC, RAW	100	113	0.0	2.6										
3½ oz		21.0	0.0											
BASS, BAKED	115	287	3.0	19.4	212	1864	2384	661	68	96	269	.07	3.5	97
1 serving (3×3×½")		23.6	0.0		897	1227	1298	1062	256		1.2	.16	0	1
BASS, BLACK SEA, RAW	100	93	0.0	1.2	192	1459	1690	557	68					
3½ oz		19.2	0.0		710	979	1018	826	256					
BASS, BLACK SEA, BAKED, STUFFED†	100	259	11.4	15.8	162	1231	1426	470						
3½ oz		16.2	0.0		599	826	859	697						
BASS, SMALLMOUTH OR LARGEMOUTH, RAW 3½ oz	100	104	0.0	2.6	189	1436	1663	548			192	.10	2.1	
		18.9	0.0		699	964	1002	813				.03		
BASS, STRIPED, RAW	100	105	0.0	2.7	189	1436	1663	548			212			
3½ oz		18.9	0.0		699	964	1002	813						
BASS, STRIPED, BROILED	100	228	7.9	12.8	231	1751	2029	669		47	230	.15	2.9	116
3½ oz		20.2	0.0		853	1176	1222	992		43	1.9	.14	0	
BASS, STRIPED, OVEN-FRIED‡	100	196	6.7	8.5	215	1634	1892	624						
3½ oz		21.5			796	1097	1140	924						
BASS, WHITE, RAW	100	98	0.0	2.3	180	1368	1584	522						
3½ oz		18.0	0.0		666	918	954	774						
BLUEFISH, RAW	100	117	0.0	3.3	205	1558	1804	594	74	23	243	.12	1.9	
3½ oz		20.5	0.0		758	1046	1086	882			0.6	.09		
BAKED FILET	125	199	0.0	6.5	328	2493	2886	951	130	36	359	.14	2.4	62
1 serving		32.8	0.0		1214	1673	1738	1410			0.9	.13		
BROILED	122	192	0.0	6.3	320	2432	2816	928	127	35	350	.13	2.3	61
1 serving; (½ fish)		32.0	0.0		1184	1632	1696	1376			0.8	.12		
BLUEFISH, FRIED§	100	205	4.7	9.8					146	35	257	.11	1.8	
3½ oz		22.7									0.9	.11		
BONITO, CND, DEL MONTE	100	257	0.0	19.1					514	8	193	.01	9.8	
3½ oz		19.8	0.0						302	28	1.0	.09		
BUFFALOFISH, RAW	100	113	0.0	4.2					52					
3½ oz		17.5	0.0						293					
BULLHEAD, BLACK, RAW	100	84	0.0	1.6	163	1239	1435	473						
3½ oz		16.3	0.0		603	831	864	701						
BURBOT, RAW	100	82	0.0	0.9	174	1322	1531	505			190	.39	1.5	
3½ oz		17.4	0.0		644	887	922	748				.14		
FRIED	100	156	0.0	1.2	370	2830	3300	1090				.54	3.7	
3½ oz		37.0	0.0		1370	1900	2020	1600				.23		
BUTTERFISH, NORTHERN, RAW	100	169	0.0	10.2	181	1376	1593	525						
3½ oz		18.1	0.0		670	923	959	778						
GULF, RAW	100	95	0.0	2.9	162	1231	1426	470						
3½ oz		16.2	0.0		599	826	859	697						
FRIED	50	211	0.0	19.1	91	692	801	264		10	104	.03	2.0	
1 fish 6¼" long		9.1			337	464	482	391				.03		
CARP, RAW	100	115	0.0	4.2	180	1368	1584	522	50	50	253	.01	1.5	170
3½ oz		18.0	0.0		666	918	954	774	286		.9	.04	1	

*Catch is almost all canned as tuna.
†Recipe calls for bacon, butter, celery, onion, and bread cubes.
‡Prepared with milk, bread crumbs, butter and salt.
§W/ egg, milk, and bread crumbs.

NOTE: Whole fish AP, 1 lb makes 2 servings; steaks, 1 lb makes 4 servings; filets, 1 lb makes 5 servings; Average EP serving is 3 to 4 ounces.

	WT g	Proximate CAL / PRO	CHO / FIB	FAT / PUFA	Amino Acids TRY / PHE mg	LEU / ISO mg	LYS / VAL mg	MET / THR mg	Minerals Na / K mg	Ca / Mg mg	P / Fe mg	Vitamins THI / RIB mg	NIA / ASC mg	A / D iu
CATFISH, RAW	100	103	0.0	3.1	176	1338	1549	510	60			.04	1.7	
3½ oz		17.6	0.0		651	898	933	757	330		.4	.03		
CAVIAR, CND, PRESSED	10	32	0.5	1.7	31	279	241	88	22					
1 round t		3.4			150	190	207	204	18					
STURGEON, GRANULAR	10	26	3.3	1.5	24	221	192	70	220	28	36			
1 round t		2.7			119	151	165	162	18		1.2			
CHUB, RAW	100	145	0.0	8.8										
3½ oz		15.3	0.0											
CLAMS, SOFT, RAW, MEAT ONLY	100	82	1.3	1.9					36		183			
4 large; 9 small		14.0							235		3.4			
HARD OR ROUND, MEAT ONLY	100	80	5.9	0.9					205	69	151			
5 large; 10 small		11.1							311		7.5			
CLAMS, CND, SOLIDS & LIQUID	100	52	2.8	0.7						55	137	.01	1.0	
½ cup		7.9							140		4.1	.11		
CND, SOLIDS ONLY	100	98	1.9	2.5										
½ cup		15.8												
CND, LIQUOR ONLY	100	19	2.1	0.1										
½ cup scant		2.3	0.0											
COD, RAW	100	78	0.0	0.3	176	1338	1549	510	70	10	194	.06	2.2	0
1 piece (3×3×¾")		17.6	0.0		651	899	933	757	382	28	0.4	.07	2	
BROILED	95	162	0.0	5.0	260	1980	2300	760	105	29	260	.08	2.8	170
4 oz before cooking		26.1	0.0		960	1330	1390	1150	386		0.9	.10		
CND	100	85	0.0	0.3	192	1459	1690	557						
3½ oz		19.2	0.0		710	979	1018	826				.08		
DEHYDRATED, LIGHTLY SALTED	100	375	0.0	2.8	818	6217	7198	2372	8100		891	.08	10.9	0
3½ oz		81.8	0.0		3027	4172	4335	3517	160		3.6	.45		
DRIED, SALTED	100	130	0.0	0.7	290	2200	2550	840		225				
3½ oz		29.0	0.0		1070	1480	1550	1280						
CRAB, STEAMED	100	93	0.5	1.9	277	1557	1540	519		43	175	.16	2.8	2170
3½ oz		17.3			830	813	865	900		34	0.8	.08		
CND OR CKD	85	86	0.9	2.1	237	1332	1317	444	850	38	155	.07	1.6	
½ cup flakes		14.8			710	696	740	770	94	34	0.7	.07		
SOFT SHELL, FRIED	65	185	8.6	12.0	171	963	952	321						
1 serving		10.7			514	503	535	556						
CRAPPIE, WHITE, RAW	100	79	0.0	0.8	168	1277	1478	487				tr	1.4	
3½ oz		16.8	0.0		622	857	890	722				.03		
CRAYFISH, FRESHWATER, RAW	100	72	1.2	0.5	131	1256	1387	467		77	201	.01	1.9	
3½ oz		14.6			686	599	657	642			1.5	.04		
CROAKER, ATLANTIC, RAW	100	96	0.0	2.2	178	1350	1560	530	87			.12	5.5	60
3½ oz		17.8	0.0		660	910	940	780	234			.08		
BAKED	100	133	0.0	3.2	243	1850	2140	710	12			.13	6.5	70
3½ oz		24.3	0.0		900	1240	1300	1080	323			.10		
CROAKER, WHITE, RAW	100	84	0.0	0.8										
3½ oz		18.0	0.0											
CROAKER, YELLOW FIN, RAW	100	89	0.0	0.8										
3½ oz		19.2	0.0											
CUSK, RAW	100	75	0.0	0.2								.03	2.3	
3½ oz		17.2	0.0									.08		
CUSK, STEAMED	100	106	0.0	0.7					74	27	283	.03	2.7	
3½ oz		23.4	0.0						386		1.0	.10		
DOGFISH, SPINY, RAW	100	156	0.0	9.0								.05		
3½ oz		17.6	0.0											
DOLLY VARDEN, RAW	100	144	0.0	6.5	199	1512	1751	577				.06		
3½ oz		19.9	0.0		736	1015	1055	856				.06		
DRUM, FRESHWATER, RAW	100	121	0.0	5.2	173	1315	1522	502	70					
3½ oz		17.3	0.0		640	882	917	744	286					
DRUM, RED (REDFISH), RAW	100	80	0.0	0.4	180	1350	1584	522	55			.15	3.5	
3½ oz		18.0	0.0		666	918	954	792	273			.05		
EEL, AMERICAN, RAW	100	233	0.0	18.3	159	1192	1399	461		18	202	.22	1.4	1610
1 serving		15.9	0.0		588	811	843	700			0.7	.36		
SMOKED	50	165	0.0	13.9	93	698	818	270						
1 serving		9.3	0.0		344	474	493	409						
EULACHON (SMELT), RAW	100	118	0.0	6.2	146	1095	1285	423				.04		
3½ oz		14.6	0.0		540	745	774	642				.04		

	WT	Proximate			Amino Acids				Minerals			Vitamins		
		CAL	CHO	FAT	TRY	LEU	LYS	MET	Na	Ca	P	THI	NIA	A
		PRO	FIB	PUFA	PHE	ISO	VAL	THR	K	Mg	Fe	RIB	ASC	D
	g	g	g	g	mg	mg	mg	mg	mg	mg	mg	mg	mg	iu
FISH, FRIED, ARTHUR TREACHER	147	354	25.3	19.7					448	15	197	.10	2.1	110
1 serving		19.1		5.6					407	9	0.6	.02	2	15
FISHLOAF, CKD*	100	124	7.3	3.7										
3½ oz		14.1												
FISHSTICKS, FRZN	100	176	6.5	8.9	170	1250	1460	480	180	11	167	.04	1.6	0
4½ sticks		16.6	0.0		620	840	880	720	390	18	0.4	.07	2	
FLATFISHES, RAW	100	79	0.0	0.8	167	1252	1470	484	78	12	195	.05	1.7	
3½ oz		16.7	0.0		618	852	885	735	342		0.8	.05		
FLOUNDER OR SOLE, RAW	100	68	0.0	0.5	148	1125	1306	434	56	61	195	.06	1.7	
1 piece (3×3×⅜″)		14.9	0.0		553	756	794	646	366	30	0.8	.05		
BAKED	100	202	0.0	8.2	300	2250	2640	870	237	23	344	.07	2.5	
1 serving		30.0	0.0		1110	1530	1590	1290	587	30	1.4	.08	2	
FLOUNDER, STFD, IN BTR SCE, FRZN, BRICKFORDS 3½ oz	100	288	9.3	19.5					281					
		18.9												
FROG LEGS, RAW†	100	73	0.0	0.3						18	147	.14	1.2	0
4 large legs, EP		16.4	0.0								1.5	.25		
FRIED	144	418	12.2	28.6						28	231	.17	1.8	0
6 large eggs		25.8	0.0								2.0	.35		
GROUPER, RAW	100	87	0.0	0.5								.17		
3½ oz		19.3	0.0											
HADDOCK, RAW	100	79	0.0	0.1	183	1391	1610	531	61	23	197	.04	3.0	
1 fillet		18.3	0.0		677	933	970	787	304	24	0.7	.07		
HADDOCK, BROILED	100	141	0.3	6.6					71	13	230	.03	2.1	276
3½ oz		20.1	0.0						356	32	0.5	.07	3	
FRIED‡	100	165	5.8	6.4	196	1491	1725	568	177	40	247	.04	3.2	
1 filet (3×3×½″)		19.6			725	1000	1039	843	348		1.2	.07		
SMOKED OR CND	100	103	0.0	0.4	232	1763	2042	673				.06	2.1	
1 piece (2½×2½×¼″)		23.2	0.0		858	1183	1230	998				.05		
HAKE (INCLUDING WHITING), RAW	100	74	0.0	0.4	165	1254	1452	478	74	41	142	.10		
FLESH ONLY 3½ oz		16.5	0.0		610	842	874	710	363			.20		
HALIBUT, ATLANTIC & PACIFIC, RAW	100	100	0.0	1.2	209	1588	1839	606	54	13	211	.07	8.3	440
1 piece (3×2×1″)		20.9	0.0		773	1066	1108	899	449		0.7	.07		
BROILED	125	214	0.0	8.8	315	2394	2772	914	168	20	310	.06	10.4	850
1 serving		31.5	0.0		1166	1606	1670	1354	656		1.0	.09		
SMOKED	100	224	0.0	15.0	208	1581	1830	603						
3½ oz		20.8	0.0		770	1061	1102	894						
HALIBUT, CALIFORNIA, RAW	100	97	0.0	1.2						13				
3½ oz		19.8	0.0							23				
HALIBUT, GREENLAND, RAW	100	174	0.0	12.0						13	210	.01		
3½ oz		16.4	0.0							23				
HERRING, ATLANTIC, RAW	100	176	0.0	11.3	173	1315	1522	502			256	.02	3.6	110
3½ oz		17.3	0.0	2.0	640	882	934	761			1.1	.15		
BROILED	85	217	0.0	14.2	208	1581	1830	603			290	.01	3.3	130
1 fish		20.8	0.0		770	1061	1123	915			1.2	.15	0	
CND, SOLIDS & LIQUID	100	208	25.3	13.6	199	1512	1751	577		147	297			
3½ oz		19.9	0.0	2.0	736	1015	1075	876			1.8	.18		
HERRING, PACIFIC, RAW	100	98	0.0	2.6	175	1312	1522	508	74		225	.02	3.5	100
3½ oz		17.5	0.0	tr	648	892	928	752	420		1.3	.16		
CND, TOMATO SCE	100	176	3.7	10.5	158	1185	1375	458			243		3.5	
3½ oz		15.8	0.0	tr	585	806	837	679				.11		
PICKLED, BISMARCK TYPE	100	223	0.0	15.1	204	1530	1775	592						
3½ oz		20.4	0.0		755	1040	1081	877						
HERRING, SMOKED, BLOATERS	100	196	0.0	12.4										
3½ oz		19.6	0.0											
KIPPERED	100	211	0.0	12.9						66	254		3.3	30
3½ oz		22.2	0.0								1.4	.28		
INCONNU (SHEEFISH), RAW	100	146	0.0	6.8										
3½ oz		19.9	0.0											
JACK MACKEREL, RAW	100	143	0.0	5.6										
3½ oz		21.6	0.0											

*Prepared w/ cnd fish, bread, egg, tomato and onion.
†Frog legs may be purchased fresh or frozen at large markets or specialty stores. A pair of frozen legs averages 80 gm AP. After thawing, legs are dipped in seasoned flour and fried in neutral fat. Edible portion averages 48 gm per pair.
‡Dipped in egg, milk, and bread crumbs.

	WT g	Proximate CAL / PRO g	CHO / FIB g	FAT / PUFA g	Amino Acids TRY / PHE mg	LEU / ISO mg	LYS / VAL mg	MET / THR mg	Minerals Na / K mg	Ca / Mg mg	P / Fe mg	Vitamins THI / RIB mg	NIA / ASC mg	A / D iu
KINGFISH, RAW	100	105	0.0	3.0					83					
3½ oz		18.3	0.0						250					
KINGFISH, CKD	100	255	11.7	13.4					101	80	287	.11	2.9	93
3½ oz		22.3	0.0						293	56	1.9	.13	0	
LAKE HERRING (CISCO), RAW	100	96	0.0	2.3	177	1545	1558	513	47	12	206	.09	3.3	
3½ oz		17.7	0.0		655	903	956	779	319		0.5	.10		
LAKE TROUT (UNDER 6½ LB), RAW	100	241	0.0	19.9										
3½ oz		14.3	0.0											
(6½ LB & OVER), RAW	100	524	0.0	54.4										
3½ oz		7.9	0.0											
LINGCOD, RAW	100	84	0.0	0.8					59			.05		0
3½ oz		17.9	0.0						433			.04		
LOBSTER, NORTHERN, RAW	100	91	0.5	1.9	152	1453	1606	541		29	183	.40	1.5	
3½ oz		16.9	0.0		794	693	760	744		22	0.6	.05		
BOILED OR BROILED*	334	308	0.8	24.9	180	1720	1900	640	†210	80	229	.11	2.3	920
1 (¾ lb), 2 T butter		20.0	0.0		940	820	900	880	180		0.7	.06	0	
CND	85	75	0.0	1.1	139	1324	1463	493		55	161	.03	1.9	
½ cup meat		15.4			724	631	693	678			0.7	.06		
PASTE	7	13	0.1	0.7						5	13	tr	0.2	
1 t		1.5									0.1	.01		
LOBSTER, SPINY, RAW	100	72	1.2	0.5						77	201	.01	1.9	
3½ oz		14.6									1.5	.04		
MACKEREL, ATLANTIC, RAW	100	191	0.0	12.2	190	1444	1672	551				.15	8.2	450
3½ oz		19.0	0.0		703	969	1026	836		28		.33		
BROILED FILET	130	300	0.0	20.5	283	2151	2490	821						
1 filet		28.3	0.0		1047	1443	1528	1245						
CND	105	192	0.0	11.7	202	1535	1778	586		194	287	.06	6.0	460
½ cup		20.2	0.0		747	1030	1091	889			2.2	.22		
MACKEREL, PACIFIC, RAW	100	159	0.0	7.3	219	1664	1927	635		8	274			120
3½ oz		21.9	0.0		810	1117	1183	964			2.1			
CND	100	180	0.0	10.0	211	1604	1857	612		260	288	.03	8.8	30
3½ oz		21.1	0.0		781	1076	1139	928			2.2	.33		
MACKEREL, SALTED	100	305	0.0	25.1	185	146	1628	536						
3½ oz		18.5	0.0		684	944	999	814						
SMOKED	100	219	0.0	13.0	238	1809	2094	690						
3½ oz		23.8	0.0		881	1214	1285	1047						
MENHADEN, ATLANTIC, CND	100	172	0.0	10.2										
3½ oz		18.7	0.0								1.3			
MULLET, STRIPED, RAW	100	146	0.0	6.9					81	26	220	.07	5.2	
3½ oz		19.6	0.0						292	32	1.8	.08		
MULLET, STRIPED, BREADED, FRIED	100	296	12.2	17.4					99	53	258	.10	5.1	95
3½ oz		22.6	0.0						341	38	2.3	.10	0	
MUSKELLUNGE, RAW	100	109	0.0	2.5							227			
3½ oz		20.2	0.0								0.6			
MUSSELS, ATLANTIC & PACIFIC MEAT & LIQUID 3½ oz	100	66	3.1	1.4						23				
		9.6												
MEAT ONLY	100	95	3.3	2.2					289	88	236	.16		
3½ oz		14.4							315	25	3.4	.21		
MUSSELS, PACIFIC, CND	100	114	1.5	3.3										
3½ oz		18.2										.13		
OCEAN PERCH, PACIFIC, RAW	100	95	0.0	1.5	190	1425	1672	557	63					
3½ oz		19.0	0.0		703	969	1007	817	390					
OCTOPUS, RAW	100	73	0.0	0.8						29	173	.02	1.8	
3½ oz		15.3	0.0									.06		
OYSTERS, EASTERN, RAW	100	66	3.4	1.8					73	94	143	.14	2.5	310
5 to 8 med oysters		8.4							121	32	5.5	.18		
FRIED‡	100	239	18.6	13.9					206	152	241	.17	3.2	440
1 serving		8.6	tr						203		8.1	.29		
CND	100	76	4.9	2.2						28	124	.02	0.8	
3½ oz		8.5	0.1						70		5.6	.20		

*Edible portion of lobster is estimated at 36 percent.
†Na and K values are for lobster boiled in tap water; when seawater is used the sodium control content would be higher.

‡Dipped in egg, milk, and breadcrumbs.

Food (serving)	WT g	CAL	CHO g	FAT g	PRO g	FIB g	PUFA g	TRY mg	LEU mg	LYS mg	MET mg	PHE mg	ISO mg	VAL mg	THR mg	Na mg	Ca mg	P mg	K mg	Mg mg	Fe mg	THI mg	NIA mg	A iu	RIB mg	ASC mg	D iu
FRZN (3½ oz)	100				6.1											380			210			.14	2.5	310	.18		
OYSTERS, PACIFIC & WESTERN, RAW (2 to 4 oysters)	100	91	6.4	2.2	10.6												85	153		24	7.2	.12	1.3			30	
OYSTERS, SCALLOPED (6 oysters)	143	356	31.6	18.0	15.9												158	232			7.0	143	1.5	894	277	0	
PERCH, ATLANTIC (REDFISH), RAW (3½ oz)	100	88	0.0	1.2	18.0	0.0		190	1425	1672	557	703	969	1007	817	79	20	207	269	32	1.0	.10	1.9		.08		
PERCH, ATLANTIC (REDFISH), BREADED FRIED (3½ oz)	100	319	16.5	18.9	18.9																						
PERCH, ATLANTIC (REDFISH), FRIED (3½ oz)	100	227	6.8	13.3	19.0	0.0										153	33	226	284		1.3	.10	1.8		.11		
PERCH, WHITE, RAW (1 med fish)	100	118	0.0	4.0	19.3	0.0												192									
FRIED, FILET (1 serving)	65	108	0.0	5.3	12.5	0.0											9	113			0.7	.04	2.7	0	.05	0	
PERCH, YELLOW, RAW (1 med fish)	100	91	0.0	0.9	19.5	0.0		195	1462	1716	566	721	994	1034	838	68		180	230		0.6	.06	1.7		.17		
PICKEREL, RAW (3½ oz)	100	84	0.0	0.5	18.7	0.0		187	1402	1646	542	692	954	991	804						0.7						
PIKE, BLUE, RAW (3½ oz)	100	90	0.0	0.9	19.1	0.0		191	1432	1681	554	707	974	1012	821												
PIKE, NORTHERN, RAW (3½ oz)	100	88	0.0	1.1	18.3	0.0		183	1372	1610	531	677	933	970	787												
PIKE, WALLEYE, RAW (3½ oz)	100	93	0.0	1.2	19.3	0.0		193	1448	1698	560	714	984	1023	830	51		214	319		0.4	.25	2.3		.16		
POLLACK, RAW (3½ oz)	100	95	0.0	0.9	20.4	0.0		204	1530	1795	592	755	1040	1081	877	48			350			.05	1.6		.10		
CKD, CREAMED (1 serving)	100	128	4.0	5.9	13.9	0.0										111			238			.03	0.7		.13	tr	
POMPANO, RAW (1 piece 3×3×¾")	100	166	0.0	9.5	18.8	0.0		188	1410	1654	545	696	959	996	808	47			191			.41		tr	.22		
POMPANO, BROIL (3½ oz)	100	284	0.0	55.4	23.0	0.0		229	1720	2018	665	849	1170	1215	986	57			223			.50			.27		
PORGY OR SCUP, RAW (3½ oz)	100	112	0.0	3.4	19.0	0.0		190	1425	1672	551	703	969	1007	817	63	54	250	287								
FRIED (1 serving)	93	279	10.8	15.5	22.7			227	1702	1998	658	840	1158	1203	976	58	24	258	266		1.5	.07	3.7	75	.08	0	
RED & GRAY SNAPPER, RAW (3½ oz)	100	93	0.0	0.9	19.8	0.0		198	1485	1742	574	733	1010	1049	851	67	16	214	323	28	0.8	.17			.02		
REDHORSE, SILVER, RAW (3½ oz)	100	98	0.0	2.3	18.0	0.0		180	1350	1584	522	666	918	954	774												
ROCKFISH, RAW (3½ oz)	100	97	0.0	1.8	18.9	0.0		189	1417	1663	548	699	964	1002	813	60			388			.06			.12		
OVEN-STEAMED (3½ oz)	100	107	1.9	2.5	18.1			181	1357	1593	525	670	923	959	778	68			446			.05			.12		
ROE, RAW (COD, HERRING, SHAD) (3½ oz)	100	130	1.5	2.3	24.4																0.6	.10	1.4		.76	14	
ROE, RAW (SALMON, STURGEON, TURBOT) (3½ oz)	100	207	1.4	10.4	25.2			227	2066	1789	655	1109	1411	1537	1512								2.3		.72	18	
ROE, BAKED OR BROILED (3½ oz)	100	126	1.9	2.8	22.0											73	13	402	132		2.3						
CND (3½ oz)	100	118	0.3	2.8	21.5												15	346			1.2					2	
SABLEFISH, RAW (3½ oz)	100	190	0.0	14.9	13.0	0.0										56			358			.11			.09		
SALMON, ATLANTIC, RAW (3½ oz)	100	217	0.0	13.4	22.5	0.0	tr	225	1688	1958	652	832	1125	1192	968		79	186			0.9		7.2		.08	9	
CND (3½ oz)	100	203	0.0	12.2	21.7	0.0	tr	217	1628	1888	629	803	1085	1150	933												
SALMON, CHINOOK, RAW (3½ oz)	100	222	0.0	15.6	19.1	0.0	tr	191	1432	1662	554	707	955	1012	821	45		301	399			.10		310	.23		

	WT	Proximate			Amino Acids				Minerals			Vitamins		
		CAL	CHO	FAT	TRY	LEU	LYS	MET	Na	Ca	P	THI	NIA	A
		PRO	FIB	PUFA	PHE	ISO	VAL	THR	K	Mg	Fe	RIB	ASC	D
	g	g	g	g	mg	mg	mg	mg	mg	mg	mg	mg	mg	iu
CND	100	210	0.0	14.0	196	1470	1705	568		*154	289	.03	7.3	230
⅖ cup		19.6	0.0	tr	725	980	1039	843	366	27	0.9	.14		
SALMON, CHUM, RAW	100								53	74		.10		
3½ oz									429	29		.06		
SALMON, CHUM, CND	100	124	0.0	4.2					53	198	214	.12	7.1	60
3½ oz		21.5	0.0						336	30	0.7	.04	0	
SALMON, COHO (SILVER), RAW	100								48	175	231	.09		
3½ oz									421	29		.11	1	
SALMON, COHO (SILVER), CND	100	153	0.0	8.2					351	185	249	.03	7.4	19
3½ oz		18.8	0.0						339	30	0.6	.06	0	
SALMON, PINK (HUMP BACK), RAW	100	119	0.0	3.7	200	1500	1740	590	64			.14		
3½ oz		20.0	0.0	tr	740	1000	1060	860	306			.05		
CND	100	141	0.0	5.9	205	1538	1784	595	387	*196	286	.03	8.0	70
⅖ cup		20.5	0.0	tr	758	1025	1086	882	361	30	0.8	.18		
SALMON, SOCKEYE (RED), RAW	100								48			.14		150
3½ oz									391			.07		
CND	100	171	0.0	9.3	203	1522	1766	589	522	*259	344	.04	7.3	230
⅖ cup		20.3	0.0		751	1015	1076	873	344	29	1.2	.16		
SALMON, BROILED/BAKED	100	182	0.0	7.4	270	2025	2349	783	116		414	.16	9.8	160
1 small serving		27.0	0.0		999	1350	1431	1161	443		1.2	.06		
SALMON, SMOKED	100	176	0.0	9.3						14	245			
3½ oz		21.6	0.0											
SALMON PATTY	100	114	16.1	12.4					96	78	104	.12	4.0	66
3½ oz		15.8	0.8	0.8					89	34	1.2	.22	4	42
SARDINES, ATLANTIC, CND IN OIL	100	311	0.6	24.4	206	1545	1792	597	510	354	434	.02	4.4	180
8 med		20.6			762	1030	1092	886	560		3.5	.16		
SARDINES, PACIFIC, RAW	100	160	0.0	8.6	192	1440	1670	557		33	215			
1½ large (Pilchards)		19.2	0.0		710	960	1018	826		24	1.8			
CND, BRINE/MUSTARD	100	196	1.7	12.0					760	303	354			30
3½ oz		18.8							260		5.2			
CND IN OIL	100											.01	7.4	
3½ oz												.30		
CND IN TOM SCE	100	197	1.7	12.2					400	449	478	.01	5.3	30
1½ large		18.7							320		4.1	.27		
SAUGER, RAW	100	84	0.0	0.8										
3½ oz		17.9	0.0											
SCALLOPS, BAY & SEA, RAW	100	81	3.3	0.2					255	26	208		1.3	
2 to 3 pieces (12/lb)		15.3							396		1.8	.06		
STEAMED	100	112		1.4					265	115	338			
3½ oz		23.2							476		3.0			
FRZN, BREADED, FRIED	100	194	10.5	8.4										
3½ oz		18.0												
SEABASS, WHITE, RAW	100	96	0.0	0.5	214	1605	1883	621						
3½ oz		21.4	0.0		792	1091	1134	920						
SHAD, AMERICAN, RAW	100	170	0.0	10.0	186	1395	1637	539	54	20	260	.15	8.4	
1 piece (3×3×¾")		18.6	0.0		688	949	686	800	330		0.5	.24		
BAKED	100	201	0.0	11.3	232	1740	2042	673	79	24	313	.13	8.6	30
1 serving (4 oz raw)		23.2	0.0		858	1183	1230	998	377		0.6	.26		
CND	100	152	0.0	8.8	169	1268	1487	490						
3½ oz		16.9	0.0		625	862	896	727			0.7	.16		
SHAD, GIZZARD SHAD, RAW	100	200	0.0	14.0										
3½ oz		17.2	0.0											
SHEEPSHEAD, ATLANTIC, RAW	100	113	0.0	2.8	206	1545	1813	597	101		197			
3½ oz		20.6	0.0		762	1051	1092	886	234					
SHRIMP, RAW	100	91	1.5	0.8	188	1429	1654	545	140	63	166	.02	3.2	
3½ oz		18.8			696	959	996	808	220	42	1.6	.03		
FRENCH-FRIED†	100	225	10.0	10.8					186	72	191	.04	2.7	
3½ oz		20.3							229	51	2.0	.08		
SHRIMP, FRIED, ARTHUR TREACHERS	115	381	27.1	24.4					537	57		.08	1.7	86
1 serving		13.1		6.2					99	29	0.6	.05	1	2

*The calcium content of each variety of canned salmon includes that in the bones. These can be crushed and eaten with the salmon. If bones are discarded, the calcium content is much lower.

†Dipped in batter, or in egg, breadcrumbs and flour

	WT	Proximate			Amino Acids				Minerals			Vitamins		
		CAL	CHO	FAT	TRY	LEU	LYS	MET	Na	Ca	P	THI	NIA	A
		PRO	FIB	PUFA	PHE	ISO	VAL	THR	K	Mg	Fe	RIB	ASC	D
	g	g	g	g	mg	mg	mg	mg	mg	mg	mg	mg	mg	iu
SHRIMP, CND, WET PACK	100	80	0.8	0.8	162	1231	1426	470		59	152	.01	1.5	50
½ cup		16.2			599	826	859	697			1.8	.03		
SHRIMP, CND, DRY PACK	100	116	0.7	1.1	242	1839	2130	702		115	263	.01	1.8	60
3½ oz		24.2			895	1234	1283	1041	122		3.1	.03		
SHRIMP, FRZN, BREADED, RAW	100	139	19.9	0.7						38	111	.03	2.0	
3½ oz		12.3	0.1								1.0	.03		
SHRIMP OR LOBSTER PASTE	100	180	1.5	9.4										
3½ oz		20.8										.26		
SKATE (RAJAH FISH), RAW	100	98	0.0	0.7								.02		
3½ oz		21.5	0.0											
SMELT, ATLANTIC, RAW	100	98	0.0	2.1							272	.01	1.4	
4 to 5 med		18.6	0.0		186	1395	1637	539			0.4	.12		
CND	100	200	0.0	13.5	688	949	986	800		358	370			
4 to 5 med		18.4			184	1380	1619	534			1.7			
SNAIL, RAW	100	90	2.0	1.4	681	938	975	791						
3½ oz		16.1									3.5			
SNAIL, GIANT AFRICAN, RAW	100	73	4.4	1.4										
3½ oz		9.9												
SOLE, RAW	100	68	0.0	0.5	148	1125	1306	434	56	61	195	.06	1.7	
3½ oz		14.9	0.0		553	756	794	646	366	30	0.8	.05		
SOLE, FILET, IDLE WILD	113	80	0.6	0.8					162					
1 serving		17.7							281					
SPANISH MACKEREL, RAW	100	177	0.0	10.4					68	71	249	.13	4.8	
1 piece (3×3×¾")		19.5	0.0		195	1462	1716	566	264		1.0	.14		
SPOT, RAW	100	219	0.0	15.9	721	994	1034	838	61			.16		
3½ oz		17.6	0.0		176	1320	1549	510				.22		
BAKED	100	295	0.0	21.9	651	898	933	757	312					
3½ oz		22.8	0.0		228	1710	2006	661						
SQUID, RAW	100	84	1.5	0.9	844	1163	1208	980		12	119	.02		
3½ oz		16.4									0.5	.12		
STURGEON, RAW	100	94	0.0	1.9										
1 piece (3×3×¾")		18.1	0.0		181	1357	1593	525						
STEAMED	100	160	0.0	5.7	670	923	959	778	108	40	263			
3½ oz		25.4	0.0		254	1905	2235	737	235		2.0			
SMOKED	100	149	0.0	1.8	940	1295	1346	1092						
3½ oz		31.2	0.0		312	2340	2746	905						
SUCKER, WHITE, RAW	100	104	0.0	1.8	1154	1591	1654	1342	56		220	tr	1.2	
3½ oz		20.6	0.0						336					
SUCKER, CARP, RAW	100	111	0.0	3.2										
3½ oz		19.2	0.0											
SWORDFISH, RAW	100	118	0.0	4.0						19	195	.05	8.0	1580
1 piece (3×3×¾")		19.2	0.0		192	1440	1690	557			0.9	.05		
BROILED*	100	174	0.0	6.0	710	979	1018	826		27	275	.04	10.9	2050
3½ oz		28.0	0.0		280	2100	2464	812			1.3	.05		
CND	100	102	0.0	3.0	1036	1428	1484	1204				.01	11.4	1580
3½ oz		17.5	0.0		175	1312	1540	508				.05		
TAUTOG (BLACKFISH), RAW	100	89	0.0	1.1	648	892	928	752			227			
3½ oz		18.6	0.0											
TERRAPIN (DIAMOND BACK), RAW	100	111	0.0	3.5										
3½ oz		18.6	0.0								3.2			
TILEFISH, RAW	100	79	0.0	0.5										
3½ oz		17.5	0.0											
BAKED	100	138	0.0	3.7										
3½ oz		24.5	0.0											
TOMCOD, ATLANTIC, RAW	100	77	0.0	0.4										
3½ oz		17.2	0.0		172	1290	1514	499				.17		
TROUT, BROOK, RAW	100	101	0.0	2.1	636	877	912	740			266			
3½ oz		19.2	0.0		192	1440	1690	557				.07		
TROUT, CKD	100	196	0.4	11.2						218	272	.12	2.5	319
3½ oz		23.5								35	1.1	.06	1	
TROUT, RAINBOW, RAW	100	195	0.0	11.4	710	979	1018	826				.08	8.4	
3½ oz		21.5	0.0	tr	215	1612	1892	624				.20		

*Prepared with butter or margarine

	WT	Proximate			Amino Acids				Minerals			Vitamins		
		CAL	CHO	FAT	TRY	LEU	LYS	MET	Na	Ca	P	THI	NIA	A
		PRO	FIB	PUFA	PHE	ISO	VAL	THR	K	Mg	Fe	RIB	ASC	D
	g	g	g	g ·	mg	mg	mg	mg	mg	mg	mg	mg	mg	iu
TROUT, RAINBOW/STEELHEAD, CND	100	209	0.0	13.4										
3½ oz		20.6	0.0											
TUNA, BLUEFIN, RAW	100	145	0.0	4.1	796	1096	1140	924						
3½ oz		25.2	0.0	tr	252	1890	2218	731			1.3			
TUNA, YELLOWFIN, RAW	100	133	0.0	3.0	932	1285	1336	1084	37					
3½ oz		24.7	0.0	tr	247	1852	2174	716						
TUNA, CND IN OIL, SOLIDS & LIQUID	100	288	0.0	20.5	914	1260	1309	1062	800	6	294	.04	10.1	90
¾ cup		24.2	0.0	8.0	242	1815	2130	702	301		1.1	.09		
IN OIL, DRAINED SOLIDS	100	197	0.0	8.2	895	1234	1283	1041		8	234	.05	11.9	80
⅝ cup		28.8	0.0	2.0	288	2160	2534	835			1.9	.12		
IN WATER, SOLIDS & LIQUID	100	127	0.0	0.8	1066	1469	1526	1238	41	16	190		13.3	
½ cup		28.0	0.0		280	2100	2464	812	279		1.6	.10		
TUNA PATTY	100	209	7.4	10.6					154	56	173	.08	7.4	78
3½ oz		19.8	tr	0.7					64	25	1.7	.16	1	15
TURTLE, GREEN, RAW	100	89	0.0	0.5										
3½ oz		19.8	0.0											
CND	100	106	0.0	0.7										
3½ oz		23.4	0.0											
WEAKFISH, RAW	100	121	0.0	5.6	165	1238	1452	478	75			.09	2.7	
3½ oz		16.5	0.0		610	842	874	710	317			.06		
BROILED	100	208	0.0	11.4	246	1845	2165	713	560			.10	3.5	
3½ oz		24.6	0.0		910	1255	1304	1058	465			.08		
WHITEFISH, LAKE, RAW	100	155	0.0	8.2	189	1417	1663	548	52		270	.14	3.0	2260
1 piece (3×3×⅞")		18.9	0.0		699	964	1002	813	299		0.4	.12	tr	
BAKED, STUFFED	100	215	5.8	14.0	152	1140	1338	441	195		246	.11	2.3	2000
3½ oz		15.2			562	775	806	654	291		0.5	.11	tr	
SMOKED	100	155	0.0	7.3	209	1568	1839	606		22	274			
3½ oz		20.9	0.0		773	1066	1108	899						
WRECKFISH, RAW	100	114	0.0	3.9	184	1380	1619	534		47	171			
3½ oz		18.4	0.0		681	938	975	791	282		3.2			
YELLOWTAIL, PACIFIC, RAW	100	138	0.0	5.4	210	1575	1848	609						
3½ oz		21.0	0.0		777	1071	1113	903						

	WT	Proximate			Amino Acids				Minerals			Vitamins		
		CAL	CHO	FAT	TRY	LEU	LYS	MET	Na	Ca	P	THI	NIA	A
		PRO	FIB	PUFA	PHE	ISO	VAL	THR	K	Mg	Fe	RIB	ASC	D
	g	g	g	g	mg	mg	mg	mg	mg	mg	mg	mg	mg	iu
ARROWROOT*	8	29	7.0	0.0					4	0	0	.00	0.0	0
1 T		0.0							1		0.0	.00	0	0
BARLEY FLOUR	8	28	6.1	0.1	9	62	19	8	tr					
1 T		0.8	0.1		48	37	42	29	13					
BARLEY FLOUR	112	401	86.1	1.9	125	878	274	114	3					
1 cup		11.4	0.8		672	524	593	410	179					
BUCKWHEAT FLOUR, DARK	100	333	72.0	2.5	199	725	690	222	1	33	347	.58	2.9	0
1 cup sifted		11.7	1.6		491	445	655	468	656		2.8	.15	0	
BUCKWHEAT FLOUR, LIGHT	100	347	79.5	1.2	109	397	378	122	1	11	88	.08	0.4	0
1 cup sifted		6.4	0.5		269	243	358	256	320		1.0	.04	0	
CAROB (ST. JOHN'S BREAD) FLOUR	140	252	113.0	2.0						493	113			
1 cup		6.3	10.8											
CHESTNUT FLOUR	100	362	76.2	3.7								.23	1.0	
		6.1	2.0									.37		
CORN FLOUR	110	405	84.5	2.9	52	1118	249	163	1	7	180	.22	1.5	374
1 cup sifted		8.6	0.8	1.1	387	396	439	344			2.0	.06	0	
CORN MEAL, ENR, DRY	45	505	109.2	1.6	65	1417	316	207	2001	420	760	.64	5.1	
1 cup		10.9	0.9		491	501	556	436	164	68	4.2	.38	0	
CORNMEAL, ENR, CKD	240	120	25.6	0.4	24	336	72	48	0	2	34	.14	1.2	144
1 cup		2.6	0.2		120	72	144	96	38	20	1.0	.10	0	
CORNMEAL, WHOLE GROUND	118	409	84.5	3.8	60	1300	290	190	1628	354	756	.45	2.4	531
1 cup		10.0	1.1	1.1	450	460	510	400	276	125	2.8	.13	0	
COTTONSEED FLOUR	100	356	33.0	6.6	674	3371	2408	770		283	1112	1.21	6.5	60
		48.1	2.0		2986	2119	2793	2022		650	12.6	.84		
FISH FLOUR FROM FILETS	100	398	0.0	0.1					40	920	610			
		93.0							80		8.0			
FISH FLOUR FROM FILET WASTE	100	305	0.0	0.2					220	6040	4060			
		71.0							540		54.0			
FISH FLOUR FROM WHOLE FISH	100	307	0.0	0.3	745	6189	7381	2019	170	4610	3100	.07	2.2	
		76.0			2845	4232	3916	4378	430		41.0	.62		
LIMA BEAN FLOUR	100	343	630	1.4										0
1 cup		21.5	2.0										0	
MANIOC (CASAVA) FLOUR	100	320	81.0	0.5	21	66	66	10	5	148	104	.08	1.6	0
		1.6	1.8		45	45	49	49	19		5.4	.07	14	
MANIOC, TAPIOCA, DRY	100	360	86.0	0.2					5	12	12	.00	0.0	0
		0.6							19	3	1.0	.00	0	
MASA HARINA, ENR	110	404	84.0	4.4	41		377	398	15	145	316	.48	5.3	
1 cup		10.2	2.6								4.8	.34		
PEANUT FLOUR	100	371	31.5	9.3	623	3353	1964	479	9	104	720	.75	27.8	
		47.9	2.7		2778	2251	2730	1485	1186	360	3.5	.22	0	
POTATO FLOUR	100	351	79.9	0.8	88	401	425	104	34	33	178	.42	3.4	
		8.0	1.6		353	353	425	313	1588		17.2	.14	19	
RICE FLOUR, GRANULATED	125	479	107.4	0.4	82	645	292	135		11	120	.52	7.2	0
1 cup		7.5	0.2		375	352	525	292		35	6.8	.14	0	
RYE FLOUR, DARK	128	419	87.2	3.3	229	1398	855	334	1	69	686	.78	3.5	0
1 cup		20.9	3.0		980	877	1085	772	1101	147	5.8	.28	0	
RYE FLOUR, LIGHT	88	314	68.6	0.9	93	556	338	130	1	19	163	.13	0.5	0
1 cup		8.3	0.4		390	392	431	306	137	64	1.0	.06	0	
RYE FLOUR, MEDIUM	88	308	65.8	1.5	114	674	409	158	1	24	231	.26	2.2	0
1 cup		10.0	0.9		473	427	523	371	179	64	2.3	.11		
SOYBEAN FLOUR, FAT-FREE	72	303	21.9	14.6	396	2244	1822	396	1	143	402	.61	1.5	79
1 cup		26.4	1.7	7.9	1426	1558	1531	1135	1195	178	6.0	.22	0	
SOYBEAN FLOUR, HIGH-FAT	88	334	29.3	10.6	543	3077	2998	543	1	211	572	.78	2.0	
1 cup		36.2	1.9	5.3	1955	2136	2100	1557	1562	239	7.9	.32	0	
SOYBEAN FLOUR, LOW-FAT	100	356	36.6	6.7	651	3689	2995	651	1	263	634	.83	2.6	80
1 cup		43.4	2.5	3.0	2344	2561	2517	1866	1859		9.1	.36	0	
SOYBEAN FLOUR, DEFATTED	138	450	52.6	1.2	974	5516	4478	974	1	366	904	.15	3.6	55
1 cup		64.9	3.2	tr	3505	3829	3764	2791	2512		15.3	.05	0	
SUNFLOWER SEED FLOUR, PART DEFATTED	100	339	37.7	3.4					56	348	898	3.60	27.3	
		45.2	4.6						1080		13.2	.46		

*Starch obtained from the rootstock of the arrowroot plant; used especially in foods prepared for children and invalids.

	WT	Proximate			Amino Acids				Minerals			Vitamins		
		CAL	CHO	FAT	TRY	LEU	LYS	MET	Na	Ca	P	THI	NIA	A
		PRO	FIB	PUFA	PHE	ISO	VAL	THR	K	Mg	Fe	RIB	ASC	D
	g	g	g	g	mg	mg	mg	mg	mg	mg	mg	mg	mg	iu
TORTILLA FLOUR, LIME-TREATED,	28	103	20.7	1.6						25	107	.10	0.5	1
WHITE CORN　1 oz		2.3	0.9								0.7	.04	tr	
TORTILLA FLOUR, UNTREATED REFINED	28	101	20.9	1.1						4	49	.07	0.5	48
YELLOW CORN　1 oz		2.7	0.4								1.0	.03	0	
WHEAT FLOUR, ALL-PURPOSE, ENR	8	29	6.1	0.1	10	62	18	10	tr	1	7	.04	0.3	0
1 T		0.8	tr		44	37	34	23	8	2	0.2	.02	0	
WHEAT FLOUR, ALL-PURPOSE, ENR	110	400	83.7	1.1	139	893	267	151	2	18	96	.48	3.9	0
1 cup sifted		11.6	0.3		638	534	499	336	105	28	3.2	.28	0	
WHEAT FLOUR, UNENR	8	29	6.1	0.1	10	62	18	10	tr	1	7	.01	.1	0
1 T		0.8	tr		44	37	34	23	8	2	tr	tr	0	
WHEAT FLOUR, UNENR	110	400	83.7	1.1	139	893	267	151	2	18	96	.07	1.0	0
1 cup sifted		11.6	0.3		638	534	499	336	105	28	0.9	.06	0	
WHEAT FLOUR, BREAD, ENR	8	29	6.0	0.1	11	69	21	12	tr	1	8	.04	0.3	0
1 T		0.9	tr		50	51	39	26	8	2	0.2	.02	0	
WHEAT FLOUR, BREAD, ENR	112	409	83.7	1.2	158	1016	304	172	2	18	106	.49	3.9	0
1 cup sifted		13.2	0.3		726	607	568	383	106	28	3.2	.29	0	
WHEAT FLOUR, BREAD, UNENR	8	29	6.0	0.1	11	69	21	12	tr	1	8	.01	0.1	0
1 T		0.9			50	41	39	26	8	2	0.1	.01	0	
WHEAT FLOUR, BREAD, UNENR	112	409	83.7	1.2	158	1016	304	172	2	18	106	.09	1.1	0
1 cup sifted		13.2	0.3		726	607	568	383	106	28	1.0	.07	0	
WHEAT FLOUR, CAKE/PASTRY	14	51	11.1	0.1	13	85	25	14	tr	2	10	tr	0.1	0
2 T		1.1	tr		61	51	47	32	13	4	0.1	tr	0	
WHEAT FLOUR, CAKE/PASTRY	100	364	79.4	0.8	90	578	172	98	2	17	73	.03	0.7	0
1 cup sifted		7.5	0.2		412	345	322	218	95	26	0.5	.03	0	
WHEAT FLOUR, GLUTEN	140	529	66.1	2.7	638	4350	1102	986	3	56	196			0
(55% patent & 45% gluten)　1 cup		58.0	0.6		3190	2668	2726	1566	84				0	
WHEAT FLOUR, SELF-RISING	100	352	74.2	1.0	112	716	214	121	1079	265	466	.44	3.5	0
⅞ cup sifted		9.3	0.4		512	428	400	270	90		2.9	.26	0	
WHEAT FLOUR, WHOLE WHEAT	120	400	85.2	2.4	192	1072	432	240	4	49	446	.66	5.2	0
1 cup		16.0	2.8		784	688	736	464	444	136	4.0	.14	0	
WHEAT FLOUR, WHOLE WHEAT	110	402	81.5	1.4	158	1016	304	172	2	26	210	.28	2.2	0
80% extraction　1 cup sifted		13.2	0.5		726	607	568	370	104	27	1.4	.08	0	
WHEAT FLOUR, WHOLE WHEAT	100	365	74.5	1.2	142	909	271	153	2	20	97	.12	1.4	0
straight hard　3½ oz		11.8	0.4		649	543	507	342	95	27	1.4	.07	0	
WHEAT FLOUR, WHOLE WHEAT	100	364	76.9	1.0	16	747	223	126	2	20	97	.08	1.2	0
straight soft　3½ oz		9.7	0.4		534	446	417	281	95	27	1.1	.05	0	
WHEAT-SOY FLOUR,	100	365	57.3	6.8					297	684	533	2.02	8.2	1660
straight grade wheat flour　3½ oz		21.4	1.3						642	169	20.5	.59	40	
WHEAT-SOY FLOUR, BULGUR FLOUR	100	360	60.1	6.2					296	685	562	1.49	0.1	361
3½ oz		19.7	1.8	0.2					624	202	20.8	.59	1	

	WT g	CAL / PRO g	CHO / FIB g	FAT / PUFA g	TRY / PHE mg	LEU / ISO mg	LYS / VAL mg	MET / THR mg	Na / K mg	Ca / Mg mg	P / Fe mg	THI / RIB mg	NIA / ASC mg	A / D iu
ADVANCE, ROSS LABS	240	130	14.9	4.8	86	564	456	144	108	192	144	.19	2.4	576
8 oz		6.7		2.4	264	271	370	276	264	19	4.3	.22	12	96
ALACTA, POWDER, MEAD JOHNSON	8	34	3.7	1.0	30	247	194	66	36	96	72			
1 T		2.6			124	130	160	110	112	6	tr			
CITROTEIN, POWDER, DOYLE	34	129	23.7	0.3					132	203	203	.61	8.1	1015
1 serving		7.8							132	81	7.3	.69	46	81
CITROTEIN, DOYLE	240	159	29.1	0.4					163	250	250	.75	10.1	1250
8 oz		9.6							163	100	9.0	.85	56	100
CONTROLYTE, POWDER, DOYLE	9	45	6.6	2.2	tr	1	tr	tr	1	tr	1			
1 T		tr		0.2	tr	tr	tr	tr	tr					
DEXTRI MALTOSE, POWDER, MEAD JOHNSON 1 T	7	27	6.9	0.0					1	1	2			
		0.0							11	1	0.0			
DIETENE, CHOC, POWDER, DOYLE	10	36	5.5	0.3					37	114	93	.14	1.5	444
		3.0	0.0						117		1.4	.14	5	
ENSURE OSMOLITE, ROSS LABS	240								130	130	130	.38	5.0	625
8 oz									210	50	2.2	.43	40	50
ENSURE PLUS, ROSS LABS	240	360	47.3	12.7					250	150	150	.03	7.4	625
8 oz		13.0							450	75	3.4	.65	37	50
ENSURE FLAVOR PACS, ROSS LABS 1 pkg	1	4												
FLEXICAL, MEAD JOHNSON	240	240	37.0	8.2					84	144	120	.46	6.0	600
8 oz		5.4							300	48	2.2	.52	36	48
ISOCAL, MEAD JOHNSON	240	250	31.2	10.6					125	150	125	.48	6.2	625
8 oz		8.2							312	50	2.3	.54	37	50
LACTOSE HYDROUS, USP POWDER, MERCK & CO. 1 T	11	40	10.5											
LIPOMUL, ORAL, UPJOHN	15	91	0.1	33.0					7	2		.00	0.0	0
1 T		tr		8.3					tr	tr		.00	0	
LOLACTENE, POWDER, VANILLA, DOYLE 1 pkg	57	229	30.2	5.4					252	536	469	.57	5.0	1257
		15.1							704	101	4.5	.64	23	100
LOLACTENE, VANILLA, DOYLE	297	227	30.0	5.4					250	533	467	.56	5.0	
1 serving		15.2							700	100	4.5	6.3	23	100
LYTREN, MEAD JOHNSON	30	8							17	2	1			
1 oz									29					
MALTSUPEX, ABBOTT LABS	15	61	10.1	0.0					21	3	30	.03	1.2	0
1 T		0.9		0.0					250	12	0.1	.03	45	
MERITENE POWDER, PLAIN, DOYLE	11	40	6.5	0.1	44	295	284	88	42	107	86	.17	1.7	309
1 T		3.3	0.0		153	317	229	175	142	21	1.5	.12	7	34
MERITENE, DOYLE	240	240	27.6	8.0					220	300	300	.45	4.0	1000
8 oz		14.4	0.0						400	80	4.1	.52	18	80
MERITENE IN WHOLE MILK, DOYLE	240	257	28.6	8.4					240	552	456	.53	4.6	1154
8 oz		16.6							720	92	4.1	.58	21	91
METRACAL, POWDER	10	40	4.9	0.9					40	88	79	.09	0.7	221
1 T		3.1							154	11	0.7	.13	4	18
METRECAL, MEAD JOHNSON	237	225	27.5	5.0					225	500	450	500	3.8	1250
8 oz		17.5									3.8	750	25	100
MILNOT, CND	240	288	23.3	14.4					655	571	442	.13	0.5	1166
8 oz		16.8		3.5						55	0.2	.91	3	233
MULL-SOY, SYNTEX	240	171	12.7	8.9					177	300	200	.18	2.3	499
8 oz		7.7	0.5						7	19	2.5	.25	13	100
NUTRAMENT, MEAD JOHNSON	240	254	29.8	7.4					372	398	372	.53	6.8	1694
8 oz		17.0							535		3.8	.60	25	134
NUTRAMENT, CND, MEAD JOHNSON	354	400	50.0	13.3					200	500	400	.50	5.0	1250
12½ oz		20.0	0.6								4.0	.60	50	125
NUTRAMIGEN, MEAD JOHNSON	240	160	21.1	6.2	55	528	478	156	77	150	113	.12	2.0	400
8 oz		5.3		3.6	271	331	407	246	163	18	3.0	.15	13	101
NUTRI-1000, CUTTER LABS	240	254	23.9	13.0					127	288	228	.26	2.5	634
8 oz		9.5							360	48	2.3	.26	11	51
NUTRI-1000, LACTOSE-FREE, CUTTER LABS 8 oz	240	249	24.0	13.1					170	125	125	.25	2.5	625
		9.5							350	50	2.3	.25	11	50
OSMALITE, ROSS LABS	240	254	12.2	9.6					138	137	137	.40	5.3	661
8 oz		9.4							234	53	2.4	.45	42	53

*Fluid and normal dilution unless otherwise indicated

	WT g	CAL PRO g	CHO FIB g	FAT PUFA g	TRY PHE mg	LEU ISO mg	LYS VAL mg	MET THR mg	Na K mg	Ca Mg mg	P Fe mg	THI RIB mg	NIA ASC mg	A D iu
		Proximate			Amino Acids				Minerals			Vitamins		
PEDIALYTE, ROSS LABS	240	48							163	19				
8 oz									185	12				
POLYCOSE, ROSS LABS	240	480	120						168	38	7			
8 oz									10					
POLYCOSE, POWDER, ROSS LABS	8	32	7.5						10	1	1			
1 T									1					
PORTAGEN, POWDER, MEAD JOHNSON	15	70	7.9	3.3					40	99	77	.09	1.1	359
1 T		2.7							435	7	1.2	.11	6	35
PORTAGEN, MEAD JOHNSON	240	240	27.6	11.5					113	225	170	.37	5.0	1875
8 oz		8.4							300	50	4.8	.45	19	187
PRECISION, HIGH-NITROGEN, POWDER DOYLE 1 serving	82	300	62.0	0.1	544	3044	2066	1341	820	293	293	.66	5.9	
		12.5			2210	2247	2681	1522	761	117	5.3	.76	26	117
PRECISION, HIGH-NITROGEN, DOYLE	323	300	62.3	0.2	284	1402	953	617	274	98	98	.22	2.0	489
8 oz		12.3			1017	1034	134	701	255	39	1.8	.26	9	13
PRECISION, ISOTONIC POWDER, DOYLE 1 pkg	56	237	35.6	7.4	1	5	3	2	190	161	161	.36	3.1	791
		7.1		0.8	3	3	4	2	237	64	2.9	.41	14	64
PRECISION, ISOTONIC, DOYLE	300	250	37.5	7.8					200	168	170	.38	3.3	833
1 serving		7.5							250	67	3.0	.43	15	67
PRECISION, LOW-RESIDUE POWDER, DOYLE 1 serving	85	320	72.0	0.2	105	690	400	260	202	169	169	.38	3.4	78
		7.6			428	435	519	295	254	68	3.1	.44	15	67
PRECISION, LOW-RESIDUE, DOYLE	325	316	70.0	0.2	143	796	540	351	197	164	164	.37	3.3	821
8 oz		7.4			579	588	702	400	246	66	3.0	.43	15	66
PRECISION, MODERATE-NITROGEN, POWDER, DOYLE 1 serving	78	331	50.0	10.2						164	164	.37	3.3	829
		10.8								66	3.0	.43	15	66
PRECISION, MODERATE-NITROGEN, DOYLE 8 oz	240	331	50.0	10.2					281	164	164	.37	2.7	833
		10.8		1.9					250	66	3.0	.44	15	66
PROBANA, POWDER, MEAD JOHNSON	9	39.6	4.5	1.2	23	237	174	60	36	57	51			300
1 T		2.4			113	132	165	105	72	7	0.2			60
PROBANA, MEAD JOHNSON	200		15.8	2.2					120	229	177	.12	1.6	1040
		8.4							239	16	0.3	.21	11	208
PROTEIN MILK, MEAD JOHNSON	30		0.8	0.8					21	66	37			
1 oz		1.1							5		tr			
PROTENUM, POWDER, MEAD JOHNSON 1 serving	8	30	3.7	0.2	39	326	251	86	29	95	72	.08	0.8	282
		3.4			168	183	222	142	92	8	0.8	.12	4	28
SEGO, CND, PET	312	225	26.0	5.0					220	500	450	.50	3.8	1250
10 oz		19.0							880		2.5	.75	25.0	100
SOYALAC, LOMA LINDA	300	204	20.0	11.4	171	954	648	102	95	169	84	.13	6.0	625
10 oz		6.6	0.0	6.0	618	579	558	417	229	54	3.1	.19	16	125
SOYAMEL, WORTHINGTON FOODS	30	150	17.1	7.3						159	130			
1 oz		5.4									0.8			
SOYAMEL, FORTIFIED, WORTHINGTON FOODS 1 oz	30	150	15.6	7.1					1059	159	159	.48	3.2	
		5.3									4.8	.60		120
SUSTACAL, POWDER, MEAD JOHNSON 1 serving	57	199	36.4	0.3					190	470	450	.43	6.9	1420
		12.7							560	97	5.9	.40	35	33
SUSTACAL, MEAD JOHNSON		120	33.1	5.5					223	240	223	.33	4.8	1110
8 oz		14.5							494	90	4.0	.40	13	89
SUSTAGEN, POWDER	8	31	5.3	0.3					22	57	42	.07	0.9	89
1 T		1.9							62	7	0.3	.08	5	7
SUSTAGEN, MEAD JOHNSON	240	360	61.4	3.1					249	655	600	.78	10.2	1028
8 oz		21.6							720	82	3.7	.88	62	82
SUSTAGEN, TUBE FEEDING, MEAD JOHNSON 1 oz	30	39	6.6	0.4								.08	1.1	
		2.3								9		.09	7	9
TRIGLYCERIDES, MEDIUM-CHAIN, MEAD JOHNSON 1 T	14	115												
TUBE FEEDING, COMPLEAT-B, DOYLE	111	111	13.3	4.4					173	69	187	.16	1.4	347
100cc		4.4							144	28	1.2	.18	6	28
TUBE FEEDING, FORMULA #2, CUTTER LABS 100cc	100	100	12.3	4.0					63	220	95	.08	1.0	50
		3.8							191	20	1.3	.09	3	24
TUBE FEEDING, PUREED, HOSP FORM	240	216	31.4	6.7					170					
8 oz		9.1							348					
TUBE FEEDING, NASO-GASTRIC, HOSP FORM 8 oz	240	269	24.0	12.2					235					
		15.8							660					

*Fluid and normal dilution unless otherwise indicated.

	WT	Proximate			Amino Acids				Minerals			Vitamins		
		CAL	CHO	FAT	TRY	LEU	LYS	MET	Na	Ca	P	THI	NIA	A
		PRO	FIB	PUFA	PHE	ISO	VAL	THR	K	Mg	Fe	RIB	ASC	D
	g	g	g	g	mg	mg	mg	mg	mg	mg	mg	mg	mg	iu
TUBE FEEDING, JEJUNOSTOMY,	240	206	20.9	9.6					139					
HOSP FORM 8 oz		9.4							298					
TUBE FEEDING (1.5 KCAL/CC), HOSP	240	369	40.1	18.2					293	342		.15		553
FORM 8 oz		15.4		5.0	340				590		2.0		22	
TUBE FEEDING (1.8 KCAL/CC), HOSP	240	468	46.6	21.4					389	348		.23		3936
FORM 8 oz		21.1		5.0	496				704		3.9		30	
VITAL POWDER, ROSS LABS	300	300	51.0	2.9					105	183	183	.27	3.6	918
1 serving		11.4							321	72	3.3	.30	17	72
VIVONEX-100, POWDER	6	23	5.1	tr	6	33	25	21	19	10	10	.02	0.2	63
1 T		0.5			23	21	23	21	26	4	0.1	.02	1	5
VIVONEX	240	240	54.2	0.2	67	348	262	226	206	106	106	.14	1.8	667
8 oz		4.8			250	221	242	221	281	47	1.3	.14	9	53
VIVONEX-HN	240	240	50.5	0.2					185	65	65	.10	1.1	401
8 oz		9.8							168	28	0.7	.10	6	

*Fluid and normal dilution unless otherwise indicated.

	WT g	CAL / PRO (g)	CHO / FIB (g)	FAT / PUFA (g)	TRY / PHE (mg)	LEU / ISO (mg)	LYS / VAL (mg)	MET / THR (mg)	Na / K (mg)	Ca / Mg (mg)	P / Fe (mg)	THI / RIB (mg)	NIA / ASC (mg)	A / D (iu)
ADVANCE	30	16	1.6	0.8	7	50	39	13	9	15	11	.02	0.3	70
1 oz		0.6		0.4	28	25	26	23	25	2	0.4	.03	1	12
CASEC, POWDER, MEAD JOHNSON	5	19	0.0	0.1	48	437	375	138	2	30	40	.00	0.0	0
1 T		4.4			236	235	34	197	2	1	0.2	.00	0	
CARBOHYDRATE-FREE, SYNTEX	30	21	1.9	1.1	7	47	37	12	11	27	19	.02	0.3	62
1 oz		0.5			31	28	29		27	2	0.2	.03	2	4
ENFAMIL, POWDER, MEAD JOHNSON	28	156	18.4	8.7	31	31	233	109	36	126	107	.12	1.9	389
1 oz		3.4	0.6		140	140	187	140	160	11	0.3	.14	12	78
ENFAMIL, (20 KCAL/OZ), MEAD JOHNSON 1 oz	30	20	2.1	1.1	8	45	37	12	8	17	14	.02	0.3	50
		0.5		0.5	22	24	29	21	21	1	0.1	.02	2	13
ENFAMIL W/ IRON (20 KCAL/OZ), MEAD JOHNSON 1 oz	30	20	2.1	1.1	8	45	37	12	8	17	14	.02	0.3	50
		0.5		0.5	22	24	29	21	21	1	0.4	.02	2	13
ENFAMIL (24 KCAL/OZ), MEAD JOHNSON 1 oz	30	24	2.4	1.3	9	53	44	14	10	19	16	.02	0.3	60
		0.5		0.6	26	29	34	24	24	2	0.4	.02	2	15
ISOMIL, ROSS LABS	30	20	2.0	1.1	6	47	35	12	9	21	15	.01	0.3	74
1 oz		0.6		0.2	31	26	27	21	21	1	0.4	.02	2	120
L-SOYALAC, LOMA LINDA	30	20	1.9	1.1					9	19	16	.01	0.3	62
1 oz		0.6		0.2					23	2	0.5	.02	2	12
LOFENALAC POWDER, MEAD JOHNSON	9	41	5.1	1.6	18	126	144	41	36	59	45	.03	0.3	97
1 T		0.5			7	68	99	75	90	8	0.9	.12	2	26
LOFENALAC, MEAD JOHNSON	30	20	2.6	0.8					18	30	21	.02	0.1	48
1 oz		0.7							44		0.1	.06	1	13
LONALAC, POWDER, MEAD JOHNSON	9	46	3.4	2.5	24			69	2	81	65	.03	0.1	72
1 T		2.4			143		174	107	90	7	0.2	.13		
LONALAC, MEAD JOHNSON	30	20	1.4	1.1					1	34	31	.13	tr	30
1 oz		1.0							38	3	tr	.05		
MEAT-BASE FORM (MBF), GERBER	30	19	1.9	1.0	9	72	67	31	5	30	20	.02	0.2	53
1 oz		0.9		0.3	41	38	42	34	11	1	0.4	.03	2	14
M-J PREMATURE (24 KCAL/OZ)	30	24	2.7	1.2	6	65	47	22	9	37	19	.02	0.3	60
1 oz		0.6		0.3	32	27	34	27	27	2	tr	.02	2	15
MODILAC, UNDILUTED, GERBER	30	40	4.9	1.7					23	50	38	.04	tr	188
1 oz		1.3							48		0.1	.06	3	38
NEOMULLSOY, SYNTEX	30	20	1.9	1.1		41	32	11	11	25	18	.02	0.2	62
1 oz		0.5		0.6	28	22	22	20	27	2	0.3	.03	2	12
NURSOY	30	20	2.0	1.1	9	57	43	14	6	19	13	.02	0.3	78
1 oz		0.7		0.2	36	31	32	23	22	2	0.4	.03	2	12
NUTRAMIGEN, MEAD JOHNSON	30	20	2.6	0.8	7	65	59	19	9	19	14	.01	0.3	50
1 oz		0.6		0.4	4	41	50	30	20	2	0.4	.02	2	12
PREGESTIMIL, MEAD JOHNSON	30	20	2.6	0.8	7	65	59	19	9	19	14	.01	0.3	50
1 oz		0.6		0.1	4	41	50	30	20	2	0.4	.02	2	12
PRO SOBEE, MEAD JOHNSON	30	20	2.0	1.0	9	54	40	14	12	23	16	.01	0.2	49
1 oz		0.7		0.6	34	31	31	25	22	2	0.4	.02	2	12
SIMILAC PM 60/40 (20 KCAL/OZ), ROSS LABS 1 oz	30	20	2.0	1.1	10	49	39	12	5	12	6	.02	0.2	74
		0.5		0.3	19	26	30	29	17	1	0.1	.03	2	12
SIMILAC (20 KCAL/OZ), ROSS LABS	30	20	2.1	1.1	6	43	36	13	7	15	11	.02	0.2	74
1 oz		0.5		0.2	22	24	30	21	23	1	tr	.03	2	12
SIMILAC W/ IRON (20 KCAL/OZ), ROSS LABS 1 oz	30	20	2.1	1.1	6	43	36	13	7	15	11	.02	0.2	74
		0.5		0.2	22	24	30	21	23	1	0.4	.03	2	12
SIMILAC LBW (24 KCAL/OZ), ROSS LABS 1 oz	30	24	2.5	1.3	8	61	48	17	11	21	16	.03	0.2	88
		0.6		0.2	29	30	38	28	29	2	0.1	.04	3	14
SIMILAC W/ IRON (24 KCAL/OZ), ROSS LABS 1 oz	30	24	2.5	1.3	8	61	48	17	10	21	17	.02	0.2	88
		0.6			29	30	38	28	31	2		.04	2	14
SIMILAC (27 KCAL/OZ), ROSS LABS	30	27	2.8	1.4	9	69	54	19	11	24	18	.03	0.3	99
1 oz		0.7		0.3	34	34	44	31	37	2	tr	.04	2	16
SMA (20 KCAL/OZ), WYETH LABS	30	20	2.1	1.1	8	54	46	10	4	13	10	.02	0.3	78
1 oz		0.4		0.2	24	30	32	31	16	2	0.4	tr	2	12
SMA (24 KCAL/OZ), WYETH LABS	30	24	2.5	1.3	9	65	26	12	5	16	11	.02	0.4	93
1 oz		0.5		0.2	28	36	9	37	20	2	0.4	.04	2	15
SMA (27 KCAL/OZ), WYETH LABS	30	27	2.9	1.4	11	74	62	14	6	18	13	.03	0.4	105
1 oz		0.6		0.2	32	41	43	42	22	2	0.5	.04	2	17
SOYALAC, LOMA LINDA	30	20	1.8	1.1					9	19	8	.01	0.2	62
1 oz		0.6		0.6					23	3	0.3	.02	2	12

*Fluid and normal dilution unless otherwise indicated.

	WT g	CAL	CHO	FAT	PRO	FIB	PUFA	TRY	LEU	LYS	MET	PHE	ISO	VAL	THR	Na	Ca	P	K	Mg	Fe	THI	NIA	A	RIB	ASC	D
					g	g	g	mg	mg	mg	mg	mg	mg	mg	mg	mg	mg	mg	mg	mg	mg	mg	mg	iu			
ACEROLA	100	28	6.8					0.3								8	12	11				.02	0.4				
3½ oz					0.4	0.4										83					0.2	.06	1300				
AMARANTH	100	36	6.5	0.5				38	206	141	25						267	67				.08	1.4	6100			
3½ oz					3.5	1.3						96	164	136	56	411					3.9	.16	80				
APPLE, RAW, WHOLE	100	58	14.5	0.6												1	7	10				.03	0.1	90			
1 small (2″ dm)					0.2	1.0										110					0.3	.02	4				
APPLE, RAW, WHOLE	150	87	21.7	0.9												1	10	15				.04	0.2	140			
1 med (2½″ dm)					0.3	1.5										165					0.5	.03	6				
APPLE, RAW, WHOLE	230	133	33.4	1.4												2	16	23				.07	0.2	200			
1 large (3″ dm)					0.5	2.3										253					0.7	.05	9				
BAKED, UNPARED	150	188	45.7	0.9												1	10	15				.04	0.2	140			
1 med, 2 T sugar					0.3											165					0.5	.03	5				
DRIED, CKD, SWEET	100	112	29.2	0.4												1	8	13				.01	0.1				
⅜ cup					0.3	0.8										144					0.4	.03	tr				
DRIED, CKD, UNSWEET	100	78	20.3	0.5												1	9	15				.01	0.1				
⅜ cup					0.3	0.9										162					0.5	.03	tr				
FRZN SLICES, SWEET	100	93	24.3	0.1												14	5	6				.01	0.2	20			
½ cup					0.2	0.7										68					0.5	.03	7				
APPLE, ESCALLOPED, FRZN, STOUFFERS	100	122	24.1	1.7													5							102			
					0.2																0.3						
APPLESAUCE, CND, SWEET	100	91	23.8	0.1												2	4	5				.02	tr	40			
⅓ cup					0.2	0.5										65					0.5	.01					
CND, UNSWEET	100	41	10.8	0.2												2	4	5				.02	tr	40			
½ cup scant					0.2	0.6										78					0.5	.01	1				
APRICOTS, RAW	100	51	12.8	0.2												1	17	23				.03	0.6	2700			
2 to 3 med					1.0	0.6										281					0.5	.04	10				
CND, WATER PACK	100	38	9.6	0.1												1	12	16				.02	0.4	1830			
3 med halves					0.7	0.4										246					0.3	.02	4				
CND, JUICE PACK	100	54	13.6	0.2												1	17	23				.03	0.5	2700			
3 med halves					1.0	0.4										362					0.5	.03	6				
CND IN HEAVY SYRUP	100	86	22.0	0.1												1	11	15				.02	0.4	1740			
3 med halves					0.6	0.4										234					0.3	.02	4				
DRIED, UNCKD	100	260	66.5	0.5												26	67	108				.01	3.3	10900			
17 large halves					5.0	3.0										979					5.5	.16	12				
DRIED, CKD, SWEET	100	122	31.4	0.1												7	19	31				tr	0.9	2600			
4 halves, 2 T juice					1.4	0.9										278					1.6	.04	2				
FRZN, SWEET	100	98	25.1	0.1												4	10	19				.02	0.8	1680			
⅓ cup					0.7	0.6										229					0.9	.04	8				
APRICOT WHIP	100	109	16.1	5.2												12	7	9				.01	0.2	52			
					0.3	0.2	1.2									132				4	0.2	.01	2	0			
AVOCACO, RAW, PEELED	100	167	6.3	16.4				23		120	19					4	10	42				.11	1.6	290			
½ (3¼ × 4″), pitted					2.1	1.6										604					0.6	.20	14				
RAW, CUBED	152	254	9.6	25.0				35		182	29					6	15	64				.17	3.4	440			
1 cup					3.2	2.4										920					0.9	.30	21				
BANANA, RAW, EP	100	85	22.2	0.2						51	10					1	8	26				.05	0.7	190			
1 small, 6″					1.1	0.5										370					0.7	.06	10				
RAW, EP	150	127	33.3	0.3				24		74	14					2	12	39				.08	1.0	270			
1 med, 1 cup sliced					1.6	0.8										550					1.0	.09	15				
RAW, EP	200	170	44.4	0.4				33		101	20					2	16	52				.10	1.4	380			
1 large					2.2	1.0										740					1.4	.12	20				
DEHYDRATED, (POWDER)	100	340	88.6	0.8				66		202	40					4	32	104				.18	2.8	760			
3½ oz					4.4	2.0										1477					2.8	.24	7				
BANANA FLAKES	50	170	44.5	0.5													16					.09	1.4	380			
½ cup					2.0																1.4	.12	4				
BLACKBERRIES, RAW	100	58	12.9	0.9												1	32	19				.03	0.4	200			
⅝ cup					1.2	4.1										170					0.9	.04	21				
RAW	144	84	18.6	1.3												1	46	27				.04	0.6	290			
1 cup					1.7	5.9										245					1.3	.06	31				
CND, WATER PACK	100	40	9.0	0.6												1	22	13				.02	0.2	140			
⅖ cup					0.8	2.8										115					0.6	.02	1				
CND, JUICE PACK	100	54	12.1	0.8												1	25	17				.02	0.3	150			
⅖ cup					0.8	2.7										170					0.9	.03	10				

Food	WT g	CAL / PRO g	CHO / FIB g	FAT / PUFA g	TRY / PHE mg	LEU / ISO mg	LYS / VAL mg	MET / THR mg	Na / K mg	Ca / Mg mg	P / Fe mg	THI / RIB mg	NIA / ASC mg	A / D iu
CND IN HEAVY SYRUP	100	91	22.2	0.6					1	21	12	.01	0.2	130
2/5 cup		0.8	2.6						109		0.6	.02	7	
FRZN, UNSWEET	100	48	11.4	0.3					1	25	24	.02	1.0	170
2/3 cup		1.2	2.7						153		1.6	.13	13	
BLUEBERRIES, RAW	100	62	15.3	0.5					1	15	13	.03	0.5	100
5/8 cup		0.7	1.5						81		1.0	.06	14	
RAW	140	87	21.4	0.7					1	21	18	.04	0.7	140
1 cup		1.0	2.1						113		1.4	.08		
CND, WATER PACK	121	47	11.9	0.2					1	12	11	.01	0.2	50
1/2 cup		0.6	1.2						73		0.8	.01	7	
CND IN HEAVY SYRUP	120	121	31.2	0.2					1	11	10	.01	0.2	48
1/2 cup		0.5	1.4						66		0.7	.01	7	
FRZN, UNSWEET	100	55	13.6	0.5					1	10	13	.03	0.5	70
5/8 cup		0.7	1.5						81		0.8	.06	7	
FRZN, SWEET	100	105	26.5	0.3					1	6	11	.04	0.4	30
2/3 cup		0.6	0.9						66		0.4	.05	8	
BOYSENBERRIES, CND WATER/JUICE PACK	244	88	22.2	0.2					2	46	46	.02	1.7	32
1 cup		1.7	4.6						207		2.9	.24	17	
BOYSENBERRIES, FRZN, UNSWEET	100	48	11.4	0.3					1	25	24	.02	1.0	170
4/5 cup		1.2	2.7						153		1.6	.13	13	
FRZN, SWEET	100	96	24.4	0.3					1	17	17	.02	0.6	140
2/3 cup		0.8	1.8						105		0.6	.10	8	
BREADFRUIT, RAW	100	103	26.2	0.3					15	33	32	.11	0.9	40
3 1/2 oz		1.7	1.2						439		1.2	.03	29	
CANTALOUPE, RAW	100	30	7.5	0.1	1		18	2	12	14	16	.04	0.6	*3400
1/4 melon (5" dm)		0.7	0.3						251		0.4	.03	33	
RAW, DICED	120	36	9.0	0.1	2		20	2	14	17	19	.05	0.7	*4100
1/2 cup		0.8	0.4						301		0.5	.04	40	
CASABA MELON, RAW	140	38	9.1	tr	8		56	10	17	20	22	.06	0.8	40
1 wedge		1.7	1.4						251	11	0.6	.04	18	
CHERRIES, RED, SOUR, RAW	100	58	14.3	0.3					2	22	19	.05	0.4	1000
1/2 cup		1.2	0.2						191		0.4	.06	10	
CND, WATER PACK	100	43	10.7	0.2					2	15	13	.03	0.2	680
1/2 cup, scant		0.8	0.1						130		0.3	.02	5	
CND IN HEAVY SYRUP	100	89	22.7	0.2					1	14	12	.03	0.2	650
1/2 cup scant		0.8	0.1						124		0.3	.02	5	
FRZN, UNSWEET	100	55	13.4	0.4					2	13	22	.04	0.3	1000
1/2 cup scant		1.0	0.3						188		0.7	.07	5	
FRZN, SUGAR ADDED	100	112	27.8	0.4					2	12	15	.03	0.3	480
1/2 cup scant		1.0	0.2						130		0.5	.06	6	
CHERRIES, SWEET, RAW	100	70	17.4	0.3					2	22	19	.05	0.4	110
15 large; 25 small		1.3	0.4						191		0.4	.06	10	
CND, WATER PACK	100	43	10.7	0.2					2	15	13	.03	0.2	680
1/2 cup, scant		0.8	0.1						130		0.3	.02	5	
CND IN HEAVY SYRUP	100	89	22.7	0.2					1	14	12	.03	0.2	650
1/2 cup, scant		0.8	0.1						124		0.3	.02	5	
CHERRIES, MARASCHINO	100	116	29.4	0.2										
12 large		0.2	0.3											
CHERRIES, WEST INDIAN	100	51	12.5	0.4						9	11	.03	0.3	1500
20 med, ripe		0.8	0.6								0.2	.04	30	
CRABAPPLES, RAW	100	68	17.8	0.3					1	6	13	.03	0.1	40
3 1/2 oz		0.4	0.6						110		0.3	.02	8	
CRANBERRIES, RAW	100	46	10.8	0.7					2	14	10	.03	0.1	40
1 cup		0.4	1.4						82		0.5	.02	11	
CRANBERRIES, FRZN, OCEAN SPRAY	100		10.7	0.2					1	8	11			
		0.4	1.2						74	5			13	
CRANBERRY SAUCE, CND	100	146	37.5	0.2					1	6	4	.01	tr	20
5 rounded T		0.1	0.2						30		0.2	.01	2	
SAUCE, HOMEMADE	100	178	45.5	0.3					1	7	5	.01	0.1	20
1/2 cup, scant		0.2	0.7						38		0.2	.01	2	

*Vitamin A values are based on deeply colored varieties; the paler kinds are lower in Vitamin A.

	WT g	CAL / PRO g	CHO / FIB g	FAT / PUFA g	TRY / PHE mg	LEU / ISO mg	LYS / VAL mg	MET / THR mg	Na / K mg	Ca / Mg mg	P / Fe mg	THI / RIB mg	NIA / ASC mg	A / D iu
CURRANTS, BLACK, RAW	100	54	13.1	0.1					3	60	40	.05	0.3	230
3½ oz		1.7	2.4						372		1.1	.05	200	
CURRANTS, RED OR WHITE, RAW	100	50	12.1	0.2					2	32	23	.04	.1	120
¾ cup		1.4	3.4						257		1.0	.05	41	
CURRANTS, ZANTE, DRIED, DEL MONTE	33	100	24.5	0.4					14	32	36	.04	0.2	0
¼ cup		1.3	.3						366	25	1.0	.02	0	
CUSTARDAPPLE, RAW	100	101	25.2	0.6						27	20	.08	0.5	tr
3½ oz		1.7	3.4								0.8	.10	22	
DATES, DOMESTIC, NATURAL	100	274	72.9	0.5	62	77	66	26	1	59	63	.09	2.2	50
10 medium, pitted		2.2	2.3		64	75	95	62	648		3.0	.10	0	
DRIED, CUT	178	488	130.0	0.9	109	136	117	47	2	105	112	.16	3.9	90
1 cup pitted		3.9	4.1		113	133	168	109	1150		5.3	.18	0	
ELDERBERRIES, RAW	100	72	16.4	0.5						38	28	.07	0.5	600
3½ oz		2.6	7.0						300		1.6	.06	36	
FIGS, FRESH, RAW	100	80	20.3	0.3					2	35	22	.06	0.4	80
2 large; 3 small		1.2	1.2						194		0.6	.05	2	
CND, WATER PACK	100	48	12.4	0.2					2	14	14	.03	0.2	30
3 med		0.5	0.7						155		0.4	.03	1	
CND IN HEAVY SYRUP	100	84	21.8	0.2					2	13	13	.03	0.2	30
3 med, 2 T syrup		0.5	0.7						149		0.4	.03	1	
DRIED	100	274	69.1	1.3					34	126	77	.10	0.7	80
5 med; ⅔ cup cut		4.3	5.6						640		3.0	.10	0	
DRIED	30	82	20.7	0.4					10	38	23	.03	0.2	24
2 small		1.3	1.6						192		0.9	.03	0	
FRUIT COCKTAIL, WATER PACK	100	37	9.7	0.1					5	9	13	.02	0.5	150
½ cup scant		0.4	0.4						168		0.4	.01	2	
CND IN HEAVY SYRUP	100	76	19.7	0.1					5	9	12	.02	0.4	140
½ cup scant		0.4	0.4						161		0.4	.01	2	
GOOSEBERRIES, RAW	100	39	9.7	0.2					1	18	15			290
⅔ cup		0.8	1.9						155		0.5		33	
CND, WATER PACK	100	26	6.6	0.9					1	12	10			200
½ cup scant		0.5	1.3						105		0.3		11	
CND IN HEAVY SYRUP	100	90	23.0	0.1					1	11	9			190
½ cup scant		0.5	1.2						98		0.3		10	
GRANADILLA, PURPLE, RAW	100	90	21.2	0.7					28	13	64	tr	1.5	700
3½ oz		2.2							348		1.6	.13	30	
GRAPEFRUIT, RAW, WHITE	100	41	10.8	0.1	1		6	0	1	16	16	.04	0.2	10
½ med (4" dm)		0.5	0.2						135		0.4	.02	38	
RAW, PINK & RED, SEEDLESS	100	40	10.4	0.1	1		6	0	1	16	16	.04	0.2	440
½ med (4" dm)		0.5	0.2						135		0.4	.02	36	
CND, WATER PACK, SEGMENTS	100	30	7.6	0.1	1		7	0	4	13	14	.03	0.2	10
½ cup		0.6	0.2						144		0.3	.02	30	
CND IN SYRUP	100	70	17.8	0.1	1		7	0	1	13	14	.03	0.2	10
½ cup		0.6	0.2						135		0.3	.02	30	
GRAPES, AMERICAN (SLIPSKIN)	100	69	15.7	1.0					3	16	12	.05	0.3	100
22 med		1.3	0.6						158		0.4	.03	4	
AMERICAN (SLIPSKIN)	153	106	24.0	1.5					5	24	18	.08	0.5	150
1 cup		2.0	0.9						242		0.6	.05	6	
EUROPEAN (ADHERENT SKIN)	100	67	17.3	0.3					3	12	20	.05	0.3	100
24 med		0.6	0.5						173		0.4	.03	4	
EUROPEAN (ADHERENT SKIN)	160	107	27.7	0.5					5	19	32	.08	0.5	160
1 cup		1.0	0.8						278		0.6	.05	6	
THOMPSON SEEDLESS, CND	100	51	13.6	0.1					4	8	13	.04	0.2	70
½ cup		0.5	0.2						110		0.3	.01	2	
GUAVA, COMMON, WHOLE, RAW	100	62	15.0	0.6	8		24	8	4	23	42	.05	1.2	280
1 med		0.8	5.6						289		0.9	.05	*242	
STRAWBERRY	100	65	15.8	0.6	10		30	10	4	23	42	.03	0.6	90
3½ oz		1.0	6.4						289		0.9	.03	37	
HAWS, SCARLET	100	87	20.8	0.7										
3½ oz		2.0	2.1											

*ASC value is average for varieties grown in the United States; they range
from 23 to 1160 mg per 100 gm.

FRUITS (RAW, COOKED AND PROCESSED) (cont'd)

		Proximate					Amino Acids							Minerals						Vitamins				
WT	CAL	CHO	FAT				TRY	LEU	LYS	MET					Na	Ca	P				THI	NIA	A		
	PRO	FIB	PUFA				PHE	ISO	VAL	THR					K	Mg	Fe				RIB	ASC	D		
g	g	g	g				mg	mg	mg	mg					mg	mg	mg				mg	mg	iu		

WT g	CAL	CHO	FAT	PRO	FIB	PUFA	TRY	LEU	LYS	MET	PHE	ISO	VAL	THR	Na	K	Ca	Mg	P	Fe	THI	RIB	NIA	ASC	A	D
HONEYDEW MELON, RAW 100	33	7.7	0.3												12		14		16		.04		0.6		40	
¼ small (5" dm)				0.8	0.6											251				0.4		.03		23		
WEDGE, RAW 150	49	11.5	0.5												18		21		24		.06		0.9		60	
1 piece 2" wide (6½" dm)				1.2	0.9											377				0.6		.05		35		
INDIAN FIG, RAW 100	67	16.6	0.4														57		32		.01		0.3			
				1.1	1.1															1.2		.02		18		
JACKFRUIT, RAW 100	98	25.4	0.3												2		22		38		.03		0.4			
				1.3	1.0											407								8		
JUJUBE (CHINESE DATE), RAW 100	105	27.6	0.2												3		29		37		.02		0.9		40	
				1.2	1.4											269				0.7		.04		69		
JUJUBE (CHINESE DATE), DRIED 100	287	73.6	1.1												3		79		100		.23		0.9		40	
				3.7	3.0											531				1.8		.38		13		
KUMQUATS, RAW 100	65	17.1	0.7												7		63		23		.08				600	
5 to 6 med				0.9	3.7											236				0.4		.10		36		
LEMON, RAW, PEELED* 100	27	8.2	0.3												2		26		16		.04		0.1		20	
1 med				1.1	0.4											138				0.6		.02		53		
RAW, INCLUDING PEEL† 100	20	10.7	0.3												3		61		15		.05		0.2		30	
1 small				1.2												145				0.7		.04		77		
LEMON PEEL, RAW 25		4.0	0.1												1		33		3		.01		0.1		12	
				0.4												40				0.2		.02		32		
LIME, ACID-TYPE, RAW 100	28	9.5	0.2				3		13	2					2		33		18		.03		0.2		10	
1 med				0.7	0.5											102				0.6		.02		37		
LOGANBERRIES, RAW 100	62	14.9	0.6												1		35		17		.03		0.4		200	
⅔ cup				1.0	3.0											170				1.2		.04		24		
CND, WATER PACK 100	40	9.4	0.4												1		24		11		.01		0.2		140	
½ cup				0.7	2.0											115				0.8		.02		8		
CND IN HEAVY SYRUP 100	89	22.2	0.4												1		22		11		.01		0.2		130	
½ cup				0.6	1.9											109				0.8		.02		8		
LOQUAT, RAW 100	48	12.4	0.2														20		36						670	
3½ oz				0.4	0.5											348				0.4				1		
LYCHEES, RAW 100	64	16.4	0.3												3		8		42							
3½ oz				0.9	0.3											170				0.4		.05		42		
DRIED 100	277	70.7	1.2												3		33		181							
3½ oz				3.8	1.4											1100				1.7						
MANGO, RAW 100	66	16.8	0.4				1		9	8					7		10		13		.05		1.1		4800	
½ med				0.7	0.9											189				.4		.03				
MELON BALLS (CANTALOUPE, HONEYDEW), FRZN IN SYRUP 230	143	36.1	0.2				2		35	5					21		33		28		.07		1.2		3540	
1 cup				1.4	0.7											432				0.7		.05		37		
MULBERRIES, BLACK, RED, WHITE 100	62	14.6	0.6												1		36		48							
⅔ to ¾ cup				1.2												200				1.6				12		
NECTARINES, RAW 100	64	17.1	tr												6		4		24						1650	
2 med				0.6	0.4											294				0.5				13		
OLIVES, GREEN, PICKLED 13	15	0.2	1.6												312		8		2							
2 med				0.2	0.2											7				0.2						
RIPE, CND 20	37	0.6	4.0												150		21		3		tr				14	
2 large				0.2	0.3											5				0.3		tr				
RIPE, SALT CURED (GREEK) 20	67	1.7	7.1												658				6							
3 med				0.4	0.8																					
ORANGE, WHOLE EP 100	49	12.2	0.2				3		27	3					1		41		20		.10		0.4		200	
1 small (2½" dm)				1.0	.5											200				0.4		.04		50		
WHOLE EP 150	73	18.3	0.3				4		40	4					2		62		30		.15		0.6		300	
1 med (3" dm)				1.5	0.8											300				0.6		.06		80		
WHOLE EP 235	115	28.7	0.5				7		59	7					2		96		47		.24		0.9		470	
1 large (3⅜" dm)				2.4	1.2											470				0.9		.10		120		
SEGMENTS 97	47	11.8	0.2				3		27	3					1		40		20		.10		0.4		194	
½ cup				1.0	0.4											194				0.4		.04		50		
ORANGE, CALIFORNIA, NAVEL, RAW 206	71	17.8	0.1				6	2	48	6					1		56		31		.14		0.6		280	
1 med				1.8	1.0	0.0	24	2	2	2					272		22		0.6		.06		85		
ORANGE, CALIF VALENCIA, RAW 161	62	15.0	0.4				5	2	39	3					1		48		27		.12		0.5		240	
1 med				1.4	0.8		19	2	2	2					230		18		1.0		.05		59		
ORANGE, FLORIDA, ALL VARIETIES, RAW 204	71	18.1	0.3				6	2	48	6					2		65		26		.15		0.6		300	
1 med				1.1	1.0		24	2	2	2					311		22		0.3		.06		68		

*ASC value applies to lemons marketed in summer.   †Value weighted for seasonal variation.

	WT	Proximate CAL · PRO	CHO · FIB	FAT · PUFA	Amino Acids TRY · PHE	LEU · ISO	LYS · VAL	MET · THR	Minerals Na · K	Ca · Mg	P · Fe	Vitamins THI · RIB	NIA · ASC	A · D
	g	g	g	g	mg	mg	mg	mg	mg	mg	mg	mg	mg	iu
ORANGE, MANDARIN, CND	100	61	16.2	0.1					6	7	10	.06	0.5	1393
1 cup		0.5		0.0					83	8	0.4	.04	20	
ORANGE PEEL, RAW	25		6.2	0.1					1	40	5	.03	0.2	105
		0.4							52	105	0.2	.02	34	
PAPAWS	100	85	16.8	0.9										
		5.2												
PAPAYAS, RAW	100	39	10.0	0.1	12		38	2	3	20	16	.04	0.3	1750
1/3 med; 2/5 cup pulp		0.6	0.9						234		0.3	.04	56	
PEACHES, RAW*	100	38	9.7	0.1					1	9	19	.02	1.0	1330
1 med		0.6	0.6						202		0.5	.05	7	
RAW, SLICED	168	64	16.3	0.2					2	15	32	.03	1.7	2240
1 cup		1.0	1.0						340		0.8	.08	12	
CND, WATER PACK	100	31	8.1	0.1					2	4	13	.01	0.6	450
1/2 cup slices		0.4	0.4						137		0.3	.03	3	
CND, JUICE PACK	100	45	11.6	0.1					2	6	19	.01	0.9	670
2 med halves, 2 T juice		0.6	0.4						205		0.5	.04	4	
CND IN HEAVY SYRUP	100	78	20.1	0.1					2	4	12	.01	0.6	430
2 med halves, 2 T syrup		0.4	0.4						130		0.3	.02	3	
DEHYDRATED, UNCKD	100	340	88.0	0.9					21	62	151	tr	7.8	5000
1/2 cup		4.8	4.0						1229		3.5	.10	14	
DRIED, UNCKD	100	262	68.3	0.7					16	48	117	.01	5.3	3900
5/8 cup		3.1	3.1						950		6.0	.19	18	
DRIED, CKD, UNSWEET	100	82	21.4	0.2					5	15	37	tr	1.5	1220
1/3 cup; 3 med halves, juice		1.0	1.0						297		1.9	.06	2	
DRIED, CKD, SWEET	100	119	30.8	0.2					4	13	32	tr	1.4	1070
1/3 cup; 3 med halves, juice		0.9	0.9						261		1.6	.05	2	
FRZN, SLICED, UNSWEET†	100	88	22.6	0.1					2	4	13	.01	0.7	650
2/5 cup		0.4	0.4						124		0.5	.04	41	
PEARS, RAW	100	61	15.3	0.4					2	8	11	.02	0.1	20
1/2 pear (3 × 2½")		0.7	1.4						130		0.3	.04	4	
CND, WATER PACK	100	32	8.3	0.2					1	5	7	.01	0.1	tr
2/5 cup		0.2	0.7						88		0.2	.02	1	
CND IN HEAVY SYRUP	100	76	19.6	0.2					1	5	7	.01	0.1	tr
2 small halves, 2 T syrup		0.2	0.6						84		0.2	.02	1	
DRIED, CKD	100	126	31.7	0.8					3	16	23	tr	0.3	30
3½ oz		1.5	2.9						269		0.6	.08	2	
PERSIMMON, HACHIYA	136	103	27.6	0.2						8	26		0.2	1340
1 med		0.7	0.5								0.2	.08	10	
PERSIMMON, JAPANESE, RAW	100	77	19.7	0.4					6	6	26	.03	0.1	2710
1 med		0.7	1.6						174		0.3	.02	11	
NATIVE, RAW	100	127	33.5	0.4					1	27	26			
1 med		0.8	1.5						310		2.5		66	
PINEAPPLE, RAW, DICED	100	52	13.7	0.2	5		9	1	1	17	8	.09	0.2	70
3/4 cup		0.4	0.4						146		0.5	.03	17	
RAW, SLICED	84	44	11.5	0.2	5		9	1	1	14	7	.08	0.2	60
1 slice (3½" dm × 3/4")		0.4	0.4						123		0.4	.03	14	
CND, WATER PACK	100	39	10.2	0.1	4		7	1	1	12	5	.08	0.2	50
1 large slice		0.3	0.3						99		0.3	.02	7	
CND, JUICE PACK	100	58	15.1	0.1	5		9	1	1	16	8	.10	0.3	60
1 large slice & juice		0.4	0.3						147		0.4	.03	10	
CND IN HEAVY SYRUP	100	74	19.4	0.1	4		7	1	1	11	5	.08	0.2	50
1 large slice & syrup		0.3	0.3						96		0.3	.02	7	
FRZN, CHUNKS, UNSWEET	100	85	22.2	0.1	5		9	1	2	9	4	.10	0.3	30
3/8 cup		0.4	0.3						100		0.4	.03	8	
PITANGA (SURINAM CHERRY), RAW	100	51	12.5	0.4						9	11	.03	0.3	1500
3½ oz		0.8	0.6								0.2	.04	30	
PLANTAIN (BAKING BANANA), RAW	100	119	31.2	0.4	10	59	50	6	5	7	30	.06	0.6	‡
1 small (5" long)		1.1	0.4		50	56	65	28	385		0.7	.04	14	
PLUMS, DAMSON, RAW	100	66	17.8	tr					2	18	17	.08	0.5	(300)
2 med		0.5	0.4						299		0.5	.03		

*Vitamin A value is based on yellow-fleshed varieties; for white-fleshed kinds the value is about 50 iu per 100 gm.

†Value when ascorbic acid is added prior to freezing. For product without added ascorbic acid the value is about 11 mg per 100 gm.

‡Vitamin A value varies with color: white varieties have about 10 iu per 100 gm, while deep yellow varieties have as much as 1200 iu per 100 gm.

	WT g	Proximate CAL	CHO	FAT	Amino Acids TRY mg	LEU mg	LYS mg	MET mg	Minerals Na mg	Ca mg	P mg	Vitamins THI mg	NIA mg	A iu
		PRO g	FIB g	PUFA g	PHE mg	ISO mg	VAL mg	THR mg	K mg	Mg mg	Fe mg	RIB mg	ASC mg	D iu
PLUMS, JAPANESE, RAW	70	32	8.1	0.1					1	8	12	.02	0.3	160
1 plum		0.3							112			.02	4	
PRUNE TYPE, RAW*	100	75	19.7	0.2					1	12	18	.03	0.5	300
3 med		0.8	0.4						170		0.5	.03	4	
GREEN GAGE, CND, WATER PACK	100	33	8.6	0.1					1	9	13	.01	0.3	160
3½ oz		0.4	0.2						82		0.2	.02	2	
PURPLE, CND, WATER PACK	100	46	11.9	0.2					2	9	10	.02	0.4	1250
2 med		0.4	0.3						148		1.0	.02	2	
PURPLE, CND IN HEAVY SYRUP	100	83	21.6	0.1					1	9	10	.02	0.4	1210
3 med, 2 T syrup		0.4	0.3						142		0.9	.02	2	
POHA (GROUND CHERRY), RAW	100	53	11.2	0.7						9	40	.11	2.8	720
⅔ cup, whole berries		1.9	2.8								1.0	.04	11	
POKEBERRY (POKE), RAW SHOOTS	100	23	3.7	0.4						53	44	.08	1.2	8700
3½ oz		2.6									1.7	.33	136	
BOILED, DRAINED	100	20	3.1	0.4						53	33	.07	1.1	8700
3½ oz		2.3									1.2	.25	82	
POMEGRANATE, RAW	100	63	16.4	0.3					3	3	8	.03	0.3	tr
1 med, pulp & seeds		0.5	0.2						259		0.3	.03	4	
PRUNES, DEHYDRATED, NOT CKD	100	344	91.3	0.5					11	90	107	.12	2.1	2170
8 large		3.3	2.2						940		4.4	.22	4	
DEHYDRATED, CKD, SWEET	100	180	47.1	0.2					4	31	37	.03	0.7	760
4 med, 2 T juice		1.2	0.8						329		1.5	.07	1	
DRIED, NOT CKD	100	255	67.4	0.6					8	51	79	.09	1.6	1600
10 large		2.1	1.6						694		3.9	.17	3	
DRIED, CKD, UNSWEET	100	119	31.4	0.3					4	24	37	.03	0.7	750
5 med, 2 T juice		1.0	0.8						327		1.8	.07	1	
DRIED, CKD, SWEET	100	172	45.1	0.2					3	19	30	.03	0.6	600
5 med, 2 T juice		0.8	0.6						262		1.5	.06	1	
DRIED, CKD, SUGAR & LEMON	90	120	31.6	0.2					4	19	29	.02	0.4	540
5 med, 2 T juice		0.8	0.8						330		1.3	.04	1	
PUMPKIN, RAW	100	26	6.5	0.1	13	52	48	9	1	21	44	.05	0.6	1600
3½ oz		1.0	1.1		26	37	37	23	340		0.8	.11	9	
CND†	100	33	7.9	0.3	13	52	48	9	2	25	26	.03	0.6	6400
⅖ cup		1.0	1.3		26	37	37	23	240		0.4	.05	5	
QUINCES, RAW	100	57	15.3	0.1					4	11	17	.02	0.2	40
3½ oz		0.4	1.7						197		0.7	.03	15	
RAISINS, DRIED, SEEDLESS	10	29	7.7	tr					3	6	10	.01	0.1	2
1 T		0.3	0.1						76		0.4	.01	tr	
DRIED, SEEDLESS	100	289	77.4	0.2					27	62	101	.11	0.5	20
⅝ cup		2.5	0.9						763		3.5	.08	1	
DRIED, SEEDED	71	205	55.0	0.2					19	44	72	.08	0.4	14
½ cup		1.8	0.7						542		2.5	.06	1	
CKD, SWEET	85	181	47.8	0.1					11	25	40	.04	0.2	8
½ cup, juice & lemon		1.0	0.3						300		1.4	.03	tr	
RASPBERRIES, BLACK, RAW	100	73	15.7	1.4					1	30	22	.03	0.9	tr
⅔ cup		1.5	5.1						199		0.9	.09	18	
CND, WATER PACK	100	51	10.7	1.1					1	20	15	.01	0.5	tr
½ cup		1.1	3.3						135		0.6	.04	6	
RASPBERRIES, RED, RAW	100	57	13.6	0.5					1	22	22	.03	0.9	130
¾ cup		1.2	3.0						168		0.9	.09	25	
CND, WATER PACK	100	35	8.8	0.1					1	15	15	.01	0.5	90
½ cup		0.7	2.6						114		0.6	.04	9	
FRZN, SWEET	123	121	30.3	0.2					1	16	21	.02	0.7	86
½ cup		0.8	2.7						123		0.7	.07	26	
FRZN, SWEET	284	278	69.9	0.6					3	37	48	.06	1.7	199
10 oz carton		2.0	6.2						284		1.7	.17	60	
RHUBARB, RAW, CUBED	100	16	3.7	0.1					2	‡96	18	.03	0.3	100
¾ to 1 cup		0.6	0.7						251		0.8	.07	9	
CKD, SWEET	100	141	36.0	0.1					2	‡78	15	.02	0.3	80
⅜ cup		0.5	0.6						203		0.6	.05	6	

*Italian and Imperial prunes are much higher in Vitamin A value, averaging 1340 iu per 100 gm.

†Canned pumpkin may be a mixture of pumpkin and winter squash.

‡Calcium content may be partially unavailable because of the presence of oxalic acid.

	WT	CAL	Proximate CHO	FAT	TRY	Amino Acids LEU	LYS	MET	Na	Minerals Ca	P	THI	Vitamins NIA	A
		PRO	FIB	PUFA	PHE	ISO	VAL	THR	K	Mg	Fe	RIB	ASC	D
	g	g	g	g	mg	mg	mg	mg	mg	mg	mg	mg	mg	iu
FRZN, SWEET	100	143	36.2	0.2					3	*78	12	.02	0.2	70
⅜ cup		0.5	0.8						176		0.7	.04	6	
SAPODILLO, RAW	100	89	21.8	1.1					12	21	12	tr	0.2	60
½ cup scant, pulp		0.5	1.4						193		0.8	.02	14	
SAPOTES (MARMALADE PLUM) RAW	100	125	31.6	0.6						39	28	.01	1.8	410
½ cup		1.8	1.9								1.0	.02	20	
STRAWBERRIES, RAW, WHOLE	100	37	8.4	0.5					1	21	21	.03	0.6	60
10 large		.7	1.3						164		1.0	.07	59	
RAW	150	56	12.6	0.8					2	32	32	.04	0.9	90
1 cup		1.0	2.0						246		1.5	.10	88	
CND, WATER PACK	100	22	5.6	0.1					1	14	14	.01	0.4	40
⅜ cup		0.4	0.6						111		0.7	.03	20	
FRZN, SLICED, SWEET	128	140	35.6	0.2					1	18	22	.02	0.6	38
½ cup		0.6	1.0						143		1.0	.07	68	
FRZN, WHOLE, SWEET	122	112	28.7	0.2					1	16	20	.02	0.6	37
½ cup		0.5	0.7						127		0.7	.07	67	
FRZN, WHOLE, SWEET	284	261	66.7	0.6					3	37	45	.06	1.4	85
10 oz carton		1.1	1.7						295		1.7	.17	156	
FRZN	454	418	106.7	0.9					5	59	73	.09	2.3	136
16 oz carton		1.8	2.7						472		2.7	.27	250	
TAMARIND, RAW	100	239	62.5	0.6					51	74	113	.34	1.2	30
3½ oz		2.8	5.1						781		2.8	.14	2	
TANGELOS, RAW	170	39	9.2	0.1					2	27	20			
1 medium		0.5							296	19	0.2		26	
TANGERINE, RAW	100	46	11.6	0.2					2	40	18	.06	0.1	420
1 large; 2 small		0.8	0.5						126		0.4	.02	31	
TOMATO, RAW	100	22	4.7	0.2	10	45	46	8	3	13	27	.06	0.7	900
1 small		1.1	0.5		31	32	31	36	244		0.5	.04	23	
TOWEL GOURD, RAW	100	18	4.1	0.2						19	33	.03	0.4	38
		0.8	0.5								0.9	.04	8	
WATERMELON, RIPE, BALLS/CUBES	100	26	6.4	0.2					1	7	10	.03	0.2	590
½ cup		0.5	0.3						100		0.5	.03	7	
SLICE	600	156	38.4	1.2					6	42	60	.18	1.2	3540
1 slice (6″ dm × 1½″)		3.0	1.8						600		3.0	.18	42	
WEDGE	900	234	57.6	1.8					9	63	90	.27	1.8	5310
¹⁄₁₆ melon (10 × 16″)		4.5	2.7						900		4.5	.27	63	
WAXGOURD (CHINESE PRESERVING	100	13	3.0	0.2	2		9	3	6	19	19	.04	0.4	
MELON), RAW		0.4	0.5						111		0.4	.11	13	

*Calcium content may be partially unavailable because of the presence of oxalic acid.

	WT	Proximate			Amino Acids				Minerals			Vitamins		
		CAL	CHO	FAT	TRY	LEU	LYS	MET	Na	Ca	P	THI	NIA	A
		PRO	FIB	PUFA	PHE	ISO	VAL	THR	K	Mg	Fe	RIB	ASC	D
	g	g	g	g	mg	mg	mg	mg	mg	mg	mg	mg	mg	iu
BARLEY, PEARLED LIGHT, RAW	100	349	78.8	1.0	100	568	279	114	3	16	189	.12	3.1	0
½ cup		8.2	0.5		429	343	411	279	160	37	2.0	.05	0	
BARLEY, PEARLED REG/QUICK, RAW, QUAKER SCOTCH ½ cup	100	360	75.4	1.2					7	22	240	.23	4.5	200
		11.4	0.7						260	74	2.4	.13	0	0
BUCKWHEAT, WHOLE GRAIN, RAW	100	360	72.9	2.4						114	282	.60	4.4	0
1 cup		11.7	9.9	0.9					448	253	3.1		0	
CORN FIELD,* WHOLE GRAIN, RAW	100	358	72.2	4.1					1	22	268	.37	2.2	490
		8.9	2.0	2.2					284	147	2.1	.12	0	
CORN, SWEET, WHITE & YELLOW, RAW	100	96	22.1	1.0	21	385	130	66	tr	3	111	.15	1.7	†400
1 med ear		3.5	0.7		196	130	220	144	280	48	0.7	.12	12	
CORN, SWEET, WHITE & YELLOW, CKD	100	100	21.0	1.0	20	363	122	63	tr	3	89	.12	1.4	†400
1 med ear		3.3	0.7		185	122	208	135	196		0.6	.10	9	
CORN, SWEET, CND	85	70	16.4	0.7	13	242	81	42	196	4	41	.02	.7	290
½ cup		2.2	0.7		123	81	14	90	81	17	0.4	.04	4	
CORN, SWEET, FRZN	100	94	21.6	1.0	21	385	130	66	1	3	96	.14	1.7	†350
1 med ear		3.5	0.7		196	130	220	144	231	34	0.8	.08	7	
CORN, CREAM STYLE, CND	256	210	51.2	1.5	41	732	246	125	604	8	143	.08	2.6	840
1 cup		5.4	1.3		374	246	420	274	248	49	1.5	.13	13	
CORN, CREAM STYLE, FRZN	256	192	41.0	1.0					512	18	128	.10	3.8	349
1 cup		5.1	1.8						384		0.8	.20	13	
CORN GRITS, ENR, DRY	40	145	31.3	0.3	21	455	101	66	tr	2	29	.18	1.4	†176
¼ cup		3.5	0.2		157	161	178	140	32	8	1.2	.10	0	
CORN GRITS, ENR, CKD	242	123	26.8	0.2	17	377	84	55		2	24	.09	1.0	†145
1 cup		2.9	0.2		131	134	147	116	27	7	.7	.07	0	
CORN, GRITS, UNENR, DRY	40	145	31.3	0.3	21	455	101	66	tr	2	29	.05	0.5	†176
¼ cup		3.5	0.2		157	161	178	140	32	8	.4	.02	0	
CORN, GRITS, UNENR, CKD	242	123	26.8	0.2	17	377	84	55		2	24	.05	0.5	†145
1 cup		2.9	0.2		131	134	147	116	27	7	.2	.02	0	
CORN, GRITS, INSTANT DRY, QUAKER	40	140	31.0	0.2					668	15	30	.22	1.8	0
¼ cup		3.5	0.2						49	8	1.4	.13	0	0
CORN, GRITS, BACON-FLAVORED, INSTANT DRY, QUAKER ¼ cup	40	142	30.5	0.6					767	10	39	.18	1.4	0
		3.8	0.2						91	13	1.1	.10	0	0
CORN GRITS, CHEESE-FLAVORED, INSTANT DRY, QUAKER ¼ cup	40	147	30.4	1.4						20		.18	1.4	0
		3.1									1.1	.10	0	
CORN GRITS, HAM-FLAVORED, INSTANT DRY, QUAKER ¼ cup	40	139	30.0	0.4					928	10	46	.18	1.4	0
		3.8	0.2						80	12	1.1	.10	0	0
CORN HOMINY‡, GRITS, ENR, DRY QUAKER ¼ cup	40	143	31.6	0.4		544	65	75	tr	1	31	.18	1.4	0
		3.4	0.2		194	147	164	112	43	8	1.1	.10	0	0
CORN HOMINY‡ GRITS, ENR, CKD	242	123	26.6	0.2	17	375	82	53		2	24	.10	1.0	145
1 cup		2.9	0.2		131	133	145	114	27	7	0.7	.07	0	
CORNMEAL														
BOLTED, ENR, DRY	36	131	28.2	0.4	17	365	81	53	tr	2	36	.16	1.3	158
¼ cup		2.8	0.2		126	129	143	112	43	17	1.0	.09	0	
BOLTED, ENR, DRY	145	530	113.6	1.7	68	1484	332	217	1	9	142	.64	5.1	640
1 cup		11.4	0.9		513	524	582	456	174	68	4.2	.38	0	
BOLTED, ENR, CKD	238	119	25.5	0.5	14	312	70	46		2	33	.14	1.2	142
1 cup		2.4	0.2		108	110	122	96	38	16	1.0	.10	0	
BOLTED, UNENR, DRY	145	530	113.6	1.7	68	1484	332	217	1	9	142	.20	1.4	640
1 cup		11.4	0.9		513	524	582	456	174	68	1.6	.07	0	
BOLTED, UNENR, CKD	238	119	25.5	0.5	14	312	70	46		2	33	.05	0.2	142
1 cup		2.4	0.2		108	110	122	96	38	16	0.5	.02	0	
CORNMEAL, WHOLE GROUND														
BOLTED, DRY	30	108	22.3	1.0	16	352	78	51	1	5	67	.01	0.6	144
¼ cup		2.7	0.3		121	124	138	108	75	32	0.5	.02	0	
NOT BOLTED, DRY	30	107	22.1	1.2	16	350	78	51	1	6	77	.11	0.6	153
¼ cup		2.8	0.5		121	123	137	109	85	32	0.7	.03	0	
MILLER PROSO (BROOMCORN, HOG MILLET), WHOLE GRAIN, RAW	100	368	72.9	4.1	109	1772	218	297		20	311	.73	2.3	0
		9.9	3.2	1.8	713	802	733	327	430	167	6.8	.38	0	

*An Indian corn (as dent corn, flint corn, or soft corn) grown for feeding stock or for market grain and having kernels that are usually white or yellow, and not sweet.

†Vitamin A values are for yellow varieties; white varieties contain only a trace.

‡Kernels of hulled corn (especially white flint corn) w/germ removed.

		Proximate			Amino Acids				Minerals			Vitamins		
	WT	CAL	CHO	FAT	TRY	LEU	LYS	MET	Na	Ca	P	THI	NIA	A
		PRO	FIB	PUFA	PHE	ISO	VAL	THR	K	Mg	Fe	RIB	ASC	D
	g	g	g	g	mg	mg	mg	mg	mg	mg	mg	mg	mg	iu
OAT FLAKES, MAPLE-FLAVORED,	48	184	34.7	2.0					tr	24	173	.17	0.5	0
INSTANT, DRY ½ cup		7.0	0.3						163	70	1.7	.07	0	
OAT FLAKES, MAPLE-FLAVORED,	240	166	31.2	1.9					257	2	156	.14		0
INSTANT, CKD 1 cup		6.2	0.2								1.4		0	
OATMEAL/ROLLED OATS, DRY	28	111	19.3	2.1	52	300	147	59	.6	15	115	.17	0.3	0
⅓ cup		4.0	0.3		214	207	238	133	96		1.3	.04	0	0
OATMEAL/ROLLED OATS, CKD	236	148	26.0	2.8	76	502	221	86	1	21	158	.22	0.4	0
1 cup		5.4	0.5		275	275	319	205	130		1.7	.05	0	0
OATMEAL, CKD, MAYPO	177	113	21.0	1.4						15	261	.12		0
¾ cup		4.0									1.1		0	0
OATMEAL W/ APPLES & CINN,	40	152	29.3	1.9					204	113	124	.34	3.4	1128
INSTANT, DRY, QUAKER ½ cup		4.4							122	34	4.1	.19	0	0
OATMEAL W/MAPLE & BROWN SUGAR,	40	154	30.3	1.8					215	94	126	.28	2.8	940
INSTANT, DRY, QUAKER ½ cup		4.4	0.3						92	80	3.4	.16	0	0
OATMEAL W/ RAISINS & SPICES, DRY,	40	150	29.5	1.7					214	94	116	.28	2.8	940
INSTANT, QUAKER ½ cup		4.1	0.4						133	54	3.4	.16	0	0
OATMEAL W/RAISINS & BRAN, DRY,	40	144	27.4	1.6					0	94	128	.28	2.8	940
INSTANT, QUAKER ½ cup		4.9	1.0								3.4	.16	0	0
OATS & WHEAT, DRY	48	175	32.8	2.4					1	25	203	.24	1.2	0
½ cup		7.1	0.7									.09	0	
OATS & WHEAT, CKD	245	159	29.6	2.2					412	27	184	.22	1.2	0
1 cup		6.4	0.7								1.7	.07	0	
OATS, TOASTED WHEAT GERM & SOY	40	153	23.4	36					3	28	236	.42	0.6	0
GRITS, DRY ½ cup		8.2	1.4								2.8	.07	0	
OATS, TOASTED WHEAT GERM & SOY	236	146	22.4	3.5					689	31	227	.38	0.5	
GRITS, CKD 1 cup		7.8									2.6	.07	0	
POPCORN, POPPED	14	54	10.7	0.7	11	234	52	34	tr	2	39		0.3	
1 cup		1.8	0.3		81	83	92	72			.4	.01	0	
POPCORN, POPPED W/ FAT & SALT	21	81	16.1	1.0	17	351	78	50	175	2	59	.09	0.5	0
1½ cups		2.7	0.5	1.0	210	124	139	107	42	36	0.6	.03	0	
POPCORN, POPPED W/ SUGAR	35	134	29.9	1.2					tr	2	47		0.4	
COATING 1 cup		2.1									0.5	.02	0	
PORRIDGE, LOW-PROTEIN, DRY,	50	170	42.6	tr		13			10					
SEMOLINO		0.2	2.0		6	6	7		5					
RICE, BROWN, RAW	49	176	37.9	0.9	41	318	144	67	4	16	108	.17	2.3	0
¼ cup		3.7	0.4	0.1	185	174	259	144	105	43	0.8	.02	0	
RICE, BROWN, CKD	150	178	38.2	0.9	42	327	148	68	423	18	110	.14	2.1	0
1 cup		3.8	0.4	0.1	190	179	266	148	105	45	0.8	.03	0	
RICE, WHITE														
BROWNED IN BEEF STOCK, FRZN,	100	118	21.3	2.5					688	14	57	.02	0.5	50
GREEN GIANT		2.6	0.2						40		0.5	.12	5	
CREAM OF RICE, DRY	10	32	7.9	tr	9	47	28	21	tr	2	14	.04	0.6	
1 T		0.7	tr		39	47	49	30	13	3	0.5	.01		
CREAM OF RICE, CKD	245	176	38.7	0.2						12	66			
1 cup		3.7	0.0						64	20				
ENR, ALL VARIETIES, RAW	49	178	39.4	0.2	36	284	129	59	3	12	46	.22	1.7	0
¼ cup		3.3	0.1		165	155	231	129	45	14	1.4	.13	0	
ENR, ALL VARIETIES, CKD	150	164	36.3	0.2	33	258	117	54	561	15	42	.16	1.5	0
1 cup		3.0	0.2	0.1	150	141	210	117	42	12	1.4	.10	0	
ENR, LONG GRAIN, PARBOILED, DRY	31	114	25.3	0.1	25	198	90	41	3	18	62	.14	1.1	0
¼ cup		2.3	0.1		115	108	161	90	47		0.9		0	
ENR, LONG GRAIN, PARBOILED, CKD	150	159	35.0	0.2	35	275	125	58	538	29	85	.16	1.8	0
1 cup		3.2	0.2	0.1	160	150	224	125	65		1.2		0	
ENR, LONG GRAIN, INSTANT, DRY	31	116	25.6	0.1	25	198	90	41	tr	2	20	.14	1.1	0
¼ cup		2.3	0.1		115	108	161	90			0.9		0	
ENR, LONG GRAIN, INSTANT, CKD	148	161	35.8	tr	36	284	129	59	404	4	28	.19	1.5	0
1 cup		3.3	0.1		165	155	231	129	tr		1.2		0	
FRIED	155	353	31.9	23.4						12	37	.14	1.2	0
1 cup		2.6								11	1.2	.02	0	
FRIED, FRZN, LA CHOY	155	203	43.1	1.1					594					
1 cup		4.2												
GRANULATED, DRY	47	180	40.4	0.1						4	45	.20	2.7	0
¼ cup		2.8	0.1								2.5	.05	0	

	WT	Proximate CAL	CHO	FAT	Amino Acids TRY	LEU	LYS	MET	Minerals Na	Ca	P	Vitamins THI	NIA	A
		PRO	FIB	PUFA	PHE	ISO	VAL	THR	K	Mg	Fe	RIB	ASC	D
	g	g	g	g	mg	mg	mg	mg	mg	mg	mg	mg	mg	iu
GRANULATED, CKD	200	100	22.4						352	4	26	.12	1.6	0
		1.6									1.4	.02	0	
LONG GRAIN & WILD BLEND,	50	178	37.4	0.4					860	42	103	.20	2.5	11
DRY, UNCLE BENS		5.2	0.3						161	25	4.1	.05	5	
LONG GRAIN & WILD BLEND, CKD,	228	221	47.9	0.5								.25	3.2	
UNCLE BENS 1 cup		6.8										.07	7	
MEDLEY, FRZN, GREEN GIANT	85	176	13.8	1.5					327	10	24	.17	1.3	123
1 serving		1.8							29		1.3	.04	3	
PILAF, FRZN, GREEN GIANT	85	94	15.6	2.7					525	10	26	.03	0.7	7
1 serving		1.9	0.1						49		0.3	.04	2	
SPANISH, HOMEMADE	150	130	24.9	2.6					475	21	58	.06	1.1	990
1 cup		2.7	0.8						347		0.9	.04	22	
SPANISH, FRZN, GREEN GIANT	200	200	36.0	4.0					840	20	40	.08	1.0	130
		4.0							190		0.6	.10	6	
SPANISH, STOKELY VAN CAMP	200	174	33.2	3.4					1240	28		.08	1.4	132
		3.6	1.0						208		1.2	.06	30	
UNENR, ALL VARIETIES, RAW	49	178	39.4	0.2	36	284	129	59	3	12	46	.03	0.8	0
¼ cup		3.3	0.1		165	155	231	129	45	14	0.4	.01	0	
UNENR, ALL VARIETIES, CKD	150	164	36.3	0.2	33	258	117	54	561	15	42	.03	0.6	0
1 cup		3.0	0.2	0.1	150	141	210	117	42	12	0.3	.01	0	
UNENR, GLUTINOUS (MOCHI GOMI),	50	180	39.9	0.4					5	18	50	.03	1.0	0
RAW ¼ cup		2.8	0.1						65		1.0	.02	0	
VERDI, FRZN, GREEN GIANT	100	108	20.5	2.0					480	10	16	.14	1.1	6
		2.0	0.6						25		1.5	.05	3	
& WILD, FRZN, GREEN GIANT	100	91	18.8	0.7					460	10	30	.14	1.4	77
		2.0	0.5						30		1.3	.05	3	
W/ ALMONDS, FRZN, GREEN	100	140	18.0	6.0					655	20	45	.04	0.5	5
GIANT		3.0							95		0.3	.05	3	
RICE, BRAN	100	276	50.8	15.8						76	1386	2.26	29.8	0
3½ oz		13.3	1.5						1495		19.4	.25	0	
RICE, POLISH	100	265	57.7	12.8						69	1106	1.84	28.2	0
3½ oz		12.1	2.4						714		16.1	.18	0	
RYE, WHOLE GRAIN, RAW	100	362	73.4	2.2	137	813	494	191	1	38	376	.43	1.6	0
		12.1	2.0	1.0	571	515	631	448	467	133	3.7	.22	0	
SORGHUM GRAIN, ALL TYPES, RAW	100	366	73.0	3.3	123	1767	299	190		28	287	.38	3.9	0
		11.0	1.7	1.6	547	598	628	394	350		4.4	.15	0	
TAPIOCA, DRY	10	35	8.6	tr					tr	1	2	.00	0.0	0
1 T		tr	tr						2	tr	tr	.00	0	0
TAPIOCA, INSTANT, GENERAL FOODS	10	36	8.9	tr					1	4	5	.00	0.0	0
1 T		tr	tr						2		0.1	.00		
WHEAT BRAN	9	32	5.6	0.4	23	85	59	17				.06	1.9	
1 T		1.4	0.8		52	58	66	40				.32		
WHEAT CEREAL														
BULGUR, DRY	28	99	21.2	0.4	173		108	75		8	95	.08	1.3	0
2 T		3.1	0.5						64		1.0	.04	0	
BULGUR, CND	28	47	9.8	0.2	95		60	41	168	6	56	.01	0.7	0
2 T		1.7	0.2						24		0.4	tr	0	
COCOA WHEATS, DRY, LITTLE CROW	50	182	37.8	0.9					3	22		.27	3.2	39
		5.6	tr						125		2.3	.32	10	
CREAM OF WHEAT, INSTANT, DRY	38	130	26.9	0.4					96	185	207	.19	1.4	0
(1 cup when ckd)		4.4	0.1								15.6	.11	0	0
CREAM OF WHEAT, REG, DRY, CKD		133	28.2	0.4					1	13	40	.02	0.3	0
1 cup		4.5	0.1								1.4	.02	0	0
FARINA, ENR, REG, DRY	38	140	29.6	0.3		262	76	46	2	6	33	.16	1.3	0
(1 cup when ckd)		3.8	0.2		198	186	167	106			2.0	.09	0	0
FARINA, ENR, INSTANT, CKD	245	135	27.9	0.2					461	189	147	.17	1.2	0
1 cup		4.2							32			.10	0	
MALTEX, DRY	30	115	24.0	0.4					1.0	18	48	.10		0
(¾ cup when ckd)		3.5									1.1		0	0
MALT-O-MEAL, DRY	28	101	21.7	0.2					1	6	51	.15	1.1	0
(¾ cup when ckd)		3.2	0.1						16		2.0	.09	0	0
PEP, DRY, KELLOGGS	28	106	23.6	0.4					226	11	108	.25	1.1	0
1 oz (1 cup ckd)		2.4	0.5								0.8	.03	0	400

	WT	Proximate			Amino Acids				Minerals			Vitamins		
		CAL	CHO	FAT	TRY	LEU	LYS	MET	Na	Ca	P	THI	NIA	A
		PRO	FIB	PUFA	PHE	ISO	VAL	THR	K	Mg	Fe	RIB	ASC	D
	g	g	g	g	mg	mg	mg	mg	mg	mg	mg	mg	mg	iu
PETTIJOHNS WHEAT, DRY	28	100	21.2	0.6					tr	12	104	.11	1.2	0
1 oz (⅔ cup ckd)		2.6	0.2						107		0.9	.02	0	0
RALSTON, DRY	28	94	20.1	0.6					4	15	115	.12	1.5	0
1 oz (⅔ cup ckd)		3.7	0.6						107	46	1.0	.05	0	0
ROLLED WHEAT, DRY	50	170	38.1	1.0					1	18	171	.18	2.0	0
		4.9	1.1						190	80	1.6	.06	0	
ROLLED WHEAT, CKD	200	150	33.8	0.8					168	16	152	.14	1.8	0
		4.4	1.0								1.4	.06	0	
WHEAT & MALTED BARLEY,	50	191	38.1	0.8					tr	20	195	.17		0
TOASTED, DRY		7.0	0.8							84	2.0	.04	0	
WHEAT & MALTED BARLEY,	200	160	32.2	0.6					204	18	164	.14		0
TOASTED, CKD		6.0	0.6							62	1.8	.04	0	
WHEATENA, DRY	28	101	21.7	0.6					tr	8	93	.03	1.1	0
1 oz (⅔ cup when ckd)		2.9	0.5								0.1	.04	0	0
WHEAT MEAL, DRY, POST	28	94	20.0	0.5						250		.13		0
1 oz (½ cup when ckd)		3.1	0.6										0	0
WHEAT GERM, CRUDE, COMMERCIAL	10	36	4.7	1.1	27	184	165	43	tr	7	112	.20	0.4	0
MILLED 1 T		2.7	0.2		97	127	146	143	83	34	0.9	.07	0)	
WHEAT GERM, CRUDE, COMMERCIAL	28	102	13.1	3.1	74	503	451	118	1	20	313	.56	1.2	0
MILLED 1 oz		7.4	0.7		266	348	400	392	232	94	2.6	.19		
WHEAT GERM, KRETSCHNER	10	37	4.5	1.1	239	151	166	40	tr	tr	115	.15	0.5	
		3.2	0.2	0.7	89	97	146	111		31	0.9	.08	tr	3
WHEAT GERM, TOASTED	10	39	4.9	1.1	27	171	153	40	tr	5	108	.17	0.5	11
		2.5	0.2		91	118	136	134	95	36	0.9	.10	10	
WHEAT, WHOLE GRAIN, RAW	100	338	72.3	2.0					2	45	398	.51	4.7	0
		13.5	2.2						370	183	3.7	.13	0	0
WHEAT, WHOLE GRAIN, CKD	245	110	23.0	0.7					519	17	127	.15	1.5	
1 cup		4.4	0.7						118		1.2	.05	0	
WHEAT, WHOLE GRAIN, DURUM, RAW	100	361	70.1	3.3	157	852	348	194	3	37	386	.66	4.4	0
		12.7	1.8	1.5	627	551	588	366	435	160	4.3	.12	0	
WHEAT, WHOLE GRAIN, HARD RED	100	357	69.1	2.7	173	939	384	214	3	36	383	.57	4.3	0
SPRING, RAW		14.0	2.3	1.3	691	607	648	403	370	160	3.1	.12	0	
WHEAT, WHOLE GRAIN, HARD RED	100	359	71.7	2.5	152	825	338	188	3	46	354	.52	4.3	0
WINTER, RAW		12.3	2.3	1.2	608	534	570	354	370	160	3.4	.12	0	
WHEAT, WHOLE GRAIN, SOFT RED	100	361	72.1	2.0	126	684	280	156	3	42	400	.43	3.6	0
WINTER, RAW		10.2	2.3	1.1	504	443	472	294	376	160	3.5	.11	0	
WHEAT, WHOLE GRAIN, WHITE, RAW	100	357	75.4	2.0	116	630	258	143	3	36	394	.53	5.3	0
		9.4	1.9	1.0	464	408	435	271	390	160	3.0	.12	0	
WILD RICE, RAW	28	99	21.1	0.2					2	5	95	.13	1.7	0
¼ cup		3.9	0.3						62	36	1.2	.18	0	

	WT	Proximate			Amino Acids				Minerals			Vitamins		
		CAL	CHO	FAT	TRY	LEU	LYS	MET	Na	Ca	P	THI	NIA	A
		PRO	FIB	PUFA	PHE	ISO	VAL	THR	K	Mg	Fe	RIB	ASC	D
	g	g	g	g	mg	mg	mg	mg	mg	mg	mg	mg	mg	iu
BAGEL MADE W/ EGG	55	165	28.0	2.0							9	.14	1.2	30
1 bagel		6.0									1.2	.10	0	
BAGEL MADE W/ WATER	55	165	30.0	2.0							8	.15	1.4	0
1 bagel		6.0									1.2	.11	0	
BISCUIT FROM MIX, DRY, ENR	100	424	68.7	12.6	92	593	177	100	1300	27	265	.44	3.0	tr
3½ oz		7.7	0.3		424	354	331	223	80		3.1	.26	tr	
BISCUIT, BAKING POWDER, ENR	35	129	16.0	6.0	31	200	60	34	219	42	61	.07	0.6	tr
1 biscuit		2.6	0.1	0.3	143	120	112	75	41	8	0.6	.07	tr	
BISCUIT, BAKING POWDER, UNENR	35	129	16.0	6.0	31	200	60	34	219	42	61	.01	0.2	tr
1 biscuit		2.6	0.1	0.3	143	120	112	75	41	8	0.2	.04	tr	
BISCUIT, BUTTERMILK	23	82	9.2	4.1					215					
1 avg		1.5												
BISCUIT, FROM CND DOUGH	35	97	16.2	2.2	31	197	59	33	304	19	174	.09	0.7	
1 avg		2.6	tr		141	118	110	74	23	8	0.6	.06	0	
BISCUIT, FROM MIX, ENR	28	91	14.6	2.6	24	153	46	26	272	19	65	.22	0.6	0
1 avg		2.0	0.1	2.0	109	92	85	58	32	7	0.6	.07	0	0
BISCUIT, W/ SELF-RISING FLOUR	35	130	16.1	6.1	30	193	58	32	231	73	111	.08	0.7	tr
1 avg		2.5	0.1	0.3	138	115	108	72	22		0.6	.08	tr	
BREAD														
BANANA TEA	49	134	22.8	3.9	33	178	92	45		8	30	.06	0.5	273
1 slice		2.4	0.1		123	117	119	81		4		.06	0	3
BOSTON BROWN	44	93	20.0	0.6	25	211	99	43	110	39	70	.05	0.5	0
1 slice		2.4	0.3		117	112	130	88	128		0.8	.03	0	
BRAN RAISIN	48	148	27.2	4.2	46	255	103	57		44	118	.08	1.4	182
1 thick slice		3.8	0.5	tr	186	163	175	110			1.6	.09	0	
CINNAMON	25	68	12.6	0.8					127	21	24	.06	0.6	1
1 slice		2.2	0.1						26		0.6	.05	0	
CINNAMON RAISIN	23	60	13.0	0.5					83	17		.02	0.2	1
1 slice		1.7									0.3	.02	tr	
CORNBREAD, FROM MIX, W/ ENR YELLOW CORNMEAL 1 piece	45	105	14.8	3.8					335	38	121	.07	0.5	122
		2.7	0.1						57		0.5	.09		
CORNBREAD, FROM MIX, W/ UNENR YELLOW CORNMEAL 1 piece	45	105	14.8	3.8					335	38	121	.02	0.1	122
		2.7	0.1						57		0.3	.06		
CORNBREAD, NORTHERN STYLE W/ ENR YELLOW CORNMEAL 1 piece	40	107	18.2	2.1	21	452	104	64	276	44	62	.08	0.6	136
		3.5	0.1	0.2	160	160	176	140	75		0.7	.12	tr	
CORNBREAD, SOUTHERN STYLE*	45	93	13.1	3.2	20	429	96	63	283	54	95	.06	0.3	68
1 piece		3.3	0.2		152	152	168	132	71		0.5	.09	tr	
CORNBREAD, SOUTHERN STYLE†	45	93	13.1	3.2	20	429	96	63	283	54	95	.06	0.3	140
1 piece		3.3	0.2		152	152	168	132	71		0.5	.09	tr	
CORNBREAD, SOUTHERN STYLE W/ WHITE ENR CORNMEAL 1 piece	45	101	15.6	2.7	132	429	96	63	266	49	70	.08	0.5	68
		3.2	0.1		152	152	168	20	71		0.6	.11	tr	
CORNBREAD, SOUTHERN STYLE W/ YELLOW ENR CORNMEAL 1 piece	45	100	15.6	2.7	132	429	96	63	266	49	70	.08	0.5	140
		3.2	0.1		152	152	168	20	71		0.6	.11	tr	
CORN & MOLASSES, PEPPERIDGE FARM 1 slice	32	89	17.9	0.9					142	15	39	.07	0.7	
		2.5	0.1						25	7	0.9	.06		
CORNPONE, WHOLE GROUND CORNMEAL 1 cake	45	92	16.3	2.4	13	265	58	40	178	28	73	.07	0.4	0
		2.0	0.4	1.0	112	90	103	81	27	9	0.5	.02	0	7
CRACKED WHEAT	23	60	12.0	0.1	24	134	54	30	122	20	29	.03	0.3	tr
1 slice		2.0	0.1		98	86	92	58	31	8	0.3	.02	tr	
CRACKED WHEAT, TOASTED	19	60	11.8	0.1	24	134	54	30	120	20	29	.02	0.3	tr
1 slice		2.0	0.1		98	86	92	58	30	8	0.2	.02	tr	
FRENCH/VIENNA, ENR	20	58	11.1	0.6	22	139	41	23	116	9	17	.06	0.5	tr
1 slice		1.8	tr	0.2	99	83	77	52	18	4	0.4	.04	tr	
FRENCH/VIENNA, ENR, TOASTED	17	58	10.9	0.6	22	139	41	23	115	9	14	.04	0.5	tr
1 slice		1.8	tr	0.2	99	83	77	52	18	4	0.4	.04	tr	
FRENCH/VIENNA, UNENR	20	58	11.1	0.6	22	139	41	23	116	9	17	.02	0.1	tr
1 slice		1.8	tr		99	83	77	52	18	4	0.4	.02	tr	
FRENCH/VIENNA, UNENR, TOASTED	17	58	10.9	0.6	22	139	41	23	115	9	14	.01	0.1	tr
1 slice		1.8	tr		99	83	77	52	18	4	0.4	.02	tr	
FRESH HORIZONS, DARK	29	54	9.6	0.6					156	51		.12	1.0	0
1 slice		2.6									0.9	.07	0	

*With white whole ground cornmeal   †With yellow whole ground cornmeal

	WT g	CAL / PRO g	CHO / FIB g	FAT / PUFA g	TRY / PHE mg	LEU / ISO mg	LYS / VAL mg	MET / THR mg	Na / K mg	Ca / Mg mg	P / Fe mg	THI / RIB mg	NIA / ASC mg	A / D iu
FRESH HORIZONS, LIGHT	29	54	9.9	0.6					162	51		.12	1.0	0
1 slice		2.3									0.9	.09	0	
HILLBILLY	23	56	10.1	0.9					122	8		.06	0.6	0
1 slice		2.1	tr								0.6	.10	0	
HOLLYWOOD, DARK	20	49	8.8	0.8					110	35		.08	0.7	0
1 slice		2.2	0.2								1.0	.09	0	
HOLLYWOOD, LIGHT	20	49	9.2	0.8					116	28		.08	0.7	0
1 slice		2.2	0.1								0.6	.06	0	
HONEY CRACKED WHEAT	28	74	13.4	1.1					148	10	30	.07		0
1 slice		2.5									0.5	.05	0	
HONEY, WHEATBERRY, PEPPERIDGE FARM 1 slice	32	78	15.4	1.1					173	9	68	.08	0.8	
		2.9	0.4						67	21	0.9	.06		
ITALIAN, ENR	20	55	11.3	0.2	22	139	41	23	117	3	15	.06	0.5	0
1 slice		1.8	tr		99	83	77	52	15		0.4	.05	0	
ITALIAN, UNENR	20	55	11.3	0.2	22	139	41	23	117	3	15	.02	0.2	0
1 slice		1.8	tr		99	83	77	52	15		0.1	.01	0	
PROFILE, DARK	24	55	11.0	0.7					129	40	46	.17	1.2	tr
1 slice		2.6									1.0	.12	0	
PROFILE, LIGHT	24	64	11.0	1.0					144	17		.10	0.8	0
1 slice		2.2	0.1								0.8	.06	0	
RAISIN	23	60	12.3	0.6	18	100	40	22	84	16	20	.01	0.2	tr
1 slice		1.5	0.2	0.2	74	64	69	30	54	6	0.3	.02	tr	
RAISIN, TOASTED	19	60	12.3	0.6	18	100	40	22	84	16	20	.01	0.2	tr
1 slice		1.5	0.2	0.2	74	64	69	30	53	6	0.3	.02		
ROMAN MEAL	23	56	11.0	0.9					138	24		.09	0.8	0
1 slice		2.5	0.3								0.7	.07	0	
RYE, AMERICAN	23	56	12.0	0.3	23	141	67	32	128	17	34	.04	0.3	0
1 slice		2.1	0.1	0.2	101	90	109	67	33	10	0.4	.02	0	
RYE, AMERICAN, TOASTED	20	56	12.1	0.3	23	141	67	32	130	17	34	.03	0.3	0
1 slice		2.1	0.1	0.2	101	90	109	67	34	10	0.4	.02		
RYE, DUTCH	19	40	5.6	1.0					129					
1 slice		2.2	0.2											
RYE, PARTY	15	40	7.8	0.4					40	3	10	.01	0.1	2
1 slice		1.2	tr						10	6	0.1	.01		
RYE, PUMPERNICKEL	32	79	17.0	0.4	32	194	119	46	182	27	73	.07	0.4	0
1 slice		2.9	0.4		136	125	151	107	145	23	0.8	.04	0	
RYE, SPROUTED, PEPPERIDGE FARM	26	67	11.8	1.4					138	8	57	.08	0.5	1
1 slice		3.1	0.3						48	7	0.7	.03		
SALT-FREE	26	61	10.9	0.7					36					
1 slice		2.9	0.1											
SALT-RISING	25	67	13.1	0.6					66	6	17	.01	0.2	3
1 slice		2.0	0.1						17		0.2	.01		
SALT-RISING, TOASTED	22	65	12.8	0.6					65	6	17	.01	0.1	2
1 slice		1.9	tr						16		0.2	.01		
SPOONBREAD, W/WHITE WHOLE GROUND CORNMEAL 1 serving	96	187	16.2	10.9	38	832	186	122	463	92	157	.09	0.4	278
		6.4	0.2	0.9	294	294	326	256	127		1.0	.17	tr	
WHEAT GERM, PEPPERIDGE FARM	32	79	14.2	0.9					178	30	81	.09	0.8	
1 slice		3.8	0.2						78	18	1.1	.06		
WHITE, ENR	23	62	11.6	0.7	24	159	60	29	117	19	22	.06	0.6	tr
1 slice		2.0	tr	tr	108	96	92	62	24	5	0.6	.05	tr	
WHITE, ENR, TOASTED	20	62	11.8	0.7	24	159	60	29	118	20	23	.05	0.6	tr
1 slice		2.0	tr	tr	108	96	92	62	24	5	0.6	.05	tr	
WHITE, UNENR	23	62	11.6	0.7	24	159	60	29	117	19	22	.02	0.2	tr
1 slice		2.0	tr	tr	108	96	92	62	24	5	0.2	.02	tr	
WHITE, UNENR, TOASTED	20	62	11.8	0.7	24	159	60	29	118	20	23	.02	0.2	tr
1 slice		2.0	tr	tr	108	96	92	62	24	5	0.2	.02	tr	
WHOLE WHEAT	23	56	11.0	0.7	29	166	71	37	121	23	52	.06	0.6	tr
1 slice		2.4	0.4	tr	117	106	113	72	63	18	0.5	.02	tr	
WHOLE WHEAT, TOASTED	19	55	10.8	0.7	29	166	71	37	119	22	52	.04	0.6	tr
1 slice		2.4	0.4	tr	117	104	113	72	62	18	0.5	.02	tr	
BREADCRUMBS, DRY, GRATED	88	345	64.6	4.0	133	871	294	150	648	107	124	.19	3.1	tr
1 cup		11.1	0.3	0.6	605	522	494	333	134	30	3.2	.26	tr	
BREADCRUMBS, SALT-FREE, DRY, GRATED 1 cup	88	346	64.6	4.0	133	871	294	150	135	107	124	.19	3.1	tr
		11.1	0.3	0.6	605	522	494	333	134	30	3.2	.26	tr	

	WT g	CAL / PRO	CHO / FIB	FAT / PUFA	TRY / PHE	LEU / ISO	LYS / VAL	MET / THR	Na / K	Ca / Mg	P / Fe	THI / RIB	NIA / ASC	A / D
		CAL g	CHO g	FAT g	TRY mg	LEU mg	LYS mg	MET mg	Na mg	Ca mg	P mg	THI mg	NIA mg	A iu
		PRO g	FIB g	PUFA g	PHE mg	ISO mg	VAL mg	THR mg	K mg	Mg mg	Fe mg	RIB mg	ASC mg	D iu
BREADSTICKS, REG	6	23	4.5	0.2					100	2	6	tr	0.1	
1 avg		0.7	tr						6		tr	tr		
BREADSTICKS, CORNSTICKS, FRZN, AUNT JEMIMA 1 oz	28	82	12.9	2.6					48	6	108	.05	0.4	16
		1.8	0.1						9		0.1	.05		
BREADSTICKS, GARLIC	6	24	4.3	0.3										
1 avg		tr												
BREADSTICKS, SESAME	10	56	4.4	3.7										
1 avg		1.1												
BREADSTICKS, VIENNA BREAD TYPE	6	18	3.5	0.2					94	3	5	tr	tr	
1 avg		0.6	tr						6		tr	tr		
CHIPS & SNACKS (GRAIN-BASED)														
BACON NIPS, FRITO-LAY	28	148	14.3	9.2					700	14	30		0.2	0
1 oz		2.5	0.1	0.8					38	3	0.1	.04	0	
BUGLES, GENERAL MILLS	28	159	14.7	10.5					272	2	13	.02	0.3	
1 oz		1.6							14		0.2	.01		
CHEESE PUFFS, CHEETOS, FRITO-LAY	28	158	14.3	9.8					293	18	29	.01	0.2	62
1 oz		2.2	0.3	0.8					23	4	0.2	.03	0	
CHEESE STRAWS	24	109	8.3	7.2					173	62	49	tr	0.1	94
4 pieces		2.7	tr						15		0.1	.04	0	
CHEESE TWISTS	28	153	14.6	9.5					329	40	52	.03	0.2	107
1 oz		2.4									0.2	.07	tr	
CORN CHIPS, WONDER	28	160	14.6	10.6					218	2	31	.03	tr	65
1 oz		1.5										.01	0	
CORN CHIPS, DORITOS, NACHO CHEESE FLAVOR, FRITO-LAY 1 oz	28	139	18.1	6.8					107	17	98	.04	0.4	127
		2.2	0.4						109	13	0.4	.03	0	
CORN CHIPS, DORITOS, TACO FLAVOR, FRITO-LAY 1 oz	28	140	17.6	6.6					191	45	91	.08	0.8	151
		2.6	0.4	1.4					72	27	0.7	.09	0	
CORN CHIPS, FRITOS	28	154	14.6	10.4					202	35	52	.01	0.3	99
1 oz		1.9	0.3	2.5					23	17	0.3	.02	0	
CORN CHIPS, FRITOS, BBQ	18	150	15.3	9.9					267	25	69	.05	0.5	181
1 oz		2.1	0.5	3.2					131	16	0.4	.04	0	
FUNYUNS, FRITO-LAY	28	136	18.7	5.6					230	12	23	.01	0.3	
1 oz		2.2	0.2	1.1					35	6	tr	.01	0	
TORTILLA CHIPS, FRITO-LAY	28	135	18.1	6.2					99	39	74	.03	0.4	203
1 oz		2.4	0.4	0.9					30	22	0.4	.02	0	
TORTILLA CHIPS, PINATA	28	128	16.8	7.5						125		.01	0.2	7
1 oz		2.2									1.0	.02	tr	
TORTILLA CHIPS, WONDER	28	143	15.9	8.0					239	40	37	.04	0.3	75
1 oz		1.7										.01	0	
CRACKER CRUMBS, GRAHAM	44	220	37.7	6.3					277	15	53	.15	1.3	
½ cup		3.1							79		1.7	.12		
CRACKERMEAL	64	269	46.5	6.1						13	61	.04	0.7	0
½ cup		6.1									0.7	.03	0	0
CRACKERS														
BACON-FLAVORED, NABISCO	8	4	5.1	2.3										
4 crackers		0.6												
BACON RINDS, NABISCO	28	145							217	1	0	.00	0.0	0
1 oz											0.0	.00	0	
BUTTER THINS	4	18	2.8	0.6					306	6	73	tr	tr	62
1 cracker		0.3							32	1	0.2	tr	0	
CHEESE	4	19	2.4	0.9					42	13	12	tr	tr	14
4 avg		0.5	tr						4	1	tr	tr	0	
CHEESE GOLDFISH, PEPPERIDGE FARM	10	49	5.8	2.3					117	13	13	.05	0.3	18
10 pieces		1.1	0.2						17	3	0.4	.02	0	
CHEESE NIPS	10	46	6.7	1.7					170	16	16	.07	0.3	
10 pieces		0.9							10		0.3	.05		
CHEESE TID-BITS, NABISCO		54	5.7	3.2					164	20	22	.07	0.3	
10 pieces		0.7							9		0.3	.03		
CHEESE W/ PNT BTR	7	34	3.9	1.7					69	4	13	tr	0.2	3
1 avg		1.1	tr						16		tr	tr	0	
CHEEZ-ITS, SUNSHINE	10	14	5.6	2.6							16			
10 pieces		1.2									tr			

	WT	Proximate			Amino Acids				Minerals			Vitamins		
		CAL	CHO	FAT	TRY	LEU	LYS	MET	Na	Ca	P	THI	NIA	A
		PRO	FIB	PUFA	PHE	ISO	VAL	THR	K	Mg	Fe	RIB	ASC	D
	g	g	g	g	mg	mg	mg	mg	mg	mg	mg	mg	mg	iu
CHICKEN IN A BISKIT, NABISCO	10	50	5.7	2.9					96	5	10	.04	0.3	
5 pieces		0.7							16		0.4	.03		
CINNAMON TOAST, SUNSHINE	10	43	7.6	1.1						2	23			
3 pieces		0.7									0.3			
ESCORT, NABISCO	8	39	5.1	1.8					71	3	6	.03	0.2	
2 crackers		0.5							5		0.2	.02		
GOLDFISH, LIGHTLY SALTED, PEPPERIDGE FARM	10	48	6.2	2.0					84	6	10	.06	0.3	1
		0.9	0.2						20	3	0.4	.02	0	
GRAHAM	14	54	10.3	1.3	14	87	25	14	94	6	21	.01	0.2	0
2 crackers		1.1	0.2	1.0	38	52	48	32	54	3	0.2	.03	0	
GRAHAM, SUGAR, HONEY-COATED	14	58	10.7	1.6					71	12	46	tr	0.1	
2 crackers		0.9	0.1						38		0.2	tr	0	
HI HO, SUNSHINE	16	82	9.5	4.4						21	40			
4 crackers		1.1									0.5			
MATZOTH	30	117	25.4	0.3					tr					
1 piece		3.0												
MELBA TOAST	4	15	2.7	0.2					3					
1 slice		0.5							9					
ONION RINGS, WONDER	28	132	19.3	5.9					435	3	0	.00	0.0	0
1 oz		0.1									0.0	.00	0	
OYSTER	10	43	7.5	1.1					145	3	8	.05	0.4	
10 crackers		1.1							9		0.4	.03		
PRETZEL GOLDFISH, PEPPERIDGE FARM	10	41	7.2	0.8					217	3	10	.05	0.3	1
		1.0	0.2						17	4	0.4	.02	0	
RICE WAFERS	10	31	6.7	0.0					8					
3 wafers		0.8							3					
RITZ, NABISCO	10	54	6.4	2.9					97	15	24	.04	0.3	
3 crackers		0.7							8		0.3	.04		
RITZ CHEESE, NABISCO	10	52	5.5	2.9					104	20	24	.08	0.3	
3 crackers		0.6							9		0.3	.05		
RUSK	10	42	7.1	0.9					25	2	12	.01	0.1	23
1 piece		1.4	tr						16		0.1	.02		
RYE WAFERS, WHOLE GRAIN	7	24	5.3	0.1					62	4	27	.02	0.1	0
1 cracker		0.9	0.2						42		0.3	.02	0	
RY-KRISP	6	22	4.9	0.1					50	2	21	.03	0.1	
1 triple cracker		0.6	0.1						32	8	0.3	.01		
RY-KRISP, SEASONED	7	26	4.7	0.5					88	2	21	.02	0.1	
1 triple cracker		0.7	0.1						29	8	0.3	.01		
RYE THINS	5	26	3.0	1.3										
2 crackers		0.4												
SALTINES	6	28	4.6	0.8	7	37	13	7	66	1	5	tr	0.1	0
2 crackers		0.6	tr	0.5	29	25	23	16	7	1	0.1	tr	0	
SALTINES W/ UNSALTED TOPS	6	26	4.3	0.6					50	10	5	.03	0.2	
2 crackers		0.4							5		0.2	.02		
SESAME (ARMENIAN CRACKER BREAD), AK-MOK 1 oz	28	117	18.9	2.3						21	0	.06	1.1	14
		4.6								41	0.5	.04	2	
SESAME	12	60	7.3	2.9					108	20	14	.06	0.3	
4 crackers		1.2									0.4	.04		
SODA	14	61	9.9	1.8					154	3	12	tr	0.1	0
2 crackers		1.3	tr						17	4	0.2	.01	0	
TOWN HOUSE, KEEBLER	10	8	1.2	2.9					128		25			
		0.8							37		tr			
TRIANGLE THINS, NABISCO	10	50	6.1	2.5					143	15	28	.06	0.3	
5 pieces		1.1							19		0.4	.03		
TRISCUITS, NABISCO	9	42	6.2	1.5										
2 pieces		0.8												
VEGETABLE THINS, NABISCO	4	18	2.3	0.7										
2 crackers		0.2												
UNEEDA, NABISCO	10	42	7.1	1.0					69	3	8	.04	0.3	
2 crackers		1.0							11		0.4	.03		
WAVERLY WAFERS, NABISCO	8	36	5.0	1.6					96	13	22	.05	0.3	
2 pieces		0.5							7		0.3	.02		
WHEAT THINS, NABISCO	7	36	5.0	1.4										
4 pieces		0.5												

	WT	Proximate			Amino Acids				Minerals			Vitamins		
		CAL	CHO	FAT	TRY	LEU	LYS	MET	Na	Ca	P	THI	NIA	A
		PRO	FIB	PUFA	PHE	ISO	VAL	THR	K	Mg	Fe	RIB	ASC	D
	g	g	g	g	mg	mg	mg	mg	mg	mg	mg	mg	mg	iu
WHEAT RYE THINS	10	48	6.6	2.0						4	16	.06	0.4	1
5 pieces		0.8								1	0.4	.04	0	
WHOLE WHEAT	8	32	5.5	1.1					44	2	15	tr	0.1	0
2 crackers		0.7	0.2								tr	tr	0	
ZWIEBACK	7	31	5.4	0.7					18	1	5	tr		3
1 piece		0.9							11		tr	tr	0	
CREPE MIX, DRY, AUNT JEMIMA	100	403	67.9	88						160		.71	5.0	0
		13.2	0.2								3.2	.59		
CROUTONS	100	359	73.5	1.2					1360	91	152	.22	1.0	148
		13.0	0.6						141	29	2.5	.17		
DUMPLINGS	28	38	6.2	1.0						29	17	.04	0.4	8
1 oz		1.0								3	0.2	.04	0	
FRENCH TOAST, HOMEMADE	65	119	16.6	3.3					75	38	68	.10	0.8	11
1 slice		5.4	0.1	0.3					60	10	1.3	.13	tr	10
FRENCH TOAST, FRZN, AUNT JEMIMA 1 slice	45	90	13.5	2.3		311	128	90	228	50	63	.08	0.8	90
		3.5	tr		173	180	205	111	51	6	0.8	.09	0	
FRENCH TOAST, CINN SWIRL, FRZN, AUNT JEMIMA 1 slice	45	102	14.3	3.4					190	50	66	.07	0.8	97
		3.3	0.1						62	8	0.7	.08	0	
FRENCH TOAST, FRZN, CAMPBELLS	67	130	15.0	5.0					330	49		.08	0.1	110
1 slice		6.0	tr								1.6	.18	0	
FRENCH TOAST, FRZN, DOWNY FLAKE 1 slice	100	140	14.5	6.5										
		3.0												
FRENCH TOAST, FRZN, EGGO	38	72	11.4	1.9					219	18	36	.13	1.8	447
1 slice		2.7	tr						66	7	1.6	.15	0	
GRANOLA CRUNCH CINN APPLE SNACK, SUN COUNTRY 1 pkg	32	155	19.0	7.0										
		4.0												
GRANOLA CRUNCH HONEY ALMOND SNACK, SUN COUNTRY 1 pkg	32	155	19.0	7.0										
		4.0												
MACARONI, ENR, DRY	110	405	82.6	1.3	164	904	438	206	2	30	178	.97	6.6	0
1 cup		13.7	0.3		712	685	781	534	217	53	3.2	.41	0	
MACARONI, ENR, CKD, FIRM	140	207	42.2	0.7	84	462	224	104	1	15	91	.25	2.0	0
1 cup		7.0	0.1		364	350	400	272	110	28	1.5	.14	0	
MACARONI, ENR, CKD, TENDER	140	151	32.2	0.6	58	317	154	72	1	11	7.0	.20	1.5	0
1 cup		4.8	0.1		250	240	274	187	85	25	1.2	.11	0	
MACARONI, UNENR, DRY	110	405	82.6	1.3	164	904	438	206	2	30	178	.10	1.9	0
1 cup		13.7	0.3		712	685	781	534	217	53	1.4	.07	0	
MACARONI, UNENR, CKD, FIRM	140	207	42.2	0.7	84	462	224	104	1	15	91	.03	0.6	0
1 cup		7.0	0.1		364	350	400	272	110	28	0.7	.03	0	
MACARONI, UNENR, CKD, TENDER	140	151	32.2	0.6	58	317	154	72	1	11	70	.01	0.4	0
1 cup		4.8	0.1		250	240	274	187	85	25	0.6	.01	0	
MACARONI, LOW-PROTEIN, DRY, ANELLINI	100	340		0.1		26			20					
		0.5	4.0		12	13	15		10					
MACARONI, LOW-PROTEIN, CKD, ANELLINI	100	91	20.7	tr					8					
		0.1	1.1		4				2					
MACARONI, LOW-PROTEIN, DRY, RIGATINI	100	340	85.3	0.1		26			20					
		0.5	4.0		12	13	15		10					
MACARONI, LOW-PROTEIN, CKD, RIGATINI	100	91	20.7	tr					8					
		0.1	1.1		4				2					
MUFFINS														
BLUEBERRY	40	112	16.8	3.7	38	243	122	53	253	34	53	.06	0.5	88
1 avg		2.9	0.1	0.4	157	155	157	106	46	10	0.6	.08	tr	
BLUEBERRY, FRZN, CHEF FRANCISCO	50	136	19.6	4.8					387	44	170	.08	0.6	66
1 avg		3.4	0.1						63		0.6	.09	tr	
BLUEBERRY, FRZN, MORTON	45	116	20.8	2.7					151	22	61	.06	0.5	29
1 avg		2.4									0.5	.07	1	
BRAN	40	104	17.2	3.9					179	57	162	.06	1.6	92
1 avg		3.1	0.7						172		1.5	.10		
CHERRY, FRZN, CHEF FRANCISCO	50	163	25.0	5.3					420	48	186	.09	0.6	69
1 avg		3.7	0.1						65		0.6	.09	tr	
CORN W/ENR CORNMEAL	45	141	21.6	4.5	35	347	153	73	216	47	76	.09	0.7	135
1 avg		3.2	0.1	0.4	151	177	195	138	61	22	0.8	.10	tr	
CORN W/ WHOLE GROUND CORNMEAL 1 avg	45	130	19.1	4.6	35	347	153	73	223	50	97	.08	0.5	140
		3.2	0.2	0.4	151	177	195	138	59	48	0.6	.08	tr	

	WT	Proximate			Amino Acids				Minerals			Vitamins		
		CAL	CHO	FAT	TRY	LEU	LYS	MET	Na	Ca	P	THI	NIA	A
		PRO	FIB	PUFA	PHE	ISO	VAL	THR	K	Mg	Fe	RIB	ASC	D
	g	g	g	g	mg	mg	mg	mg	mg	mg	mg	mg	mg	iu
CORN, FROM MIX	40	130	20.0	4.2					192	96	152	.07	0.6	96
1 avg		2.8	tr						44		0.6	.08		
CORN, FRZN, MORTON	50	140	21.7	4.5					290	29	169	.06	0.4	89
1 avg		2.9									0.4	.07	tr	
ENGLISH, NEWLY WED	71	153	29.3	0.9					1044	153		.30	2.2	0
1 whole		7.2	0.2	0.3					103	5	1.9	.14	tr	
ENGLISH, THOMAS	56	138	28.3	1.4					203					
1 whole		5.3												
ENGLISH, WONDER	57	131	26.2	1.1					251	80	64	.23	2.0	0
1 whole		4.0									1.5	.14	0	
ENGLISH, BUTTER, McDONALDS	62	186	28.3	5.6					446	87	94	.22	6.4	106
1 whole		5.6	0.1						66	13	1.6	.14	1	8
HONEY BRAN, GENERAL MILLS	45	119	20.7	3.3					31	32		.07	0.6	0
1 avg		1.7							61	tr	0.6	.08	0	
PLAIN, ENR	40	118	16.9	4.0	41	260	131	57	176	42	60	.07	0.6	40
1 avg		3.1	0.1	0.4	168	166	168	113	50	11	0.6	.09	tr	
PLAIN, UNENR	40	118	16.9	4.0	41	260	131	.57	176	42	60	.02	0.2	40
1 avg		3.1	0.1	0.4	168	166	168	113	50	11	0.2	.06	tr	
PLAIN, FROM MIX, PILLSBURY	42	124	20.9	3.9					253	14	95	.07	0.5	
1 avg		1.5							17		0.5	.07		
RAISIN ROUND, WONDER	61	159	30.0	2.4					226	86	139	.24	1.7	0
1 avg		4.9									1.6	.18	0	
SOY*	40	119	16.7	4.4	54	332	199	69		35	56	.08	0.5	196
1 med		3.9	tr	0.4	211	217	219	152		52	0.9	.10	0	
WHOLE WHEAT	40	103	20.9	1.1	52	320	165	74	226	42	112	.14	1.2	tr
1 avg		4.0	0.6		209	206	215	142	117	45	1.0	.05	tr	
NOODLES, CHOW MEIN, CND	50	248	29.0	11.7										
1 cup		6.6												
NOODLES, CHOW MEIN, FRZN, LA CHOY　1 cup	58	129	14.4	6.6					228					
		2.8												
NOODLES, EGG, ENR, DRY	73	283	52.6	3.4	102	614	307	158	4	23	134	.64	4.4	161
1 cup		9.3	0.3		446	456	549	391	99	39	2.1	.28	0	
NOODLES, EGG, ENR, CKD	160	200	37.3	2.4	73	436	218	112	3	16	94	.22	1.9	112
1 cup		6.6	0.2	1.6	317	323	389	277	70	37	1.4	.13	0	
NOODLES, EGG, UNENR, DRY	73	283	52.6	3.4	102	614	307	158	4	23	134	.12	1.5	161
1 cup		9.3	0.3		446	456	549	391	99	39	1.4	.07	0	
NOODLES, EGG, UNENR, CKD	160	200	37.3	2.4	73	436	218	112	3	16	94	.05	0.6	112
1 cup		6.6	0.2		317	323	389	277	70		1.0	.03	0	
NOODLES, LOW-PROTEIN, DRY, TAGLIATELLE	100	340	85.3	0.1		26			20					
		0.5	4.0		12	13	15		10					
NOODLES, LOW-PROTEIN, CKD, TAGLIATELLE	100	91	20.7	tr					8					
		0.1	1.1		4				2					
PANCAKE MIX, COMPLETE, AUNT JEMIMA	100	367	69.7	5.0					1192	210	680	.46	3.0	68
		10.2	0.4						261	44	3.2	.35	0	0
PANCAKE & WAFFLE MIX, ORIGINAL, AUNT JEMIMA	100	348	72.5	2.0		892	295	269	1555	463	780	.48	2.9	0
		9.8	0.5		383	406	489	316	193	62	3.7	.17	0	0
PANCAKE & WAFFLE MIX, BUCKWHEAT AUNT JEMIMA	100	346	68.7	2.7		997	513	282	1384	460	880	.55	3.7	0
		11.1	1.9		543	483	609	382	298	135	5.4	.32	0	0
PANCAKES														
APPLE, FRZN, CHEF FRANCISCO	75	125	21.5	2.9					240	50	59	.08	0.5	56
1 avg		3.2	0.2						98		0.6	.11	1	
BLUEBERRY, FRZN, CHEF FRANCISCO	75	132	23.3	3.0					235	50	60	.08	0.6	56
1 avg		3.3	0.3						98		0.7	.11	2	
BUCKWHEAT, FROM MIX	45	90	10.7	4.1	46	258	189	68	209	99	152	.05	0.3	104
1 avg		3.1	0.2	tr	153	168	192	131	110	22	0.6	.07	tr	
CORNMEAL	48	68	11.4	1.3	26	251	126	55		23	36	.05	0.4	95
1 avg		2.3	0.1	tr	114	132	146	103		23	0.5	.07	0	
FRZN, CAMPBELLS	28	70	9.3	2.7					131	20	28	.04	0.3	28
1 avg		1.8	0.0						36		0.4	.06		
FRZN, DOWNY FLAKE	44	90	16.0	2.5								.14	1.0	
1 avg		1.5									1.0	.10		

*Soy flour replaces ¼ of the white flour.

	WT g	CAL / PRO	CHO / FIB	FAT / PUFA	TRY / PHE	LEU / ISO	LYS / VAL	MET / THR	Na / K	Ca / Mg	P / Fe	THI / RIB	NIA / ASC	A / D
	g	PRO g	FIB g	PUFA g	PHE mg	ISO mg	VAL mg	THR mg	K mg	Mg mg	Fe mg	RIB mg	ASC mg	D iu
HOMEMADE, ENR	45	104	15.3	3.2	43	274	151	66	191	45	63	.08	0.6	54
1 avg		3.2	0.1	0.5	173	178	184	124	55	11	0.6	.10	tr	
HOMEMADE, UNENR	45	104	15.3	3.2	43	274	151	66	191	45	63	.02	0.2	54
1 avg		3.2	0.1	0.5	173	178	184	124	55	11	0.3	.06	tr	
PLAIN & BUTTERMILK, FROM MIX, ENR	45	101	14.6	3.3	45	275	153	68	254	97	117	.07	0.4	113
1 avg		3.2	tr	2.3	176	180	185	126	69	6	0.5	.11	0	
PLAIN & BUTTERMILK, FROM MIX, UNENR	45	101	14.6	3.3	45	275	153	68	254	97	117	.03	0.2	113
1 avg		3.2	tr	2.3	176	180	185	126	69	6	0.3	.08	0	
SOUTHERN	45	68	10.0	2.2	29	284	143	62		33	63	.05	0.3	162
1 avg		2.6	0.2	tr	129	150	165	116		24	0.4	.07	0	
SOY*	45	68	10.2	1.9	40	247	148	51		26	42	.07	0.4	90
1 avg		2.9	tr	tr	157	162	163	113		55	0.6	.07	0	
W/BUTTER & SYRUP, McDONALDS	206	472	89.0	9.3					107	154	404	.31	4.0	255
1 serving		8.0	0.2						264	30	2.4	.43	2	12
PATTY SHELL, FRZN, PEPPERIDGE FARM	47	249	18.3	18.3					247	tr	12	.20	0.4	5
1 shell		3.3	tr						27	5	0.9	.08	0	
POPOVER, ENR, HOMEMADE	50	112	12.9	4.6	62	383	234	100	110	48	70	.07	0.5	165
1 avg		4.4	0.1	0.5	239	257	271	183	75	12	0.8	.12	tr	
PRETZELS	13	51	9.9	0.6					218	3	17	tr	0.3	0
1 avg		1.3	tr	0.1					17	5	0.2	tr	0	
3-RING, NABISCO	3	11	2.3	0.1					54	1	2	.02	tr	
1 avg		0.3							3		0.1	.01		
GEMS, PEPPERIDGE FARM	28	109	22.1	1.1					364	8	43	.03	0.4	0
1 oz (5 pieces)		2.2	0.5						44		0.4	.01	0	
GEMS, CHEESE, PEPPERIDGE FARM	28	109	21.3	1.4					386	25	63	.03	tr	0
1 oz (5 pieces)		2.8	0.5						47		0.3	.02	0	
RODS, FRITO-LAY	28	108	21.7	1.1					390	7	30	.03	0.3	
1 oz		3.0	0.2	0.2					33	10	0.5	.01	0	
STICKS, VERY THIN, NABISCO	28	100	21.0	1.0					882	13	23	.14	0.9	25
1 oz		3.0									1.2	.14		
TWISTS, FRITO-LAY	28	110	22.3	1.1					500	7	26	.09	1.4	0
1 oz		2.5	0.2	0.1					92	10	1.1	.07	0	
ROLLS														
BROWN & SERVE	28	92	15.3	2.2					157	14	25	.07	0.6	0
1 avg		2.4	tr						28	6	0.6	.06	1	
BUTTERFLAKE, FROM MIX, PILLSBURY	20	58	8.9	1.6					261					
1 avg		1.6												
BUTTERFLY, FRZN, SARA LEE	28	87	12.5	2.9					152	19	30	.08	0.6	51
1 avg		2.3	0.1						34	7	0.5	.07		
BUTTERMILK, FROM MIX, PILLSBURY	35	113	14.8	4.9					343					
1 avg		2.0												
CLOVERLEAF, FRZN, SARA LEE	27	90	13.8	2.6					186	17	27	.09	0.8	13
1 avg		2.2	0.1						21	7	0.6	.08	0	
CROISSANT, FRZN, SARA LEE	26	99	10.1	5.6					119	11	23	.07		40
1 avg		1.9	0.1						25		0.4	.05	0	
DINNER, KENTUCKY FRIED CHICKEN	17	52	9.2	1.1					83	17		.08	0.8	15
1 avg		1.4									0.4	.07	tr	
FROM MIX	35	105	19.1	1.6					110	20	34	.09	0.8	
1 avg		3.2	tr						43		0.7	.09		
FRZN, ENR	24	75	13.4	1.3					134	9	21	.06	0.6	8
1 avg		2.0	tr						23		0.5	.05	tr	
FRZN, UNENR	24	75	13.4	1.3					134	9	21	.02	0.3	8
1 avg		2.0	tr						23		0.2	.02	tr	
HAMBURGER	30	89	15.9	1.7	31	200	82	38	152	22	26	.08	0.7	tr
1 whole		2.5	0.1	0.2	134	124	121	82	28	11	0.6	.05	tr	
HARD, ENR, HOMEMADE	35	109	20.8	1.1	42	273	112	52	219	16	32	.09	0.9	tr
1 avg		3.4	0.1	0.1	182	168	164	112	34	8	0.8	.08	tr	
HARD, UNENR, HOMEMADE	35	109	20.8	1.1	42	273	112	52	219	16	32	.02	0.3	tr
1 avg		3.4	0.1	0.1	182	168	164	112	34	8	0.3	.03	tr	
HOMEMADE, ENR	35	119	19.6	3.1	36	234	96	44	98	16	36	.09	0.8	28
1 avg		2.9	0.1	0.1	156	145	141	96	41	13	0.7	.09	tr	

*Soy flour replaces ¼ of the white flour.

	WT	Proximate			Amino Acids				Minerals			Vitamins		
		CAL	CHO	FAT	TRY	LEU	LYS	MET	Na	Ca	P	THI	NIA	A
		PRO	FIB	PUFA	PHE	ISO	VAL	THR	K	Mg	Fe	RIB	ASC	D
	g	g	g	g	mg	mg	mg	mg	mg	mg	mg	mg	mg	iu
HOT DOG	36	108	18.9	1.8					263	27		.10	0.8	0
1 whole		2.7	0.1								0.7	.06	0	
PANROLL, COMMERCIAL, ENR	38	113	20.1	2.1	38	249	102	47	192	28	32	.11	0.8	tr
1 avg		3.1	0.1	0.2	166	153	149	102	36	14	0.7	.07	tr	
PANROLL, COMMERCIAL, UNENR	38	113	20.1	2.1	38	249	102	47	192	28	32	.02	0.3	tr
1 avg		3.1	0.1	0.2	166	153	149	102	36	14	0.3	.03	tr	
PARKER HOUSE, FROM MIX, PILLSBURY 1 avg	23	61	11.2	1.0					294					
		1.5												
PARKER HOUSE, FRZN, SARA LEE	18	61	9.1	1.9					123	12	18	.06	0.5	9
1 avg		1.5	tr						14	5	0.4	.06	0	
RAISIN	60	165	33.8	1.7	49	316	94	53	230	45	55	.04	0.4	
1 avg		4.1	0.5		226	189	176	119	147	12	0.8	.06		
RYE	57	165	26.2	3.9					359					
1 avg		6.3	0.2											
SESAME, FRZN, SARA LEE	18	61	9.0	2.0					120	15	21	.06	0.5	8
1 avg		1.5	0.1						13	7	0.4	.05	0	
WHOLE WHEAT	35	90	18.3	1.0					197	34	98	.12	1.1	tr
1 avg		3.5	0.6	tr					102	40	0.8	.05	tr	
SHAKE 'N BAKE, GENERAL FOODS														
FOR CHICKEN	5	21	3.1	0.8					172	4	2	.01	0.1	2
1 T		0.5	tr						3.0	1	0.1	tr	tr	
FOR CHICKEN, BARBEQUE FLAVOR	5	17	3.9	0.2					221	2	tr		tr	0
1 T		0.2	0.1						tr	1	tr			
FOR CHICKEN, CRISPY COUNTRY, MILD 1 T	5	24	3.1	1.2					151	3	2	.01	0.1	0
		0.2	tr						2	1	tr	tr		
FOR CHICKEN, ITALIAN FLAVOR	5	21	3.1	0.8					144	2	2	.01	0.1	1
1 T		0.6	tr						tr	tr	tr	tr	tr	
FOR FISH	5	20	3.0	0.7					222	3	9	.01		tr
1 T		0.6	tr						9		0.1	.11		
FOR HAMBURGER	5	14	2.9	0.1					318	3	2	tr	tr	0
1 T		0.6	tr						2	1	tr	tr	0	
FOR PORK	5	19	3.5	0.3					180	2	tr		tr	1
1 T		0.5	tr						1	tr	tr		0	
FOR PORK W/RIBS, BBQ STYLE	5	18	3.7	0.4					209	2	1	.01	tr	0
1 T		0.2	0.1						2	1	tr	tr		
PLAIN	5	30	3.1	0.6					166	3	2	tr	tr	1
1 T		0.5	tr						3		0.1	tr	0	
SPAGHETTI, ENR, DRY	38	140	28.7	0.5	58	317	154	72	1	10	62	.33	2.3	0
¼ cup		3.8	0.1		250	240	274	187	75	18	1.1	.14	0	
SPAGHETTI, ENR, CKD FIRM	146	216	44.0	0.7	88	482	234	110	1	16	95	.26	2.0	0
1 cup		7.3	0.1		380	365	416	285	115	29	1.6	.15	0	
SPAGHETTI, ENR, CKD TENDER	150	166	34.5	0.6	61	337	163	76	2	12	75	.21	1.6	0
1 cup		5.1	0.2		265	255	291	199	92	27	1.4	.12	0	
SPAGHETTI, UNENR, DRY	38	140	28.7	0.5	58	317	154	72	1	10	62	.03	0.6	0
¼ cup		4.8	0.1		250	240	274	187	75	18	0.5	.02	0	
SPAGHETTI, UNENR, CKD FIRM	146	216	44.0	0.7	88	482	234	110	1	16	95	.03	0.6	0
1 cup		7.3	0.1		380	365	416	285	115	29	0.7	.03	0	
SPAGHETTI, UNENR, CKD TENDER	150	166	34.5	0.6	61	337	163	76	2	12	75	.02	0.4	0
1 cup		5.1	0.2		265	255	291	179	92	27	0.6	.02	0	
STUFFING														
BREAD, FROM MIX, DRY CRUMBLY	70	251	24.9	15.3					627	46	68	.06	1.1	455
½ cup		4.6	0.3						63		1.1	.08		
BREAD, FROM MIX, MOIST W/EGG	95	198	18.7	12.2					479	38	63	.05	0.8	399
½ cup		4.2	0.2						55		1.0	.09	1	
CORNBREAD	95	352	68.7	5.4	146			170	1303	134	231	.31	3.0	68
½ cup		12.2	1.0		665		543	374	186		3.0	.25		
STUFF 'N SUCH, CKD, UNCLE BENS	83	118	24.2	0.6					647	14	39	.06	0.7	10
½ cup		4.2	0.2						64	30	1.1	.03	1	
TURKEY STUFFING W/GVY	207	196	21.1	6.0					188	10		.05	0.6	72
1 serving		14.7							257		0.5		14	

	WT	Proximate			Amino Acids				Minerals			Vitamins		
		CAL	CHO	FAT	TRY	LEU	LYS	MET	Na	Ca	P	THI	NIA	A
		PRO	FIB	PUFA	PHE	ISO	VAL	THR	K	Mg	Fe	RIB	ASC	D
	g	g	g	g	mg	mg	mg	mg	mg	mg	mg	mg	mg	iu
STUFFING, DRY MIX														
BREAD, COARSE CRUMBS	76	263	2.7	2.7	109	717	241	124	945	88	134	.14	2.3	
1 cup		9.2	0.6	0.3	498	429	407	274	122		2.3	.18		
CHICKEN FLAVOR, STOVE TOP	28	105	18.5	2.1					516	31		.14	0.9	59
1 oz		2.8	0.2						99		0.7	.10		
CORNBREAD, FRITO-LAY	28	81	19.5	0.6					480	28	46	.17	2.1	
1 oz		3.9	0.5	0.1					52	10	0.7	.07	0	
CORNBREAD, STOVE TOP	28	100	19.7	1.0					345	86		.29	2.2	62
1 oz		2.7	0.3						76		1.5	.22		
HERB-SEASONED, PEPPERIDGE FARM	28	103	22.0	0.4					485	31	57	.08	0.6	
1 oz		3.1	0.3						62	16	0.7	.06		
POULTRY, FRITO-LAY	28	98	19.9	0.5					680	34	48	.06	0.6	
1 oz		3.5	0.5	0.1					66	10	0.4	.03	0	
RICE, STOVE TOP	28	106	20.9	1.0					491	27	28	.16	1.3	0
1 oz		3.1	0.1						41		1.2	.06		
TORTILLA, CORN, YELLOW	30	63	13.5	0.6	8	242	38	28	33	60	42	.04	0.3	6
1 avg (6" diameter)		1.5	0.3	0.3	64	88	78	60	5	32	0.9	.02	0	
TORTILLA, CORN, WHITE	30	63	13.5	0.4	8	242	38	28	33	60	42	.04	0.3	1
1 avg (6" diameter)		1.5	0.3	0.3	64	88	78	60	5	32	0.8	.02	0	
TORTILLA, CORN FLOUR, REFINED, YELLOW 1 avg (6" diameter)	30	108	22.4	1.2	17	376	84	54		4	5.2	.08	0.5	21
		2.9	0.4		132	134	148	116		14	1.0	.04	0	
TORTILLA, CORN FLOUR, REFINED, WHITE 1 avg (6" diameter)	30	108	22.4	1.2	17	376	84	54		4	52	.08	0.5	0
		2.9	0.4		132	134	148	116		14	1.0	.04	0	
TORTILLA, CORN, FRIED	45	142	19.2	8.1						139		.01	0.3	7
1 avg		2.2									0.9	.02	tr	
TORTILLA, CORN & SOY	30	69	13.2	1.2	24	202	110	33		38	52	.08	0.4	
1 avg (6" diameter)		2.0	0.7		101	109	111	84		55	1.4	.02		
TORTILLA, FLOUR	30	95	17.3	1.8						46	25	.01	1.0	2
1 avg		2.5								7	1.1	.08	tr	
TORTILLA, FLOUR, WHOLE WHEAT	30	91	15.0	2.9						43		.26	1.4	3
1 avg		2.9									1.0	.05	tr	
TORTILLA, TACO SHELL	11	48	7.6	2.2						38		tr	0.2	5
1 avg		0.9									0.3	.01	tr	
WAFFLES, HOMEMADE, ENR	75	209	28.1	7.4	94	598	331	144	356	85	130	.13	1.0	248
1 med		7.0	0.1	0.3	379	391	404	272	109	19	1.3	.19	tr	
WAFFLES, HOMEMADE, UNENR	75	209	28.1	7.4	94	598	331	144	356	85	130	.04	0.3	248
1 med		7.0	0.1	0.3	379	391	404	272	109	19	0.7	.14	tr	
WAFFLES, FROM MIX	75	206	27.2	8.0					515	179	257	.11	0.7	173
1 med		6.6	0.2						146		1.0	.17		
WAFFLES, FRZN	37	89	14.2	2.7		215	60	47	273	89	141	.16	1.6	48
1 med		2.3	0.1		104	106	130	64	41	7	1.8	.20		
WAFFLES, BLUEBERRY, FRZN, AUNT JEMIMA 1 med	37	91	14.4	2.7					522	90	141	.17	1.7	
		2.3	0.1						41	0	1.8	.20	2	
WAFFLES, BUTTERMILK, FRZN	37	90	14.2	2.7					279	89	141	.17	1.7	
1 med		2.3	0.1						43	8	1.8	.20		

	WT	Proximate			Amino Acids				Minerals			Vitamins		
		CAL	CHO	FAT	TRY	LEU	LYS	MET	Na	Ca	P	THI	NIA	A
		PRO	FIB	PUFA	PHE	ISO	VAL	THR	K	Mg	Fe	RIB	ASC	D
	g	g	g	g	mg	mg	mg	mg	mg	mg	mg	mg	mg	iu
ACEROLA JUICE	100	23	4.8	0.3					3	10	9	.02	0.4	
⅖ cup		0.4	0.3								0.5	.06	1600	
APPLE JUICE, CND	30	15	3.7	tr					1	2	3	tr	tr	
1 fl oz; 2 T		tr	tr						30		0.2	.01	tr	
CND	100	47	11.9	tr					1	6	9	.01	0.1	
⅖ cup		0.1	0.1						101		0.6	.02	1	
CND	185	87	22.0	tr					2	11	17	.02	0.2	
¾ cup; 6 fl oz		0.2	0.2						187		1.1	.04	2	
APRICOT JUICE, UNSWEET	100	49	11.7	0.1					3					
⅖ cup		0.4												
NECTAR	100	57	14.6	0.1					tr	9	12	.01	0.2	950
⅖ cup		0.3	0.2						151		0.2	.01	3	
NECTAR	185	105	27.0	0.2					1	17	22	.02	0.4	1760
¾ cup; 6 fl oz		0.6							279		0.4	.02	6	
BLACKBERRY JUICE, UNSWEET	100	37	7.8	0.6					1	12	12	.02	0.3	
⅖ cup		0.3	tr						170		0.9	.03	10	
BLUEBERRY JUICE	100	54	13.7	0.0					1					
⅖ cup		0.1							111					
CARROT JUICE, CND	240	96	21.4	0.2										11,520
1 cup		1.9	0.5											
CRANBERRY JUICE COCKTAIL*	100	65	16.5	0.1					1	5	3	.01	tr	tr
⅖ cup		0.1	tr						10		0.3	.01		
CURRANT JUICE, BLACK	100	55	13.7	0.0						20	16			
⅖ cup		0.5											162	
GRAPEFRUIT JUICE, FRESH†	100	39	9.2	0.1					1	9	15	.04	0.2	80
⅖ cup		0.5	tr						162		0.2	.02	38	
FRESH	250	98	23.0	0.2					2	22	38	.10	0.5	200
1 cup		1.2	tr						405		0.5	.05	95	
CND, UNSWEET	100	30	7.6	0.1					4	13	14	.03	0.2	10
⅖ cup		0.6	0.2						144		0.3	.02	30	
CND, SWEET	100	70	17.8	0.1					1	13	14	.03	0.2	10
⅖ cup		0.6	0.2						135		0.3	.02	30	
FROZEN CONC, UNDILUTED	100	145	34.6	0.4					4	34	60	.14	0.7	30
⅖ cup		1.9	0.1						604		0.4	.06	138	
DILUTED, SWEET	100	41	9.8	0.1					1	10	17	.04	0.2	10
⅖ cup		0.5	tr						170		0.1	.02	39	
FROZEN CONC, SWEET, UNDILUTED	100	165	40.2	0.3					3	28	50	.12	0.6	20
⅖ cup		1.6	0.1						508		0.3	.05	116	
DILUTED 1:4	100	47	11.4	0.1					1	8	14	.03	0.2	10
⅖ cup		0.4	tr						144		0.1	.01	33	
GRAPEFRUIT-ORANGE, CND, UNSWEET ⅖ cup	100	43	10.1	0.2					1	10	15	.05	0.2	100
		0.6	0.1						184		0.3	.02	34	
CND, SWEET	100	50	12.2	0.1					1	10	15	.05	0.2	100
⅖ cup		0.5	0.1						184		0.3	.02	34	
FRZN CONC, UNSWEET, UNDILUTED ⅖ cup	100	157	37.1	0.5					2	29	47	.23	1.1	380
		2.1	0.1						623		0.4	.03	144	
DILUTED 1:4	100	44	10.5	0.1					tr	8	13	.06	0.3	110
⅖ cup		0.6	tr						177		0.1	.01	41	
GRAPE JUICE, BOTTLED	100	66	16.6	tr					2	11	12	.04	0.2	
⅖ cup		0.2	tr						116		0.3	.02	tr	
BOTTLED	180	119	29.9	tr					4	20	22	.07	0.4	
¾ cup		0.4	tr						209		0.5	.04	tr	
FRZN CONC, SWEET, UNDILUTED	100	183	46.3	tr					3	10	15	.06	0.7	20
3½ ounces		0.6	0.1						118		0.4	.10	15	
DILUTED 1:4	100	53	13.3	tr					1	3	4	.02	0.2	tr
⅖ cup		0.2	tr						34		0.1	.03	4	
GRAPEJUICE DRINK‡	100	54	13.8	tr					1	3	4	.01	0.1	
⅖ cup		0.1	tr						35		0.1	.01	16	
GUAVA JUICE, CND	125	86	23.9	0.0						8	8			
½ cup		1.0									0.1		100	

*Approximately 33 percent cranberry juice. Ascorbic acid is usually added.

†Juice of white-fleshed varieties have about 10 IU of Vitamin A per 100 g; red-fleshed, about 440 IU of Vitamin A per 100 g.

‡Fruit juice content may range from 10 to 50 percent, with ascorbic acid added as a preservative or nutrient.

	WT	Proximate			Amino Acids				Minerals			Vitamins		
		CAL	CHO	FAT	TRY	LEU	LYS	MET	Na	Ca	P	THI	NIA	A
		PRO	FIB	PUFA	PHE	ISO	VAL	THR	K	Mg	Fe	RIB	ASC	D
	g	g	g	g	mg	mg	mg	mg	mg	mg	mg	mg	mg	iu
LEMON JUICE, FRESH	15	4	1.2	tr					tr	1	2	.01	tr	3
1 T		0.1	tr								tr	tr	7	
FRESH	60	15	4.8	0.1					1	4	6	.02	0.1	12
2 fl oz		0.3	tr						85		0.1	.01	28	
FRESH	100	25	8.0	0.2					1	7	10	.03	0.1	20
⅖ cup		0.5	tr						141		0.2	.01	46	
FRESH	240	60	19.2	0.5					2	17	24	.07	0.2	48
1 cup		1.2	tr						338		0.5	.02	110	
CND, UNSWEET	15	4	1.1	tr					tr	1	2	.01	tr	3
1 T		0.1	tr						85		tr	tr	6	
CND, UNSWEET	100	23	7.6	0.1					1	7	10	.03	0.1	20
⅖ cup		0.4	tr						141		0.2	.01	42	
FRZN, UNSWEET	100	22	7.2	0.2					1	7	9	.03	0.1	20
⅖ cup		0.4	tr						141		0.3	.01	44	
FRZN CONC, UNSWEET	100	116	37.4	0.9					5	33	47	.14	0.3	80
⅖ cup		2.3	tr						658		0.9	.06	230	
LEMONADE, FRZN, CONC,	100	195	51.1	0.1					2	4	6	.02	0.3	20
UNDILUTED ⅖ cup		0.2	0.1						70		0.2	.03	30	
DILUTED WITH WATER (4⅓ ×)	100	44	11.4	tr					tr	1	1	tr	0.1	tr
⅖ cup		0.1	tr						16		tr	.01	7	
LIME JUICE, FRESH	15	4	1.3	tr					tr	1	2	tr	tr	2
1 T		0.1	tr						16		tr	tr	3	
FRESH	60	16	5.4	0.1					1	5	7	.01	0.1	6
2 fl oz		0.2	tr						62		0.2	.01	19	
FRESH	100	26	9.0	0.1					1	9	11	.02	0.1	10
⅖ cup		0.3	tr						104		0.2	.01	32	
CND, UNSWEET	15	4	1.3	tr					tr	1	2	tr	tr	2
1 T		0.1	tr						16		tr	tr	3	
CND, UNSWEET	60	16	5.4	0.1					1	5	7	.01	0.1	6
2 fl oz		0.2	tr						62		0.1	.01	13	
CND, UNSWEET	100	26	9.0	0.1					1	9	11	.02	0.1	10
⅖ cup		0.3	tr						104		0.2	.01	21	
LIMEADE FRZN, CONC,	100	187	49.5	0.1					tr	5	6	.01	0.1	tr
UNDILUTED ⅖ cup		0.2	tr						59		0.1	.01	12	
DILUTED (4⅓ ×)	100	41	11.0	tr					tr	1	1	tr	tr	tr
⅖ cup		tr	tr						13		tr	tr	2	
LOGANBERRY JUICE	123	50	12.4	0.0						8	4			
½ cup		0.7	0								0.1			
ORANGE JUICE, FRESH	31	14	3.4	0.1					.3	6	5	.03	0.1	(59)
2 T; 1 fl oz		0.2	tr						52		0.1	.01	15	
FRESH	100	45	10.4	0.2					1	11	17	.09	0.4	200
⅖ cup		0.7	0.1						200		0.2	.03	*50	
FRESH	248	111	25.8	0.6					3	27	42	.22	1.0	496
1 cup		1.7	0.3						496		0.5	.07	124	
CND, UNSWEET	100	48	11.2	0.2					1	10	18	.07	0.3	200
⅖ cup		0.8	0.1						199		0.4	.02	40	
CND, SWEET	100	52	12.2	0.2					1	10	18	.07	0.3	200
⅖ cup		0.7	0.1						199		0.4	.02	40	
CND CONC, UNSWEET, UNDILUTED	100	223	50.7	1.3					5	51	86	.39	1.7	960
⅖ cup		4.1	0.5						942		1.3	.12	229	
DILUTED 1:6	100	46	10.3	0.3					1	10	18	.08	0.3	200
⅖ cup		0.8	0.1						192		0.3	.02	47	
FRZN CONC, UNSWEET	100	158	38.0	0.2					2	33	55	.30	1.2	710
⅖ cup		2.3	0.2						657		0.4	.05	158	
DILUTED 1:4	100	45	10.7	0.1					1	9	16	.09	0.3	200
⅖ cup		0.7	tr						186		0.1	.01	45	
ORANGE JUICE CRYSTALS,	100	380	88.9	1.7					8	84	134	670	2.9	1680
DEHYDRATED 3½ oz		5.0	.8						1728		1.7	210	359	
WATER ADDED (4 oz makes 1 qt)	240	110	26.0	.5					2	24	38	190	1.0	480
1 cup		1.4	.2						500		.5	70	105	
ORANGE JCE, INSTANT, RECON,	270	143	33.5	0.3					16	226	173	.00	0.0	3213
GENERAL FOODS 1 cup		0.0							108		0.2	.00	162	

*Subject to seasonal variation in ascorbic acid.

	WT	Proximate			Amino Acids				Minerals			Vitamins		
		CAL	CHO	FAT	TRY	LEU	LYS	MET	Na	Ca	P	THI	NIA	A
		PRO	FIB	PUFA	PHE	ISO	VAL	THR	K	Mg	Fe	RIB	ASC	D
	g	g	g	g	mg	mg	mg	mg	mg	mg	mg	mg	mg	iu
ORANGE JCE DRINK, CND	240	131	29.6	0.0					77	2	5	.02	0.1	0
1 cup		1.1	0.2						46	3	0.0	.00	80	
ORANGE–APRICOT JUICE DRINK	100	50	12.7	0.1					tr	5	8	20	0.2	580
⅖ cup		0.3	0.2						94		0.1	10	16	
ORANGE–GRAPEFRUIT JCE, FRZN, DILUTED 1 cup	248	109	26.0									.15		270
		1.0									0.3			
ORANGE–GRAPEFRUIT JCE, CND, SWEET, DEL MONTE 1 cup	250	110	28.5	0.3					3	28	35	.13	1.5	100
		1.8							560	45	1.3	.08	50	
ORANGE–GRAPEFRUIT JCE, CND, UNSWEET, DEL MONTE 1 cup	250	95	24.8	0.3					3	20	35	.15	0.8	295
		1.5							335	25	1.3	.08	68	
PAPAYA JUICE, CND	125	60	15.1	0.0						22	12	19	0.1	2500
½ cup		0.5									0.4	12	51	
PEACH NECTAR, CND	100	48	12.4	tr					1	4	11	10	0.4	430
⅖ cup		0.2	0.1						78		0.2	20	tr	
CND	185	89	23.0	tr					2	7	20	18	0.7	795
¾ cup		0.4	0.2						144		0.4	40	tr	
PEAR NECTAR, CND	100	52	13.2	0.2					1	3	5	tr	tr	tr
⅖ cup		0.3	0.3						39		0.1	20	tr	
CND	185	96	24.4	0.4					2	6	9	tr	tr	tr
¾ cup		0.6	0.6						72		0.2	40	tr	
PINEAPPLE JUICE, CND, UNSWEET	30	16	4.1	tr					tr	4	3	15	0.1	15
2 T; 1 fl oz		0.1							45		0.1	6	3	
CND	100	55	13.5	0.1					1	15	9	50	0.2	50
⅖ cup		0.4	0.1						149		0.3	20	9	
CND	190	104	25.6	0.2					2	29	17	100	0.4	95
¾ cup		0.8	0.2						284		0.6	40	17	
FRZN UNSWEET, UNDILUTED	100	179	44.3	0.1					3	39	28	230	0.9	50
3½ oz		1.3	0.3						472		0.9	60	42	
DILUTED 1:4	100	52	12.8	tr					1	11	8	70	0.2	10
⅖ cup		0.4	0.1						136		0.3	20	12	
PINEAPPLE–GRAPEFRUIT JCE DRINK, CND, DEL MONTE 1 cup	250	123	12.7	0.3					58	13	10	.19	0.8	
		0.5							140	15	0.8	.05	43	
PINEAPPLE–PINK GRAPEFRUIT JCE DRINK, CND, DEL MONTE 1 cup	250	123	32.0						60	18	10	.02	0.5	
		0.5							133	15	0.5	.05	43	
PINEAPPLE–ORANGE JCE DRINK, CND, DEL MONTE 1 cup	250	123	32.0						8	13	10	.08	0.5	135
		0.5							115	15	0.5	.05	43	
PRUNE JUICE, CND	100	77	19.0	0.1					2	14	20	10	0.4	
⅖ cup		0.4	tr						235		4.1	10	2	
CND	180	138	34.2	0.2					4	25	36	20	0.7	
¾ cup		0.7	tr						423		7.4	20	4	
RASPBERRY JUICE	120	49	12.8	0.0						29	14	20		120
½ cup		0.2	tr								1.0		18	
TANGELO JUICE	100	41	9.7	0.1										
⅖ cup		0.5	tr										27	
TANGERINE JCE, RAW	100	43	10.1	0.2					1	18	14	60	0.1	420
⅖ cup		0.5	0.1						178		0.2	20	.31	
CND, UNSWEET	100	43	10.2	0.2					1	18	14	60	0.1	420
⅖ cup		0.5	0.1						178		0.2	20	22	
FRZN CONC, UNDILUTED, UNSWEET	100	162	38.3	0.7					2	62	48	200	0.4	1460
3½ oz		1.7	0.3						613		0.7	60	96	
DILUTED 1:4	100	46	10.8	0.2					1	18	14	60	0.1	410
⅖ cup		0.5	0.1						174		0.2	20	27	
TOMATO JUICE, CND	30	6	1.3	tr					60	2	5	15	0.2	240
2 T; 1 fl oz		0.3	0.1						67		0.3	9	5	
CND	100	19	4.3	0.1					200	7	18	50	0.8	800
⅖ cup		0.9	0.2						227		0.9	30	16	
CND	180	34	7.7	0.2					360	13	32	90	1.4	1440
¾ cup		1.6	0.4						408		1.6	50	29	
TOMATO JCE COCKTAIL, CND	100	21	5.0	0.1					200	10	18	50	0.6	800
⅖ cup		0.7	0.2						221		0.9	20	16	

	WT	Proximate			Amino Acids				Minerals			Vitamins		
		CAL	CHO	FAT	TRY	LEU	LYS	MET	Na	Ca	P	THI	NIA	A
		PRO	FIB	PUFA	PHE	ISO	VAL	THR	K	Mg	Fe	RIB	ASC	D
	g	g	g	g	mg	mg	mg	mg	mg	mg	mg	mg	mg	iu

Beef

	WT	CAL/PRO	CHO/FIB	FAT/PUFA	TRY/PHE	LEU/ISO	LYS/VAL	MET/THR	Na/K	Ca/Mg	P/Fe	THI/RIB	NIA/ASC	A/D
ARM/BLADE, LEAN & MARB,	83	218	0.0	11.2	319	2243	2391	679	45	7	158	.08	3.6	0
POT-ROASTED 2 slices		27.4	0.0	0.6	1126	1433	1519	1208	320	16.4	4.1	.19	0	0
ARM/BLADE, LEAN ONLY,	41	90	0.0	3.1	169	1181	1259	357	20	3	80	.04	1.8	0
POT-ROASTED 2 slices		14.4	0.0	0.2	593	755	801	637	169	8.8	2.2	.10	0	0
BEEF BURGUNDY	100	104	2.0	3.9					679	16	71	.04	2.1	196
3½ oz									89		2.0	.11	0	
BEEF IN BARBEQUE SCE, FRZN,	100	138		5.9					237					
CONTINENTAL 3½ oz		7.6												
BEEF JERKY	10	38	1.4	1.7							4	.02	0.8	
		4.2									0.6	.08		
BEEF SNACK, LANCE	10	57	0.2	5.1							4	.09	0.5	
1 pkg		2.5									0.5	.03		
BEEF TIPS W/MADEIRA SCE, FRZN,	100	72	3.6	1.8					320	9	93	.03	1.9	
SARA LEE 3½ oz		9.4	0.3						169	21	1.3	.09	1	
BRISKET, LEAN, MARB & FAT, CKD	95	411	0.0	37.4	201	1415	1509	428	52	11	123	.04	3.7	0
3 slices		17.3	0.0	1.9	710	904	959	763	285	12.7	2.6	.21	0	0
CHUCK, TOTAL EDIBLE, BRAISED	100	327	0.0	23.9					60	11	140	.05	4.0	40
3½ oz		26.0	0.0						370	15	3.3	.20		
CHUCK, SEPARABLE, LEAN, CKD	100	214	0.0	9.5					60	13	160	.05	4.6	20
3½ oz		30.0	0.0						370	18	3.8	.23		
CHUCK, GROUND, CKD	100	327	0.0	23.9					60	11	140	.05	4.0	40
3½ oz		26.0	0.0						370	15	3.3	.20		
CHUCK, STEW MEAT, RAW	113	476	0.0	40.7	297	2088	2226	632	68	15	190	.11	5.8	0
¼ lb		25.4	0.0	2.0	1048	1333	1415	1125	563	20.9	3.7	.28	0	0
CLUB STEAK, LEAN, MARB & FAT, CKD	93	260	0.0	17.5	278	1948	2079	590	48	10	151	.09	5.7	0
1 steak		23.9	0.0	0.9	978	1244	1321	1050	370	17.9	3.6	.11	0	0
CLUB STEAK, LEAN ONLY	55	108	0.0	4.5	187	1310	1397	397	27	4	100	.07	3.7	0
1 steak		16.0	0.0	0.2	658	836	888	706	236	11.4	2.4	.05	0	0
CORNED, MED FAT, CKD	100	372	0.0	30.4					1740	9	93	.02	1.5	
3½ oz		22.9	0.0						150	29	2.9	.18	0	
CORNED, MED FAT, CND	28	60	0.0	3.4	74	520	555	157	268	6	30	tr	1.0	0
1 slice		7.1	0.0	0.2	261	332	353	280	17		1.2	.06	0	0
CORNED, MED FAT, CND	85	184	0.0	10.2	251	1761	1878	533	803	17	90	.01	2.9	0
3 slices		21.5	0.0	0.5	884	1125	1194	949	51		3.7	.20	0	0
CUBED STEAK, CKD	100	261	0.0	15.4	379	2661	2838	806	60	12	250	.08	5.6	30
3½ oz		28.6	0.0	0.7	1336	1700	1804	1434	370	21	3.5	.22	0	
DRIED, CHIPPED	28	57	0.0	1.8	113	788	841	239	1253	6	113	.02	1.1	0
2 thin slices		9.6	0.0	0.1	396	504	534	425	56		1.4	.09	0	0
DRIED, CHIPPED	85	173	0.0	5.4	341	2388	2547	723	3660	17	343	.06	3.2	0
3 oz		29.1	0.0	0.3	1199	1526	1618	1288	170		4.3	.27	0	0
DRIED, CHIPPED, CREAMED	120	209	6.1	13.1	191	1312	1212	365	1161	106	232	.06	1.6	437
½ cup		16.1	0.0	0.7	709	828	861	667	184		2.1	.26	0	2
DRIED, CHIPPED, CREAMED, FRZN,	114	175	7.5	13.0					875	57	66	.04	0.8	0
CAMPBELLS ½ pkg		7.5	0.1						146		0.9	.12	1	
DRIED, CHIPPED, CREAMED, FRZN,	100	151	6.7	10.3					546	83		.05	1.5	225
STOUFFERS		8.0									0.9	.19	2	
FLANK STEAK, LEAN & MARB, CKD	141	331	0.0	14.4	551	3859	4117	1169	67	20	299	.07	3.9	0
1 serving		47.1	0.0	0.7	1938	2465	2617	2080	344	30	7.1	.22	0	0
FLANK STEAK, LEAN ONLY, CKD	75	158	0.0	5.2	301	2106	2249	638	47	6	176	.04	3.8	0
1 serving		25.7	0.0	0.3	1057	1345	1428	1135	281	16	3.9	.02	0	0
HAMBURGER, MED FAT, CKD	85	224	0.0	14.5	255	1786	1906	541	40	6	187	.13	4.8	0
1 patty		21.8	0.0	0.7	897	1141	1211	963	382	18	3.3	.15	0	0
HAMBURGER, LEAN, CKD	86	140	0.0	3.4	303	2122	2264	643	41	14	233	.18	7.1	0
1 patty		25.9	0.0	0.2	1066	1355	1439	1144	480	22	3.9	.11	0	0
HINDSHANK, SEPARABLE, LEAN, CKD	100	176	0.0	4.8					60	14	151	.06	4.7	10
3½ oz		31.0	0.0						370	18	3.9	.23		
NECK, LEAN & MARB, POT-ROASTED	132	325	0.0	16.9	472	3302	3523	1001	72	15	230	.06	5.1	0
2 slices		40.3	0.0	0.8	1658	2109	2239	1780	396	27	6.1	.03	0	0
NECK, LEAN ONLY, POT-ROASTED	30	60	0.0	1.8	121	844	900	256	19	3	57	.01	1.5	0
2 slices		10.3	0.0	0.1	424	539	572	455	112	7	1.6	.09	0	0
PORTERHOUSE STK, LEAN & MARB,	100	242	0.0	14.7	297	2081	2220	631	52	11	183	.10	6.1	0
BROILED 1 steak		25.4	0.0	0.7	1045	1329	1411	1122	398	20	3.8	.12	0	0
PORTERHOUSE STK, LEAN ONLY	54	102	0.0	4.4	172	1204	1285	365	26	4	106	.07	3.6	0
1 steak		14.7	0.0	0.2	605	769	817	649	232	11.5	2.2	.04	0	0

Food	WT g	CAL / PRO	CHO / FIB	FAT / PUFA	TRY / PHE	LEU / ISO	LYS / VAL	MET / THR	Na / K	Ca / Mg	P / Fe	THI / RIB	NIA / ASC	A / D
POT ROAST, FRZN, BRAISED, SARA LEE 1 serving	180	162	7.2	6.5					830	40	283	.09	5.6	140
		18.7	0.4						472	36	3.2	.23	1	
POT ROAST, INVISIBLE CHEF 3½ oz	100	168	4.0	11.9					507	12	93	.04	2.1	1154
		10.7							96		1.8	.07		
RIB ROAST, LEAN & MARB, CKD 2 slices	106	302	0.0	6.7	332	2327	2483	705	57	8	217	64	4.2	0
		28.4	0.0	0.3	1168	1486	1578	1254	438	24	3.5	223	0	0
RIB ROAST, LEAN ONLY, CKD 2 slices	41	70	0.0	2.6	129	901	961	273	17	4	84	33	2.3	0
		11.0	0.0	0.1	453	576	611	486	169	9	1.4	94	0	0
RIB STEAK, LEAN & MARB, CKD 1 serving	94	246	0.0	16.0	281	1966	2098	596	50	7	159	56	3.8	0
		24.0	0.0	0.8	987	1256	1333	1060	388	22	3.6	197	0	0
RIB STEAK, LEAN ONLY, CKD 1 serving	43	73	0.0	2.2	144	1008	1075	305	23	4	68	47	1.9	(0)
		12.3	0.0	0.1	506	644	683	543	205	12	1.8	103	0	0
RIBEYE STEAK, CKD 3½ oz	100	440	0.0	39.4					60	9	186	.05	3.6	80
		19.9	0.0						370	20	2.6	1.5	0	
ROUND, BOTTOM, LEAN & MARB, BROILED 1 slice	114	271	0	10.8	474	3318	3541	1006	51	14	260	148	6.5	0
		40.5	0	0.5	1666	2119	2250	1789	552	28	6.1	376	0	0
ROUND, BOTTOM, LEAN ONLY, BROILED 1 slice	86	205	0.0	8.2	357	2499	2666	757	38	12	196	112	4.9	0
		30.5	0.0	0.4	1255	1596	1695	1341	437	21	4.6	284	0	0
ROUND, TOP, LEAN & MARB, BROILED 1 slice	111	254	0.0	5.9	504	3531	3768	1070	46	9	268	111	6.3	0
		43.1	0.0	0.3	1773	2255	2395	1904	547	28	6.5	344	0	0
ROUND, TOP, LEAN ONLY, BROILED 1 slice	80	173	0.0	4.2	367	2573	2745	780	34	8	193	88	4.5	0
		31.4	0.0	0.2	1292	1643	1745	1387	369	29	4.7	264	0	0
ROUND HEEL, LEAN & MARB, POT-ROASTED 1 slice	118	261	0.0	11.4	431	3015	3217	914	64	10	204	118	5.1	0
		36.8	0.0	0.6	1514	1926	2045	1625	455	28	5.5	271	0	0
ROUND HEEL, LEAN ONLY 1 slice	53	100	0.0	2.9	204	1426	1521	432	26	5	109	58	2.3	0
		17.4	0.0	0.1	716	911	967	769	218	13	2.6	127	0	0
ROUND, STEW MEAT, RAW ¼ lb	113	294	0.0	15.3	429	3007	3208	911	68	15	215	113	5.8	0
		36.7	0.0	0.8	1510	1921	2039	1621	563	24	5.5	283	0	0
RUMP, LEAN & MARB, POT-ROASTED 2 slices	80	188	0.0	8.7	300	2097	2238	636	43	7	158	80	3.4	0
		25.6	0.0	0.4	1053	1340	1422	1131	309	16	3.8	184	0	0
RUMP, LEAN ONLY, POT-ROASTED 2 slices	37	77	0.0	2.8	140	983	1049	297	18	3	68	41	1.6	0
		12.0	0.0	0.1	494	628	667	530	152	8	1.8	89	0	0
SALISBURY STK, GREEN GIANT 1 serving	200	274	16.0	16.2					1036	36	150	.08	2.8	544
		15.8	0.2						378		2.2	.24	6	
SALISBURY STK W/ GVY & ONIONS, INVISIBLE CHEF 1 serving	181	397	5.5	31.7					1860	27	160	.12	5.8	238
		22.3							152		4.2	.22	1	
SALISBURY STK W/ MUSHROOM SCE, HORMEL	200	276	12.2	18.2					1276	84	232	.14	3.8	166
		16.0	1.0							42	2.7	.19	6	
SALISBURY STK W/ MUSHROOM GVY, FRZN, CAMPBELLS 1 serving	185	220	9.0	11.0					1050	18	204	.20	4.4	0
		22.0	0.4						370		3.3	.20	0	
SAUERBRATEN (SOUR POT ROAST), FRZN, SARA LEE	200	202	7.8	6.8					794	22	226	.08	4.8	164
		25.2	0.2						378	30	3.2	.20	2	
SHORT PLATE, SEPARABLE, LEAN, SIMMERED 3½ oz	100	199	0.0	7.7					60	13	149	.05	4.6	10
		30.3	0.0						370	18	3.8	.23		
SHORT RIBS, LEAN & MARB, CKD 1 serving	72	290	0.0	23.9	206	1442	1539	437	39	5	106	43	2.9	0
		17.6	0.0	1.2	724	921	978	777	297	14	2.6	151	0	0
SHORT RIBS, LEAN ONLY, CKD 1 serving	17	46	0.0	2.5	64	451	481	137	9	1	32	14	0.9	0
		5.5	0.0	0.1	226	288	306	243	81	4	0.8	39	0	0
SHORT RIBS, BRAISED, FRZN, CAMPBELLS 1 package	200	420	2.9	34.7					522	12	79	.13	5.9	514
		23.1	0.3						514		4.0	.13	2	
SHORT RIBS, BONELESS W/ CARROTS FRZN, INVISIBLE CHEF 1 serving	151	391	2.6	34.4					868	21	536	.06	3.3	3210
		17.7									2.7	.12	0	
SHORT RIBS W/ BBQ SCE, FRZN, INVISIBLE CHEF 1 serving	188	372	6.0	32.1					1104	22	488	.04	2.7	324
		14.7							670		2.4	.12	4	
SIRLOIN, LEAN & MARB, BROILED 1 slice	125	260	0.0	13.8	373	2614	2789	792	57	18	282	125	4.1	0
		31.9	0.0	0.7	1312	1669	1772	1409	545	26	4.8	575	0	0
SIRLOIN, LEAN ONLY, BROILED 1 slice	77	128	0.0	4.4	243	1704	1818	516	34	12	178	85	2.2	0
		20.8	0.0	0.2	856	1088	1156	919	349	18	3.1	462	0	0
SIRLOIN TIP, LEAN & MARB, ROASTED 1 slice	76	141	0.0	5.1	261	1827	1949	554	35	11	180	76	2.5	0
		22.3	0.0	0.3	917	1167	1239	985	331	17	3.5	350	0	0
SIRLOIN TIP, LEAN ONLY, ROASTED 1 slice	44	74	0.0	1.8	157	1098	1171	333	19	7	108	48	1.3	0
		13.4	0.0	0.1	551	701	745	592	199	10	2.0	264	0	0
SIRLOIN, GROUND, CKD 3½ oz	100	408	0.0	34.7					60	10	186	.06	4.6	60
		22.2	0.0						370	21	2.9	.18	0	

	WT	Proximate CAL	CHO	FAT	Amino Acids TRY	LEU	LYS	MET	Minerals Na	Ca	P	Vitamins THI	NIA	A
		PRO	FIB	PUFA	PHE	ISO	VAL	THR	K	Mg	Fe	RIB	ASC	D
	g	g	g	g	mg	mg	mg	mg	mg	mg	mg	mg	mg	iu
STEAK, CHICKEN-FRIED	100	389	12.3	30.0					815	11	110	.11	2.7	26
3½ oz		17.9							126		2.3	.14		
STEAK, TERIYAKI	100	453	0.2	40.3					85	9	175	.06	4.3	70
3½ oz		20.4	0.0						371		2.6	.16	0	
SWISS STK W/ GVY, FRZN, CAMPBELLS	150	200	5.0	9.0					465	10	146	.11	3.5	258
1 pkg		24.0	0.3						209		3.4	.16	3	
SWISS STK W/ GVY & ONIONS,	181	213	2.9	8.4					1249	27	159	.08	4.9	209
INVISIBLE CHEF 1 serving		31.4									4.3	.26	0	
SWISS STK W/ TOMATOES,	181	213	2.8	8.4					1321	26	174	.09	4.8	403
INVISIBLE CHEF 1 serving		31.4							94		4.4	.26	8	
T-BONE, LEAN & MARB, BROILED	95	235	0.0	14.7	281	1966	2098	596	49	10	172	95	5.8	0
1 steak		24.0	0.0	0.7	987	1256	1333	1060	378	19	3.6	114	0	0
T-BONE, LEAN ONLY, BROILED	59	116	0.0	5.4	183	1278	1364	387	29	4	105	77	4.0	0
1 steak		15.6	0.0	0.3	642	816	867	689	253	12	2.3	53	0	0
TENDERLOIN, LEAN & MARB, BROILED	66	148	0.0	8.3	201	1409	1504	427	30	10	137	66	2.2	0
1 steak		17.2	0.0	0.4	708	900	956	760	288	15	2.6	304	0	0
TENDERLOIN, LEAN ONLY, BROILED	52	107	0.0	4.6	178	1245	1329	377	23	8	112	57	1.5	0
1 steak		15.2	0.0	0.2	625	795	845	671	236	12	2.3	312	0	0
Chicken														
ALL CLASSES, DARK MEAT, NO SKIN,	100	130	0.0	4.7					67	13	188	.08	5.2	150
RAW 3½ oz		20.6	0.0						250		1.5	.20		
ALL CLASSES, DARK MEAT, NO SKIN,	100	176	0.0	6.3					86	13	229	.07	5.6	150
ROASTED 3½ oz		28.0	0.0						321	23	1.7	.23		
ALL CLASSES, LIGHT MEAT, NO SKIN,	100	117	0.0	1.9					50	11	218	.05	10.7	60
RAW 3½ oz		23.4	0.0						320	23	1.1	.09		
ALL CLASSES, LIGHT MEAT, NO SKIN,	100	166	0.0	3.4					64	11	265	.04	11.6	60
ROASTED 3½ oz		31.6	0.0						411	19	1.3	.10		
BAKED BREAST, IDLE WILD	100	199	0.4	11.5					58					
3½ oz		23.6							143					
BAKED LEG, IDLE WILD	100	233	0.3	15.0					68					
3½ oz		24.3							213					
BROILER, RAW	100	151	0.0	7.2	250	1490	1810	537	78	14	200	.08	10.2	0
3½ oz		20.2	0.0	1.9	811	1088	1012	877	320		1.5	.16	0	
BROILER, FLESH ONLY, BROILED	100	136	0.0	3.8					66	9	201	.05	8.8	90
3½ oz		23.8	0.0						274		1.7	.19		
BROILER, FRIED	85	232	3.1	13.6					80	18	225	.07	9.7	230
¼ bird w/o bone		22.4	0.0	3.5					242		1.8	.17	0	
CAPON, TOTAL EDIBLE, RAW	100	283	0.0	21.2										
3½ oz		21.4	0.0											
CAPON, FLESH, SKIN, RAW	100	291	0.0	22.0										
3½ oz		21.6	0.0											
CND, BONED	100	178	0.2	10.0	253	1507	1830	543	543	14	149	.04	5.5	230
3½ oz		20.4	0.0	2.6	820	1100	1023	887	138	19	1.8	.11	4	
CREAMED	118	208	6.6	12.1						83	138	.04	3.8	328
½ cup		17.6									1.1	.18	tr	
FRICASSEE, INVISIBLE CHEF	100	164	3.3	9.4						6		.21	0.7	71
		15.5									0.9	.07	0	
FRIED FILET, ARTHUR TREACHERS	136	369	16.5	21.6					326	11	192	.07	11.0	102
1 serving		27.1		5.6					326	27	0.8	.15	1	4
FRIED PARTS, BANQUET	100	285	13.1	17.1					717	167	242	.05	4.0	20
3½ oz		19.7							386		2.2	.12	0	
FRIED PARTS, KENTUCKY FRIED,	100	290	9.0	17.8					535	47		.08	6.4	65
ORIG RECIPE 1 serving		23.4									1.3	.19	1	
FRIED PARTS, KENTUCKY FRIED,	100	323	12.0	20.8					440	34		.07	6.0	74
CRISPY . 1 serving			22.1								1.2	.14	1	
FRYER														
BACK, RAW	100	157	0.0	9.6					54	12	185	.05	4.3	310
3½ oz		16.5	0.0						402	47	1.7	.23		
BACK, FRIED	50	173	3.4	10.6						7	131	.03	3.4	195
		15.0	0.0							11	1.3	.25		
BREAST, RAW	100	104	0.0	0.5	289	1721	2091	620	90	14	212	.07	10.5	0
3½ oz		23.3	0.0	0.1	937	1257	1169	1013	370	47	1.1	.09	0	

	WT g	CAL / PRO (g)	CHO / FIB (g)	FAT / PUFA (g)	TRY / PHE (mg)	LEU / ISO (mg)	LYS / VAL (mg)	MET / THR (mg)	Na / K (mg)	Ca / Mg (mg)	P / Fe (mg)	THI / RIB (mg)	NIA / ASC (mg)	A / D (iu)
BREAST, FRIED	96	232	3.1	11.9						19	245	.07	10.2	460
½ breast		26.8		3.1							1.3	.10	0	
DARK MEAT, W/ SKIN, RAW	100	132	0.0	6.3					54	12	185	.06	4.7	200
3½ oz		17.7							402	47	1.7	.30		
DARK MEAT W/ SKIN, FRIED	100	263	3.1	13.6						12	228	.07	6.7	210
3½ oz		29.9								23	2.0	.45		
DARK MEAT W/O SKIN, RAW	100	112	0.0	3.8					67	13	188	.06	5.3	120
3½ oz		18.1	0.0						250		1.5	.34		
DARK MEAT W/O SKIN, FRIED	100	220	1.5	9.3					88	14	235	.07	6.8	130
3½ oz		30.4							330	23	1.8	.45		
DRUMSTICK, RAW	100	112	0.0	2.7	254	1514	1839	546	79	15	188	.10	5.6	0
3½ oz		20.5	0.0	0.7	824	1105	1028	891	325	26	1.8	.24	0	
DRUMSTICK, FRIED	80	126	3.0	10.7						17	193	.09	4.8	322
2 small legs w/o bone		20.9		2.8						18	1.9	.23	0	
FLESH, RAW	100	107	0.0	2.7					58	12	203	.06	6.4	90
3½ oz		19.3	0.0						285		1.3	.25		
FLESH, FRIED	100	209	1.2	7.8					78	13	257	.06	9.7	90
3½ oz		31.2							381	23	1.6	.35		
FLESH & SKIN, RAW	100	126	0.0	5.1					54	11	198	.05	5.6	170
3½ oz		18.8	0.0						402	47	1.5	.23		
FLESH & SKIN, FRIED	100	250	2.8	11.9						12	243	.06	9.2	170
3½ oz		30.6								23	1.8	.36		
LIGHT MEAT, W/ SKIN, RAW	100	120	0.0	3.9	205	1472	1590	502	54	11	211	.05	6.7	130
3½ oz		19.9	0.0		800	1069	1018	794	402	47	1.3	.16		
LIGHT MEAT, W/ SKIN, FRIED	100	234	2.4	9.9						11	260	.05	11.9	130
3½ oz		31.5								23	1.5	.27		
LIGHT MEAT, W/O SKIN, RAW	100	101	0.0	1.5	250	1490	1810	537	50	11	218	.05	7.6	50
3½ oz		20.6	0.0		811	1088	1012	877	320	23	1.1	.17		
LIGHT MEAT W/O SKIN, FRIED	100	197	1.1	6.1					68	12	280	.05	12.9	50
3½ oz		32.1							434	19	1.3	.25		
NECK, RAW	100	151	0.0	9.4					54	11	182	.05	3.0	310
3½ oz		15.5	0.0						402	47	1.9	.25		
NECK, FRIED	50	144	2.2	8.7						6	117	.04	2.8	175
		13.3								11	1.3	.20		
RIB, RAW	100	124	0.0	5.4					54	11	212	.04	5.1	170
3½ oz		17.7	0.0						402	47	1.3	1.8		
RIB, FRIED	50	149	2.9	7.7						6	145	.02	4.7	105
		15.7								11	1.0	.23		
THIGH, RAW	100	128	0.0	5.6					54	12	186	.06	5.7	180
3½ oz		18.1	0.0						402	47	1.6	.33		
THIGH, FRIED	50	118	1.2	5.7						6	118	.03	3.4	100
		14.5								11	1.1	.24		
WING, RAW	100	146	0.0	7.4					54	10	203	.04	4.1	290
3½ oz		18.5	0.0						402	20	1.5	.14		
WING, FRIED	20	54	0.5	3.0						2	47	.01	1.4	50
1 piece		5.8								5	0.4	.05		
HEN & COCK														
DARK MEAT W/O SKIN, RAW	100	154	0.0	7.5					67	13	188	.10	8.7	240
3½ oz		20.2	0.0						250		1.5	.18		
DARK MEAT W/O SKIN, STEWED	100	207	0.0	9.5					64	13	138	.06	8.3	270
3½ oz		28.5	0.0						239		1.8	.20		
FLESH, RAW	100	155	0.0	7.0	259	1540	1871	556	58	12	203	.08	10.1	230
3½ oz		21.3	0.0		838	1125	1046	907	285		1.3	.14		
FRESH, STEWED	100	208	0.0	8.9					55	12	149	.04	9.6	250
3½ oz		30.0	0.0						272		1.5	.15		
FLESH & SKIN, RAW	100	251	0.0	18.8					54	11	182	.06	9.2	610
3½ oz		19.0	0.0						402	47	1.3	.13		
FLESH & SKIN, STEWED	100	317	0.0	22.8						11	134	.04	8.8	670
3½ oz		26.1	0.0								1.5	.14		
FLESH & SKIN, GIBLETS, RAW	100	246	0.0	18.3						11	185	.07	9.1	900
3½ oz		19.0	0.0								1.5	.20		
FLESH & SKIN, GIBLETS, STEWED	100	312	0.0	22.2						11	136	.04	8.6	990
3½ oz		26.2	0.0								1.8	.23		

	WT (g)	CAL / PRO (g)	CHO / FIB (g)	FAT / PUFA (g)	TRY / PHE (mg)	LEU / ISO (mg)	LYS / VAL (mg)	MET / THR (mg)	Na / K (mg)	Ca / Mg (mg)	P / Fe (mg)	THI / RIB (mg)	NIA / ASC (mg)	A / D (iu)
LIGHT MEAT W/O SKIN, RAW	100	133	0.0	3.7					50	11	21.8	.05	11.5	120
3½ oz		23.4	0.0						320	23	1.1	.09		
LIGHT MEAT W/O SKIN, STEWED	100	180	0.0	4.7					48	11	160	.03	11.0	130
3½ oz		32.2	0.0						306	19	1.3	.09		
TOTAL ED, RAW	100	298	0.0	24.8					54	10	167	.06	8.2	1080
3½ oz		17.4	0.0						402	47	1.4	.19		
TOTAL ED, STEWED	100	369	0.0	29.5						10	123	.04	7.8	1190
3½ oz		24.0	0.0								1.6	.21		

ROASTER

	WT (g)	CAL / PRO (g)	CHO / FIB (g)	FAT / PUFA (g)	TRY / PHE (mg)	LEU / ISO (mg)	LYS / VAL (mg)	MET / THR (mg)	Na / K (mg)	Ca / Mg (mg)	P / Fe (mg)	THI / RIB (mg)	NIA / ASC (mg)	A / D (iu)
DARK MEAT W/O SKIN, RAW	100	132	0.0	4.7					67	13	188	.13	4.7	150
3½ oz		21.0	0.0						250	16	1.5	.16		
DARK MEAT W/O SKIN, ROASTED	100	184	0.0	6.5					88	14	235	.12	5.3	160
3½ oz		29.3	0.0						330	23	1.8	.19		
FLESH, RAW	100	131	0.0	4.5					58	12	203	.10	7.7	150
3½ oz		21.1	0.0						285	27	1.3	.12		
FLESH, ROASTED	100	183	0.0	6.3					77	12	254	.10	8.5	150
3½ oz		29.5	0.0						376	23	1.5	.15		
FLESH & SKIN, RAW	100	197	0.0	12.6					75	11	191	.08	7.4	410
3½ oz		19.5	0.0						372	47	1.5	.12		
FLESH & SKIN, ROASTED	100	248	0.0	14.7						11	239	.08	8.2	420
3½ oz		27.1	0.0							23	1.8	.14		
FLESH & SKIN, GIBLETS, RAW	100	191	0.0	11.9	205	1472	1590	502	75	12	194	.08	7.3	760
3½ oz		19.6	0.0	800	1069	1018	794	372	27	1.7	.21		
FLESH & SKIN, GIBLETS, ROASTED	100	242	0.0	14.0						12	242	.08	8.1	790
3½ oz		27.2	0.0							23	2.0	.25		
LIGHT MEAT W/O SKIN, RAW	100	128	0.0	3.2					50	11	218	.08	10.6	100
3½ oz		23.3	0.0						320	17	1.1	.08		
LIGHT MEAT W/O SKIN, ROASTED	100	182	0.0	4.9					66	11	272	.08	11.8	110
3½ oz		32.3	0.0						422	19	1.3	.10		
TOTAL EDIBLE, RAW	100	239	0.0	17.9					75	10	176	.08	6.7	920
3½ oz		18.2	0.0						372	27	1.6	.19		
TOTAL EDIBLE, ROASTED	100	290	0.0	20.2						10	220	.07	7.4	960
3½ oz		25.2	0.0							23	1.9	.22		

Game & specialty meats

	WT (g)	CAL / PRO (g)	CHO / FIB (g)	FAT / PUFA (g)	TRY / PHE (mg)	LEU / ISO (mg)	LYS / VAL (mg)	MET / THR (mg)	Na / K (mg)	Ca / Mg (mg)	P / Fe (mg)	THI / RIB (mg)	NIA / ASC (mg)	A / D (iu)
ALLIGATOR MEAT	100	232	0.0	4.2						1231	767	.10	6.3	45
3½ oz		45.6	0.0								11.4	.30	0	
ARMADILLO MEAT	100	172	0.0	5.4						30	208	.10	6.0	0
3½ oz		29.0	0.0								10.9	.40	0	
BEAVER, ROASTED	100	248	0.0	13.7								.08		
3½ oz		29.2	0.0									.38		

DUCK

	WT (g)	CAL / PRO (g)	CHO / FIB (g)	FAT / PUFA (g)	TRY / PHE (mg)	LEU / ISO (mg)	LYS / VAL (mg)	MET / THR (mg)	Na / K (mg)	Ca / Mg (mg)	P / Fe (mg)	THI / RIB (mg)	NIA / ASC (mg)	A / D (iu)
DOMESTIC, TOTAL ED, RAW	100	326	0.0	28.6		1243	1382	398	82	10	176	.08	6.7	0
3½ oz		16.0	0.0	632	832	770	701	285		1.6	.19	8	
DOMESTIC, FLESH ONLY, RAW	100	165	0.0	8.2		1657	1842	531	74	12	203	.10	7.7	0
3½ oz		21.4	0.0	842	1109	1027	935	285		1.3	.12		
DOMESTIC, ROASTED	100	310	0.0	23.6						19	230			0
3½ oz		22.8	0.0								5.8			
WILD, TOT ED, RAW	100	233	0.0	15.8						12	200			0
3½ oz		21.1	0.0								3.0			
WILD, FLESH ONLY, RAW	100	138	0.0	5.2		1657	1842	531						
3½ oz		21.4	0.0	842	1109	1027	935						
FROG LEGS, RAW	100	73	0.0	0.2					55	18	147	.14	1.2	0
3½ oz		16.4	0.0						308	24	1.5	.25		
FROG LEGS, FRIED	100	290	8.5	19.8						19	160	.12	1.3	0
3½ oz		17.9	0.0								1.4	.24		
GOAT MEAT	100	165	0.0	9.4						11		.17	5.6	0
3½ oz		18.7									2.2	.32		
GOOSE, TOTAL ED, ROASTED	100	426	0.0	36.0					86	11	240	.08	8.1	0
3½ oz		23.7	0.0						420	31	2.1	.24	13	
GOOSE, FLESH ONLY, ROASTED	100	233	0.0	9.8					124	14	277	.11	9.3	
3½ oz		33.9	0.0						605		1.7	.16		

	WT	Proximate			Amino Acids				Minerals			Vitamins		
		CAL	CHO	FAT	TRY	LEU	LYS	MET	Na	Ca	P	THI	NIA	A
		PRO	FIB	PUFA	PHE	ISO	VAL	THR	K	Mg	Fe	RIB	ASC	D
	g	g	g	g	mg	mg	mg	mg	mg	mg	mg	mg	mg	iu
GOOSE, FLESH & SKIN, ROASTED	100	441	0.0	38.1						13	260	.09	8.9	
3½ oz		22.9	0.0						406	31	1.9	.16		
GUINEA HEN, TOTAL ED, RAW	100	156	0.0	6.4										
3½ oz		23.1	0.0											
GUINEA HEN, FLESH & SKIN, RAW	100	158	0.0	6.4										
3½ oz		23.4	0.0											
GUINEA PIG MEAT, RAW	100	96	0.0	1.6						29	253	.06	6.5	
3½ oz		19.0									1.9	.14		
HARE, FLESH ONLY	100	135	0.0	5.0						12	157	.09	5.0	
3½ oz		21.0	0.0								3.2	.19		
HORSE MEAT, RAW	100	118	0.9	4.1						10	150	.07	4.3	
3½ oz		18.1	0.0	1.2							2.7	.12		
MUSKRAT, ROASTED	100	153	0.0	4.1								.16		
3½ oz		27.2	0.0									.21		
OPOSSUM, ROASTED	100	221	0.0	10.2								.12		
3½ oz		30.2	0.0									.38		
PHEASANT, TOTAL ED, RAW	100	151	0.0	5.2						14	262			
3½ oz		24.3	0.0								3.7			
PHEASANT, FLESH & SKIN, RAW	100	152	0.0	5.2										
3½ oz		24.7	0.0											
PHEASANT, FLESH ONLY, RAW	100	162	0.0	6.8										
3½ oz		23.6	0.0											
QUAIL, TOTAL ED, RAW	100	168	0.0	6.8					40	15	270			
3½ oz		25.0	0.0						175		3.8			
QUAIL, FLESH, SKIN, RAW	100	172	0.0	7.0					40					
3½ oz		25.4	0.0						175					
RABBIT, DOMESTIC, FLESH ONLY, RAW	100	162	0.0	8.0		1636	1818	541	43	20	352	.08	12.8	
3½ oz		21.0	0.0		793	1082	1021	1021	385		1.3	.06	4	
RABBIT, DOMESTIC, BAKED	100	177	0.0	5.0								.11	7.0	
3½ oz		30.9										.15	0	
RABBIT, DOMESTIC, STEWED	100	216	0.0	10.1		1636	1818	541	41	21	259	.05	11.3	
3½ oz		29.3	0.0		793	1082	1021	1021	368		1.5	.07	0	
RABBIT, WILD, FLESH ONLY, RAW	100	135	0.0	5.0					47					
3½ oz		21.0	0.0						415	29			0	
RACCOON, ROASTED	100	255	0.0	14.5								.59		
3½ oz		29.2	0.0									.52	0	
REINDEER, TOTAL ED SIDE, RAW	100	217	0.0	14.4										
3½ oz		20.5	0.0											
REINDEER, FOREQUARTER, RAW	100	178	0.0	9.4										
3½ oz		21.8	0.0											
REINDEER, HINDQUARTER, RAW	100	256	0.0	19.2										
3½ oz		19.4	0.0											
SNAIL, RAW	100	90	2.0	1.4										
3½ oz		16.1									3.5			
SNAIL, GIANT AFRICAN, RAW	100	73	4.4	1.4										
3½ oz		9.9												
SQUAB (PIGEON), TOTAL ED, RAW	100	279	0.0	22.1						17	411	.10	5.6	
3½ oz		18.6	0.0								1.8	.24	0	
SQUAB (PIGEON), FLESH & SKIN, RAW	100	294	0.0	23.0										
3½ oz		18.5	0.0											
SQUAB (PIGEON), FLESH ONLY	100	142	0.0	7.5									6.6	
3½ oz		17.5	0.0											
SQUAB (PIGEON), LIGHT MEAT	100	125	0.0	4.2									7.6	
3½ oz		20.4	0.0											
TURTLE, GREEN, RAW	100	89	0.0	0.5										
3½ oz		19.8	0.0											
TURTLE, GREEN, CND	100	106	0.0	0.7										
3½ oz		23.4	0.0											
VENISON, RAW	100	124	0.0	3.6					70	19	183	.37	7.4	0
3½ oz		21.4	0.0	0.3					336	29	5.0	.28	0	
VENISON, ROASTED	100	146	0.0	2.2						20	264	.37	7.4	
3½ oz		29.5	0.0	0.2						29	3.5	.28	0	0
VENISON, SALTED, DRIED	100	142	0.0	0.9						60	298	.09	10.0	
3½ oz		31.4		0.1							1.9	.34		

	WT g	CAL / PRO	CHO / FIB	FAT / PUFA	TRY / PHE	LEU / ISO	LYS / VAL	MET / THR	Na / K	Ca / Mg	P / Fe	THI / RIB	NIA / ASC	A / D
		g	g	g	mg	mg	mg	mg	mg	mg	mg	mg	mg	iu
WHALE MEAT, RAW	100	156	0.0	7.5	220		1880	520	78	12	144	.09		1860
3½ oz		20.6	0.0		1260			800	22			.08	6	
Lamb														
ARM CHOP, LEAN, MARB & FAT, CKD	100	339	0.0	27.0	290	1734	1813	538	66	7	210	.16	6.3	0
3½ oz		22.4	0.0	1.1	911	1161	1104	1025	388	2.2	2.2	.26	0	0
ARM CHOP, LEAN & MARB, CKD	57	144	0.0	8.9	189	1130	1182	350	43	4	135	.10	4.1	0
1 chop		14.6	0.0	0.4	593	757	719	668	252	14	1.3	.17	0	0
ARM CHOP, LEAN, MARB, CKD	114	287	0.0	17.9	379	2269	2372	703	87	9	269	.22	8.3	0
2 chops		29.3	0.0	0.7	1191	1519	1444	1341	509	28	3.4	.34	0	0
ARM CHOP, LEAN ONLY	59	110	0.0	4.4	214	1278	1336	396	49	5	151	.12	4.7	0
2 chops		16.5	0.0	0.2	671	855	813	755	286	16	1.8	.19	0	0
BLADE CHOP, LEAN, MARB & FAT, CKD	100	340	0.0	26.1	319	1905	1991	590	70	7	227	.17	6.7	0
CKD 3½ oz		24.6	0.0	1.0	1000	1275	1212	1126	412	22	3.0	.28	0	0
BLADE CHOP, LEAN & MARB, CKD	93	260	0.0	16.8	329	1967	2056	610	72	11	226	.18	6.9	0
1 chop		25.4	0.0	0.7	1033	1316	1251	1163	422	22	2.8	.28	0	0
BLADE CHOP, LEAN ONLY	56	128	0.0	6.4	211	1262	1319	391	46	5	149	.11	4.4	0
1 chop		16.3	0.0	0.3	663	845	803	746	272	14	1.7	.18	0	0
LEG, LEAN, MARB & FAT, ROASTED	100	242	0.0	14.5	267	1595	1667	494	61	6	199	.16	5.3	0
3½ oz		20.6	0.0	0.6	837	1068	1015	943	369	22	3.1	.22	0	0
LEG, LEAN & MARB, ROASTED	49	96	0.0	4.1	179	1069	1117	331	41	4	105	.11	3.5	0
1 slice		13.8	0.0	0.2	561	715	680	632	246	12	1.5	.14	0	0
LEG, LEAN & MARB, ROASTED	98	192	0.0	8.2	357	2137	2234	662	82	8	210	.22	7.0	0
2 slices		27.6	0.0	0.3	1122	1431	1360	1263	492	24	3.0	.29	0	0
LEG, LEAN ONLY, ROASTED	61	107	0.0	3.5	228	1363	1425	422	52	5	134	.14	4.5	0
2 slices		17.6	0.0	0.2	715	912	867	806	312	14	1.9	.18	0	0
LOIN CHOP, LEAN, MARB & FAT, CKD	100	302	0.0	22.5	298	1781	1862	552	49	10	193	.17	6.5	0
3½ oz		23.0	0.0	0.9	935	1192	1133	1053	466	20	3.0	.27	0	0
LOIN CHOP, LEAN & MARB, CKD	46	103	0.0	5.5	162	968	1012	300	37	4	98	.09	3.6	0
1 chop		12.5	0.0	0.2	508	648	616	572	218	11	1.4	.14	0	0
LOIN CHOP, LEAN & MARB, CKD	92	205	0.0	10.9	322	1928	2015	598	75	7	197	.18	7.1	0
2 chops		24.9	0.0	0.4	1012	1291	1227	1140	437	22	2.8	.29		0
LOIN CHOP, LEAN ONLY, CKD	70	130	0.0	5.2	254	1518	1586	470	58	6	160	.14	5.5	0
2 chops		19.6	0.0	0.2	797	1016	966	897	340	15	2.2	.23	0	0
RIB CHOP, LEAN, MARB & FAT, CKD	100	423	0.0	37.2	263	1572	1643	487	83	8	182	.21	7.9	0
3½ oz		20.3	0.0	1.5	825	1052	1000	929	485	17	3.0	.33	0	0
RIB CHOP, LEAN & MARB, CKD	41	119	0.0	8.3	136	813	850	252	34	3	90	.08	3.2	0
1 chop		10.5	0.0	0.3	427	544	517	481	199	8	1.2	.13	0	0
RIB CHOP, LEAN & MARB, CKD	82	238	0.0	16.6	272	1626	1700	504	68	6	180	.17	6.4	0
2 chops		21.0	0.0	0.7	854	1088	1035	961	398	16	2.4	.27	0	0
RIB CHOP, LEAN ONLY, CKD	52	102	0.0	4.3	193	1154	1206	358	43	4	125	.10	4.1	0
2 chops		14.9	0.0	0.2	606	772	734	682	252	11	1.6	.17	0	0
Luncheon meats														
BARBEQUE LOAF, OSCAR MAYER	28	50	1.7	2.8					374	15	37	.10	0.6	
1 oz slice		4.5		0.2					92	5	0.3	.07	5	10
BOLOGNA	28	88	0.3	8.2	35	297	333	52	364	2	36	.05	0.8	0
1 slice		3.4		0.3	151	201	208	170	64	5	0.5	.06	0	
BOLOGNA, ALL BEEF, OSCAR MAYER	28	88	0.8	8.1					292	3	22	.01	0.7	
1 oz slice		3.1		2.8					40	3	0.4	.03	6	9
CANADIAN BACON, OSCAR MAYER	28	45		2.0					384	2	46	.20	1.6	tr
1 oz slice		6.2		0.1					87	4	0.1	.04	8	10
CAPICOLA/CAPACOLA	28	140	0.0	12.8										
1 oz slice		5.7	0.0	1.3										
HAM, BOILED	28	66	0.0	4.8						3	46	.12	0.7	0
1 oz slice		5.3	0.0	0.5						3	0.8	.04		
HAM, CHOPPED	28	64	0.8	4.8	40	322	351	101	387	2	40	.21	1.1	0
1 oz slice		4.5	0.0	0.6	160	207	217	171	89	4	0.2	.06	5	
HAM, DEVILED, CND	13	46	0.0	4.2					117	1	12	.02	0.2	0
1 T		1.8	0.0	0.5					27		0.3	.01	tr	
HAM & CHEESE LOAF, OSCAR MAYER	28	70	0.3	5.6					372	15	62	.17	1.0	
1 slice		4.8		0.8					80	4	0.2	.06		13
HEADCHEESE	28	55	0.3	4.1	22	265	254	70		3	48	.01	0.3	0
1 oz slice		4.2	0.0	0.5	159	143	173	117			0.6	.03	0	0

	WT	Proximate			Amino Acids				Minerals			Vitamins		
		CAL	CHO	FAT	TRY	LEU	LYS	MET	Na	Ca	P	THI	NIA	A
		PRO	FIB	PUFA	PHE	ISO	VAL	THR	K	Mg	Fe	RIB	ASC	D
	g	g	g	g	mg	mg	mg	mg	mg	mg	mg	mg	mg	iu
HONEY LOAF, OSCAR MAYER	28	39	0.9	1.7					377	4	38	.19	0.8	
1 oz slice		5.0		0.1					94	4	0.2	.08	5	
LIVERCHEESE, OSCAR MAYER	28	84	0.4	7.3					345	2	56	.06	3.4	490
1 oz slice		4.2							63	3	2.9	.60	12	12
LIVERWURST	28	139	0.5	9.1	52	392	364	97	81	3	67	.06	1.6	1778
1 slice		4.7	0.0	1.1	213	229	290	203	65	6	1.5	.36	0	4
LIVERWURST, SMOKED	28	102	0.6	9.1	48	361	336	90		3	69	.05	2.3	1828
(BRAUNSCHWEIGER) 1 oz slice		4.3		1.1	196	211	268	187		4	1.7	.40	0	4
LIVERWURST SPREAD, UNDERWOOD	28	91	1.1	7.7					211	5	72	.17	1.5	1736
1 oz		4.2							69		2.2	.34	1	
MEATLOAF	28	56	0.9	3.7						3	50	.04	0.7	
1 oz		4.5	0.0								0.5	.06		
OLD-FASHIONED LOAF, OSCAR MAYER	28	62	2.3	3.9					366	27	42	.11	0.8	tr
1 oz slice		4.2		0.4					105	5	0.2	.08	3	
OLIVE LOAF, OSCAR MAYER	28	64	2.9	4.5					416	28	35	.08	0.5	28
1 oz slice		3.4		0.6					81	5	0.2	.08	2	13
PICKLE & PIMENTO LOAF	28	60	3.5	3.6					406	30	36	.10	0.6	
1 oz slice		3.4		0.5					96	5	0.1	.08	3	
POTTED MEAT (BEEF, CHICKEN,	28	78	0.0	5.4	42	337	297	101				.01	0.3	
TURKEY) 1 oz		4.5	0.0		179	179	264	185				.06		
PREM	83	274		24.4					1012					
1 slice		9.6												
SALAMI, CKD	28	73	0.4	5.8						3	56	.07	1.1	
1 oz		4.9	0.0	0.2						4	0.7	.07		
SALAMI, BEEF, COTTO	28	60	0.7	4.5					356	2	28	.03	1.0	
1 oz slice		4.2		0.2					70	4	0.5	.07	5	
SALAMI, COTTO	28	66	0.5	5.3					297	tr	29	.08	1.0	
1 oz slice		4.2	0.0	0.5					51	4	0.6	.10	5	8
SALAMI, DRY	28	112	0.2	9.8	57	480	538	141	540	2	36	.17	1.3	
1 oz slice		5.6	0.0	0.8	244	325	336	274	102	4	0.3	.08	6	
SANDWICH SPREAD, OSCAR MAYER	28	63	3.4	4.5					298	3	16	.05	0.5	28
1 slice		2.2		0.8					32	2	0.2	.04	0	
SPAM, HORMEL	28	87	1.1	7.4					336	3	35	.11	0.6	
1 oz		3.9	tr						59		0.5	.04		
TREET	60	167	0.7	14.7					780	10	71	.20	1.6	
2 oz		7.9									0.7	.09	0	
Meats, miscellaneous														
MEAT BALLS, HOMEMADE, CKD	28	78	1.9	5.5						11	42	.02	0.8	14
1 oz		5.0								6	0.6	.04	0	
MEAT BALLS, ITALIAN, INVISIBLE	28	42	1.5	3.1					25	3	27	.02	0.4	141
CHEF 1 oz		2.0							10		0.4	.03	2	
MEATBALLS, SWEDISH, INVISIBLE	28	16	0.5	1.3					66	6	10	.01	0.1	22
CHEF 1 oz		0.7							1		tr	0.1		
MEATBALLS, SWEDISH, FRZN, PRONTO	28	58	1.8	4.5					235	20	35	.02	0.3	80
FOODS 1 oz		2.5							3		0.2	.05	0	
MEATLOAF, HOMEMADE	100	160	4.6	7.6					653	38	162	.07	8.0	179
3½ oz		17.0	0.1						374		2.3	.19	2	
MEATLOAF, FRZN, CAMPBELLS	100	171	8.8	9.6					671	24	128	.07	4.0	0
3½ oz		12.3	0.1						249		2.0	.02	0	
MEATLOAF W/ TVP*	100	285	15.3	15.9										
3½ oz		20.2												
MEATLOAF W/ BROWN GVY, FRZN,	100	165	10.3	9.2					543	26	84	.04	3.0	0
CAMPBELLS 3½ oz		11.4	0.1						140		1.0	.18	0	
MEATLOAF W/ BROWN GVY,	100	161	3.1	12.2					1054	12	125	.11	1.6	154
INVISIBLE CHEF 3½ oz		9.5							246		1.4	.14	0	
MEATLOAF W/ TOM SCE, FRZN,	100	122	7.6	5.9					478	30	100	.07	3.7	246
CAMPBELLS 3½ oz		9.7	0.2						250		1.6	.10	4	
MEAT, POTTED, CND, STOKELY	28	69		5.4								.01	0.3	
VAN CAMP 1 oz		4.9										.06		

*TVP = texturized vegetable protein.

	WT g	Proximate CAL	CHO	FAT	Amino Acids TRY mg	LEU mg	LYS mg	MET mg	Minerals Na mg	Ca mg	P mg	Vitamins THI mg	NIA mg	A iu
		PRO g	FIB g	PUFA g	PHE mg	ISO mg	VAL mg	THR mg	K mg	Mg mg	Fe mg	RIB mg	ASC mg	D iu

Meat substitutes

	WT g	CAL	CHO	FAT	TRY	LEU	LYS	MET	Na	Ca	P	THI	NIA	A
BRKFST LINKS, MORNING STAR	23	61		4.0					175					
FARMS 1 piece		3.7		1.5										
BRKFST PATTIES, MORNING STAR	38	100		7.0					291					
FARMS 1 piece		6.0		2.5										
BRKFST SLICES, MORNING STAR	28	59		4.4					413					
FARMS 1 slice		4.4		1.5										

Organ meats

	WT g	CAL	CHO	FAT	TRY	LEU	LYS	MET	Na	Ca	P	THI	NIA	A
BRAINS, ALL KINDS, RAW	85	106	0.7	7.3	117	718	646	187	106	9	265	.20	3.7	0
3 oz		8.8	0.0		430	428	456	420	186	11	2.0	.22	15	0
CHITTERLINGS*	100	335		25.7	94	457	670	193						
3½ oz		8.6			359	308	462	398						

GIBLETS

	WT g	CAL	CHO	FAT	TRY	LEU	LYS	MET	Na	Ca	P	THI	NIA	A
CHICKEN, CAPON, RAW	28	62	0.1	4.1										
1 oz		5.7	0.0											
CHICKEN, FRYER, RAW	28	29	tr	0.9						4	62	.04	1.4	1268
1 oz		4.9	0.0								1.3	.38		
CHICKEN, FRYER, FRIED	28	71	1.3	3.1						5	94	.05	2.2	1613
1 oz		8.6								6	1.8	.61		
CHICKEN, HEN & COCK, RAW	28	53	0.5	3.2						4	60	.03	1.9	1204
1 oz		5.2									1.2	.31	2	
GOOSE, RAW	28	44	0.2	2.0										
1 oz		5.9	0.0											
GUINEA HEN, RAW	28	44	0.3	2.0										
1 oz		5.8	0.0											
PHEASANT, RAW	28	39	0.4	1.4										
1 oz		5.8	0.0											
QUAIL, RAW	28	49	1.9	1.7										
1 oz		6.1	0.0											
ROASTER, RAW	28	38	0.5	1.3	57	413	445	141		4	61	.03	1.9	1201
1 oz		5.5	0.0		224	299	285	222			1.2	.30	2	
SQUAB (PIGEON), RAW	28	43	0.3	2.0										
1 oz		5.5	0.0											
TURKEY, SIMMERED	28	65	0.4	4.3										
1 oz		5.8	0.0											
GIZZARD, CHICKEN, SIMMERED	28	41	0.2	0.9					16	3	20	tr	1.4	
1 oz		7.6	0.0						59		0.9	.06		
GIZZARD, GOOSE, RAW	28	39	0.0	1.5										
1 oz		6.0	0.0											
GIZZARD, TURKEY, SIMMERED	28	55	0.3	2.4					14			.01	1.6	
1 oz		7.5	0.0						42			.04		

HEART

	WT g	CAL	CHO	FAT	TRY	LEU	LYS	MET	Na	Ca	P	THI	NIA	A
BEEF, LEAN, BRAISED	100	179	0.7	5.7	250	1725	1585	461	104	6	181	.25	7.6	30
3½ oz		31.3	0.0	1.1	874	979	1112	887	232	18	5.9	122	1	0
BEEF, LEAN W/ FAT, BRAISED	100	372	0.1	29.0					102		169			
3½ oz		25.8	0.0						329	35				
CALF, BRAISED	100	208	1.8	9.1					113	4	148	.29	8.1	40
3½ oz		27.8	0.0						250	35	4.4	1.44	1	
CHICKEN, RAW	100	134	0.1	6.0	266	1830	1683	489	79	4	158	.06	4.6	30
3½ oz		18.6	0.0		928	1040	1181	941	159	16	3.3	.80	4	0
CHICKEN, SIMMERED	100	166	0.1	7.2					69	4	107	.06	5.3	30
3½ oz		25.3	0.0						140		3.6	.92	4	
HOG, RAW	100	113	0.4	4.4	219	1509	1387	403	54	3	131	.32	6.6	30
3½ oz		16.8	0.0	0.9	765	857	973	776	106	13	3.3	.93	3	0
HOG, BRAISED	100	195	0.3	6.9					65	4	121	.64	6.7	40
3½ oz		30.8	0.0	1.5					128		4.9	1.89	1	
LAMB, BRAISED	100	260	1.0	14.4					102	14	231	.93	6.4	100
3½ oz		29.5	0.0						329	35		1.62	1	
TURKEY, SIMMERED	100	210	0.2	13.2					61			.25	5.7	30
3½ oz		22.6	0.0						211	35		.98	4	

*hog intestine.

	WT g	CAL / PRO	CHO / FIB	FAT / PUFA	TRY / PHE	LEU / ISO	LYS / VAL	MET / THR	Na / K	Ca / Mg	P / Fe	THI / RIB	NIA / ASC	A / D
		Proximate			**Amino Acids**				**Minerals**			**Vitamins**		
		g	g	g	mg	mg	mg	mg	mg	mg	mg	mg	mg	iu
KIDNEYS														
BEEF, BRAISED	100	252	0.8	12.0	285	1678	1402	396	253	18	244	.67	10.7	1150
3½ oz		33.0	0.0		911	942	1130	858	324	18	13.1	4.58	0	0
CALF, RAW	100	113	0.1	4.6					238	9	171	.26	7.4	1200
3½ oz		16.6	0.0						240	10	4.0	2.40	6	
HOG, RAW	100	106	1.1	3.6	240	1414	1181	334	115	11	218	.58	9.8	130
3½ oz		16.3	0.0	0.5	767	793	952	722	178	17	6.7	1.73	12	0
LAMB, RAW	100	105	0.9	3.3	244	1440	1203	340	200	13	218	.51	7.4	690
3½ oz		16.8	0.0		781	807	969	736	230	16	7.6	2.42	15	0
LIVER														
BEEF, RAW	100	135	5.3	3.8	296	1819	1475	463	136	8	352	.25	13.6	43900
3½ oz		19.9	0.0	0.6	993	1031	1239	936	281	8	6.5	3.26	31	45
CALF, RAW	100	140	4.1	4.7	286	1754	1423	447	73	8	333	.20	11.4	22500
3½ oz		19.0	0.0	0.1	958	994	1195	903	281	16	8.8	2.72	36	15
CALF, FRIED	100	261	4.0	13.2	2078	2078	1686	531	118	13	537	.24	16.5	32700
3½ oz		29.5	0.0	0.3	1133	1178	1414	1069	453	26	14.2	4.17	37	14
CHICKEN, RAW	100	129	2.9	3.7	332	2040	1655	520	70	12	236	.19	10.8	12100
3½ oz		19.7	0.0	1.0	1114	1156	1390	1050	172	17	7.9	2.49	17	50
CHICKEN, SIMMERED	100	165	3.1	4.4	440	2707	2197	690	61	11	159	.17	11.7	12300
3½ oz		26.5	0.0		1480	1533	1847	1393	151		8.5	2.69	16	67
GOOSE, RAW	100	182	5.4	10.0					140					
½ oz		16.5	0.0						230	21				
HOG, RAW	100	131	2.6	3.7	296	1819	1475	463	73	10	356	.30	16.4	10900
3½ oz		20.6	0.0	0.6	993	1031	1239	936	261	16	19.2	3.03	23	45
HOG, FRIED	100	241	2.5	11.5	357	2197	1781	559	111	15	539	.34	22.3	14900
3½ oz		29.9	0.0	1.8	1200	1246	1495	1230	395	24	29.1	4.36	22	51
LAMB, RAW	100	136	2.9	3.9	316	1939	1572	494	52	10	349	.40	16.9	50500
3½ oz		21.0	0.0		1058	1099	1320	998	202	23	10.9	3.28	33	20
LAMB, BROIL	100	261	2.8	12.4	378	2322	1881	592	85	16	572	.49	24.9	74500
3½ oz		32.3	0.0	0.3	1268	1316	1581	1195	331	23	17.9	5.11	36	23
TURKEY, RAW	100	138	2.9	4.0					63			.18	13.2	17700
3½ oz		21.2	0.0						160	21		1.93		
TURKEY, SIMMERED	100	174	3.1	4.8					55			.16	14.3	17500
3½ oz		27.9	0.0						141			2.09		
LIVER PASTE (PATE DE FOIS GRAS)*	15	69	0.7	6.6								.01	0.4	
1 T		1.7	0.0									.05	0	
LUNGS														
BEEF, RAW	100	96	0.0	2.3							216		6.2	
3½ oz		17.6	0.0											
CALF, RAW	100	106	0.0	3.8										
3½ oz		16.8	0.0											
LAMB, RAW	100	103	0.0	2.3							180			
3½ oz		19.3	0.0											
PANCREAS														
BEEF, MED-FAT, RAW	100	283	0.0	25.0	175	1054	996	244	67	9	270		5.8	
3½ oz		13.5	0.0		562	683	724	626	276	17	1.2	.55		
BEEF, LEAN ONLY, RAW	100	141	0.0	7.3					67	8	330			
3½ oz		17.6	0.0						276		2.8			
CALF, RAW	100	161	0.0	8.8	250	1500	1423	346			326			
3½ oz		19.2	0.0		808	981	1038	885						
HOG (SWEETBREAD), RAW	100	287	0.0	19.9	188	1132	1070	262	44	11	282			
3½ oz		14.5	0.0		603	733	777	673	217	17	1.0			
SPLEEN														
BEEF & CALF, RAW	100	104	0.0	3.0							272		8.2	
3½ oz		18.1	0.0								10.6	.37		
HOG, RAW	100	107	0.0	3.8							298			
3½ oz		17.1	0.0								29.4			
LAMB, RAW	100	115	0.0	3.9										
3½ oz		18.8	0.0											

*A paste made of the livers of fattened geese.

	WT	Proximate			Amino Acids				Minerals			Vitamins		
		CAL	CHO	FAT	TRY	LEU	LYS	MET	Na	Ca	P	THI	NIA	A
		PRO	FIB	PUFA	PHE	ISO	VAL	THR	K	Mg	Fe	RIB	ASC	D
	g	g	g	g	mg	mg	mg	mg	mg	mg	mg	mg	mg	iu
STOMACH, PORK, SCALDED	100	152	0.0	9.0							118			
3½ oz		16.5	0.0											
SWEETBREAD (THYMUS)														
BEEF (YEARLING), BRAISED	100	320	0.0	23.2					116		364	.04	3.1	
3½ oz		25.9	0.0						433			.40	44	0
CALF, BRAISED	100	168	0.0	3.2					98		400	.09	2.9	17
3½ oz		32.6	0.0						360		1.2	.27	44	0
LAMB, BRAISED	100	175	0.0	6.1					96		204	.04	3.1	17
3½ oz		28.1	0.0						360			.40	44	
TONGUE														
BEEF, LAMB, ETC, CND/PICKLED	28	75	tr	5.7										
1 oz		5.4	0.0											
BEEF, LAMB, ETC, POTTED/DEVILED	28	81	0.2	6.4								.01	0.4	
1 oz		5.2	0.0									.03		
BEEF, VERY FAT, RAW	100	271	0.4	23.0	173	1139	1197	317						
3½ oz		14.4	0.0		577	692	735	634		16				
BEEF, FAT, RAW	100	231	0.4	18.0										
3½ oz		15.7	0.0											
BEEF, MED FAT, RAW	100	207	0.4	15.0	197	1286	1364	357	73	8	182	.12	5.0	
3½ oz		16.4	0.0		661	792	840	708	197	16	2.1	.29		
BEEF, MED FAT, BRAISED	100	244	0.4	16.7	196	1286	1364	356	61	7	117	.05	3.5	0
3½ oz		21.5	0.0	0.7	661	792	840	708	164	16	2.2	.29	0	0
BEEF, LEAN, RAW	100	175	0.4	11.0										
3½ oz		17.4	0.0											
BEEF, SMOKED	100			28.8								.04	3.0	
3½ oz		17.2	0.0									.21		
CALF, RAW	100	130	0.9	5.3										
3½ oz		18.5	0.0											
CALF, BRAISED	100	160	1.0	6.0										
3½ oz		23.9	0.0											
HOG, RAW	100	215	0.5	15.6	202	1317	1398	366		29	186	.17	5.0	
3½ oz		16.8	0.0	2.0	677	812	860	726			1.4	.29		
HOG, BRAISED	100	247	0.5	17.4						26	119	.07	3.5	
3½ oz		22.0	0.0	2.3							1.4	.29		
LAMB, RAW	100	199	0.5	15.3							147			
3½ oz		13.9	0.0											
LAMB, BRAISED	100	254	0.5	18.2							102			
3½ oz		20.5	0.0											
SHEEP, RAW	100	265	2.4	21.8										
3½ oz		13.7	0.0											
SHEEP, BRAISED	100	323	2.4	25.3										
3½ oz		19.8	0.0								3.4			
TRIPE, BEEF	100	100	0.0	2.0					72	127	86		1.6	0
3½ oz		19.1	0.0						9		1.6	.15	0	0
TRIPE, BEEF, PICKLED	28	17	0.0	0.4					13					
1 oz		3.3	0.0						5					
Pork														
BACON														
CANADIAN, RAW	28	65	0.1	4.2	60	479	521	151	440	4	59	.25	1.5	0
1 slice		6.2	0.0	0.6	237	309	323	254	90	5.3	0.9	.07	0	0
CANADIAN, RAW	113	261	0.3	17.0	240	1932	2100	608	2140	16	238	1.03	5.9	0
4 slices		25.0	0.0	2.6	956	1244	1301	1024	362	21.5	3.7	.28	0	0
CANADIAN, BROIL/FRIED	21	65	3.0	4.2	60	479	521	151	442	4	59	.18	1.1	0
1 slice		6.2	0.0	0.6	237	309	323	254	91	5.3	0.9	.03	0	0
CANADIAN, UNHEATED, HORMEL	28	45	tr	2.7					371					
1 oz		5.1												
CANADIAN, UNHEATED, OSCAR	28	39	0.2	1.7					384	2	52	.20	1.6	tr
MAYER 1 oz		5.9		0.1					91	4	0.1	.04	8	10
CURED, RAW	28	156	0.2	16.2	24	184	148	36	71	1	22	.17	0.9	0
1 slice (1 oz)		2.3	0.0	2.4	110	101	110	77	16	1.6	0.2	.05	0	0
CURED, RAW	113	712	1.2	73.6	108	824	664	160	283	15	122	.43	2.2	0
4 slices (¼ lb)		10.3	0.0	11.4	491	452	491	346	63	6.4	0.9	.13	0	0

	WT g	CAL	CHO	FAT	TRY	LEU	LYS	MET	Na	Ca	P	THI	NIA	A
		PRO	FIB	PUFA	PHE	ISO	VAL	THR	K	Mg	Fe	RIB	ASC	D
	g	g	g	g	mg	mg	mg	mg	mg	mg	mg	mg	mg	iu
CURED, BROILED/FRIED CRISP	7	48	0.2	4.4	19	144	116	28	76	1	22	.04	0.3	0
1 strip (1 oz raw)		1.8	0.0	0.7	86	79	86	61	17	1.7	0.2	.02	0	0
CURED, BROILED/FRIED CRISP	23	147	0.5	13.5	56	432	348	84	250	4	66	.13	1.0	0
3 strips (3 oz raw)		5.4	0.0	2.0	258	237	258	182	56	4.9	0.6	.06	0	0
CURED, BROILED, HORMEL	7	37	0.0	3.2					116					
1 piece		2.0												
CURED, BROILED, OSCAR MAYER	6	35	0.1	3.1					113	tr	21	.04	0.4	
1 piece		1.4		0.4					24	1	0.1	.02	2	3
CURED, CND, CKD	100	692	0.6	72.4	91	696	561	135	1087	15	94	.24	1.5	0
4-5 slices		8.7	0.0	10.9	415	381	415	293	241	24.7	1.4	.10	0	0
FRESH, FAT CLASS (75% FAT), RAW	28	177	0.0	18.6					20	1	17	.10	0.5	0
1 oz		2.0	0.0						80		0.3	.02	0	0
FRESH, MED FAT (67% FAT), RAW	28	165	0.0	14.3					20	1	21	.11	0.6	0
1 oz		2.3	0.0	1.6					80		0.3	.03	0	0
FRESH, THIN (60% FAT), RAW	28	153	0.0	15.7					20	1	26	.13	0.7	0
1 oz		2.6	0.0						80		0.4	.03	0	0
BACON RINDS, WONDER	28	144	0.0	7.7					217	1	0	.00	0.0	0
1 oz		19.1									0.0	.00	0	
BACON SUBSTITUTES														
BAC-O-BITS, GENERAL MILLS	9	38	2.7	1.6					473	11	45	.01	tr	0
1 T		3.2							141		0.6	.05	0	
BACOS, GENERAL MILLS	9	36	1.1	1.6					432	7	23	.08	tr	
1 T		4.8									0.7	.04		
LEAN STRIPS W/ BACON FLAVOR,	28	144	2.5	11.9					580	40	55	1.68	2.7	
GENERAL FOODS 1 oz		6.2							37		tr	.10		
BLADE, LEAN, MARB & FAT, CKD	100	366	0.0	28.8	323	1833	2044	621	50	5	199	.46	3.7	0
3½ oz		24.9	0.0	4.9	980	1279	1295	1155	352	20	3.7	.22	0	0
BLADE, LEAN & MARB, CKD	54	150	0.0	9.2	204	1156	1289	392	39	4	123	.36	2.9	0
1 slice		15.7	0.0	1.6	618	806	816	728	275	13	2.4	.17	0	0
BLADE, LEAN & MARB, CKD	108	299	0.0	18.4	406	2304	2570	781	78	8	245	.73	5.7	0
2 slices		31.3	0.0	3.1	1232	1608	1628	1452	551	26	4.7	.34	0	0
BLADE, LEAN ONLY, CKD	61	134	0.0	5.9	244	1384	1543	469	44	5	159	.41	3.2	0
2 slices		18.8	0.0	1.0	740	966	978	872	311	15	2.8	.19	0	0
BOSTON BUTT, LEAN, MARB & FAT,	100	348	0.0	28.0	292	1656	1847	561	37	4	210	.34	2.7	0
CKD 3½ oz		22.5	0.0	4.8	886	1156	1170	1044	260	21	3.4	.16	0	0
BOSTON BUTT, LEAN & MARB,	58	164	0.0	11.2	191	1082	1207	367	42	4	131	.39	3.1	0
ROASTED 1 slice		14.7	0.0	1.9	578	755	764	682	296	13	2.2	.18	0	0
BOSTON BUTT, LEAN & MARB,	116	328	0.0	22.4	383	2172	2422	736	83	8	261	.78	6.1	0
ROASTED 2 slices		29.5	0.0	3.8	1161	1515	1534	1369	592	25	4.4	.37	0	0
BOSTON BUTT, LEAN ONLY,	41	82	0.0	3.5	153	869	969	294	29	3	102	.27	2.2	0
ROASTED 2 slices		11.8	0.0	0.6	464	606	614	548	209	10	1.8	.13	0	0
HAM, FRESH, LEAN, MARB & FAT,	100	306	0.0	18.3	427	2422	2701	821	61	6	263	.57	4.5	0
CKD 3½ oz		32.9	0.0	3.1	1295	1690	1711	1527	434	26	2.3	.27	0	0
HAM, FRESH, LEAN & MARB, CKD	53	126	0.0	4.7	254	1443	1609	489	37	4	146	.34	2.7	0
1 slice		19.6	0.0	0.8	771	1007	1019	909	260	15	1.2	.16	0	0
HAM, FRESH, LEAN & MARB, CKD	107	254	0.0	9.4	514	2915	3251	988	73	7	295	.69	5.4	0
2 slices		39.6	0.0	1.6	1559	2034	2059	1837	520	30	2.5	.32	0	0
HAM, FRESH, LEAN ONLY, CKD	75	167	0.0	4.8	376	2135	2381	723	54	5	202	.51	4.0	0
2 slices		29.0	0.0	0.8	1141	1489	1508	1346	382	21	1.7	.24	0	0
HAM, CURED BUTT, LEAN, MARB &	100	348	0.3	28.0	216	1739	1890	547	718	5	210	.70	3.6	0
FAT, CKD 3½ oz		22.5	0.0	4.8	860	1120	1170	922	332	21	2.5	.22	0	0
HAM, CURED BUTT, LEAN & MARB,	60	123	0.2	6.5	145	1167	1268	367	518	5	123	.46	2.5	0
CKD 1 slice		15.1	0.0	1.1	577	751	786	618	239	13	1.5	.14	0	0
HAM, CURED BUTT, LEAN & MARB,	120	246	0.4	13.0	290	2334	2537	734	1036	10	246	.93	5.0	0
CKD 2 slices		30.2	0.0	2.2	1155	1503	1571	1237	478	26	3.0	.28	0	0
HAM, CURED BUTT, LEAN ONLY, CKD	64	102	0.2	2.9	170	1368	1487	430	589	4	140	.57	2.9	0
2 slices		17.7	0.0	0.5	677	881	921	725	272	14	1.6	.18		0
HAM, CURED SHANK, LEAN, MARB &	100	371	0.3	31.4	196	1577	1714	496	672	5	157	.65	3.4	0
FAT, CKD 3½ oz		20.4	0.0	5.3	780	1015	1061	836	310	16	2.5	.21		0
HAM, CURED SHANK, LEAN & MARB,	39	91	0.1	5.4	96	773	840	243	336	3	73	.30	1.6	0
CKD 1 slice		10.0	0.0	0.9	382	498	520	410	155	8	1.0	.09		0
HAM, CURED SHANK, LEAN & MARB,	79	185	0.2	10.9	194	1561	1697	491	682	6	147	.61	3.2	0
CKD 2 slices		20.2	0.0	1.9	772	1015	1050	827	314	15	2.0	.19		0

		Proximate			Amino Acids				Minerals			Vitamins		
	WT	CAL	CHO	FAT	TRY	LEU	LYS	MET	Na	Ca	P	THI	NIA	A
		PRO	FIB	PUFA	PHE	ISO	VAL	THR	K	Mg	Fe	RIB	ASC	D
	g	g	g	g	mg	mg	mg	mg	mg	mg	mg	mg	mg	iu
HAM, CURED SHANK, LEAN ONLY,	43	76	0.1	2.7	116	935	1016	294	396	3	89	.38	2.0	0
CKD 2 slices		12.1	0.0	0.5	463	602	629	496	183	9	1.1	.12		0
HAM, CURED, CND	85	142	0.4	8.5	151	1213	1319	382	837					
1 slice (3 oz)		15.7	0.0	1.4	600	781	817	643						
HAM, CURED, CND, OSCAR MAYER	100	120	0.3	5.0					1527	5	164	.78	4.6	
3½ oz		8.0		0.4					330	15	0.6	.25	27	
HAM, CURED, CND, SWIFTS	100	224	0.3	16.0					1078	5	168	.77	3.3	
3½ oz		18.3							250	14	1.0	.22		
HAM, CURED, CND, CHOPPED,	100	257	2.0	20.5					1020	10	145	.62	3.4	
HORMEL 3½ oz		16.1							236		0.9	.17		
HAMLOAF, GLAZED	100	247	14.3	14.7					326	49	148	.27	4.0	153
3½ oz		13.3	tr						262		2.1	.18	1	
HAM PATTIES, CND, SWIFTS	62	209	0.6	19.3					735	3	87	.26	1.6	
1 patty		7.7								6	0.4	.06		
LOIN CHOP, LEAN & FAT, CKD	100	357	0.0	25.6	382	2164	2414	733	60	12	229	1.18	5.5	0
3½ oz		29.4	0.0	4.4	1157	1510	1529	1364	568	22	4.4	.19	0	0
LOIN CHOP, LEAN & FAT, CKD	88	314	0.0	22.5	336	1907	2126	646	52	10	202	1.03	4.8	0
1 chop		25.9	0.0	3.8	1019	1330	1347	1202	500	19	3.9	.16	0	0
LOIN CHOP, LEAN ONLY, CKD	68	170	0.0	7.7	305	1730	1929	586	41	8	179	.80	3.7	0
1 chop		23.5	0.0	1.3	925	1207	1222	1090	386	17	3.5	.12	0	0
LOIN ROAST, SLICED, FRZN, PRONTO	100	203	tr	13.1					396	12	135	1.46	5.1	tr
FOODS 3½ oz		27.4							150			.25	0	
LOIN W/ GVY & DRESS, FRZN,	228	280	17.0	14.0					830	50	193	.48	5.9	0
CAMPBELLS 1 serving		22.0	2.2						300		1.5	.16	1	
PICNIC, FRESH, MED FAT, SEPARABLE	100	150	0.0	7.4					70	11	225	.94	5.0	0
LEAN, SIMMERED 3½ oz		19.4	0.0						285	22	2.9	.23	0	
PICNIC, CURED, MED FAT, SEPARABLE	100	211	0.0	9.9					930	13	220	.65	5.0	0
LEAN, ROASTED 3½ oz		28.4	0.0	1.0					326		3.7	.26		
PICNIC SHOULDER, LEAN, MARB &	100	312	0.0	23.5	302	1715	1913	581	30	3	197	.28	2.2	0
FAT, CKD 3½ oz		23.3	0.0	4.0	917	1197	1212	1081	214	20	3.5	.13	0	0
PICNIC SHOULDER, LEAN & MARB,	47	116	0.0	7.0	158	898	1002	304	34	3	102	.31	2.5	0
ROASTED 1 slice		12.2	0.0	1.2	480	627	634	566	240	10	1.8	.15	0	0
PICNIC SHOULDER, LEAN & MARB,	95	234	0.0	14.3	319	1811	2020	614	68	7	207	.64	5.0	0
ROASTED 2 slices		24.6	0.0	2.4	968	1263	1279	1141	485	21	3.7	.30	0	0
PICNIC SHOULDER, LEAN ONLY,	45	81	0.0	2.7	170	964	1076	327	32	3	108	.30	2.4	0
ROASTED 2 slices		13.1	0.0	0.5	516	673	681	608	229	11	2.0	.14	0	0
PIGS FEET, PICKLED	28	56	0.0	4.1										
1 oz		4.7	0.0											
PORK & GVY, CND	100	256	6.3	17.8						13	183	.49	3.5	0
3½ oz		16.4	0.0								2.4	.17		
PORK RINDS, FRITO-LAY	28	137	0.5	7.3	tr	tr	1	tr	242	7	13	tr	0.6	65
1 oz		18.1	0.2	0.4	tr	tr	tr	tr	48	4	0.7	.03	0	
PORK SKINS, LANCE	14	77	0.1	4.6					250			tr	0.2	
1 pkg		8.8							18		0.1			
PORK, SLICED, FRZN, INVISIBLE	100	204	6.4	13.1					396	12	135	1.46	5.1	7
CHEF 3½ oz		27.4							150			.25	0	
PORK, SWEET & SOUR, HOMEMADE	200	386	28.7	21.7					856	25	209	.71	8.8	49
1 serving		19.1	0.5						400		2.9	.24	10	
PORK, SWEET & SOUR, FRZN,	100	99	22.0	0.2					716					
LA CHOY 3½ oz		4.4												
PORK W/ DRESS & GVY, FRZN,	207	180	20.0	6.0					192	11		.44	2.9	72
INVISIBLE CHEF 1 serving		13.6							259		1.6	.15	14	
SHOULDER, MED FAT, RAW	100	401	0.0	38.5					70	7	136	.62	3.3	0
3½ oz		12.7	0.0						285	20	1.9	.15	0	0
SHOULDER BUTT, CURED, LEAN,	100	385	0.0	33.4	188	1515	1647	477	699	6	168	.63	3.3	0
MARB & FAT, CKD 3½ oz		19.6	0.0	5.7	749	976	1020	803	322	19	3.0	.19		0
SHOULDER BUTT, CURED, LEAN &	75	238	0.0	18.4	164	1318	1433	415	647	6	142	.58	3.1	0
MARB, ROASTED 1 slice		17.0	0.0	3.1	652	849	887	699	298	15	2.6	.18		0
SHOULDER BUTT, CURED, LEAN &	150	477	0.0	36.8	327	2632	2860	829	1294	12	285	1.17	6.2	0
MARB, ROASTED 2 slices		33.9	0.0	6.3	1301	1695	1771	1395	597	30	5.2	.36		0
SHOULDER BUTT, CURED, LEAN ONLY	55	109	0.0	5.1	141	1136	1235	358	507	4	115	.49	2.5	0
ROASTED 2 slices		14.7	0.0	0.9	562	732	765	602	234	12	2.2	.16		0
SIRLOIN, LEAN & MARB, ROASTED	100	297	0.0	20.0	356	2016	2248	683	32	3	244	.73	2.6	0
3½ oz		27.3	0.0	3.4	1079	1406	1425	1271	295	25	4.2	.19		0

	WT	Proximate			Amino Acids				Minerals			Vitamins		
		CAL	CHO	FAT	TRY	LEU	LYS	MET	Na	Ca	P	THI	NIA	A
		PRO	FIB	PUFA	PHE	ISO	VAL	THR	K	Mg	Fe	RIB	ASC	D
	g	g	g	g	mg	mg	mg	mg	mg	mg	mg	mg	mg	iu
SIRLOIN, LEAN, MARB, ROASTED	71	161	0.0	7.8	278	1572	1753	533	39	4	186	.90	3.3	0
3 slices		21.3	0.0	1.3	842	1097	1112	991	361	17	3.2	.24		0
SIRLOIN, LEAN ONLY, ROASTED	46	91	0.0	3.3	186	1053	1174	357	25	2	123	.58	2.1	0
3 slices		14.3	0.0	0.6	564	735	745	664	234	11	2.2	.15		0
SPARERIBS, RAW, AP	100	209	0.0	19.0	114	649	724	220	51	5	92	.55	1.9	0
3½ oz (3-4 small ribs)		8.8	0.0	3.2	347	453	459	410	360		1.3	.10	0	0
SPARERIBS, ROASTED	45	123	0.0	10.0	99	565	630	191		4	80	.20	1.4	0
3 med ribs		7.7	0.0	1.7	302	394	399	356			1.1	.07	0	0
SPARERIBS, ROASTED	90	246	0.0	20.0	198	1130	1260	382		8	160	.40	2.8	0
6 med ribs		15.4	0.0	3.4	604	788	798	712			2.3	.15	0	0
TENDERLOIN, LEAN ONLY, ROASTED	100	239	0.0	12.1	398	2257	2517	765	55	5	301	1.27	4.6	0
3½ oz		30.6	0.0	2.1	1208	1575	1595	1423	509	26	4.7	.34		0
Sausages														
BLOOD SAUSAGE/BLOOD PUDDING	60	226	0.0	20.8						4	73			0
1 link or 1 slice		8.9	0.0								1.0		0	0
BOCKWURST	28	74	0.2	6.6										
1 oz slice		3.2	0.0											
CERVELAT, DRY	28	126	0.5	10.5						4	82	.08	1.5	0
1 oz slice		6.9	0.0								0.8	.06	0	0
CERVELAT, SOFT	28	86	0.5	6.9						3	60	.03	1.2	
1 oz slice		5.2	0.0								0.8	.07	0	
FRANKFURTER, RAW	50	128	1.3	10.2	60	509	571	150	550	4	50	90	1.4	0
1 avg		7.1	0.0	0.6	259	344	356	291	110	8	0.7	95	0	0
FRANKFURTER, CKD	50	124	1.0	10.0	59	502	563	148	542	3	25	.08	1.2	0
1 avg		7.0	0.0		255	339	351	286	108	5	0.6	.09	0	0
FRANKFURTER, ALL BEEF, CKD, OSCAR MAYER	45	142	1.4	13.5					464	4	34	.02	1.1	
1 avg		5.0		0.5					71	4	0.6	.04	12	10
FRANKFURTER, CND	50	110	0.1	9.0						5	72	.01	1.2	
1 avg		6.7	0.0								1.1	.06		
FRANKFURTER, CHICKEN, TYSON	34	100	0.6	8.5						12		.04	1.0	14
1 avg		4.2									0.6	.07	0	
FRANKFURTER, CHICKEN, WEAVER	45	120	2.0	10.0					630					
1 avg		6.0												
ITALIAN, HOLLOWAY HOUSE	28	39	1.5	2.8					161	4	17	.04	0.6	135
1 oz		2.1	0.1						81		0.4	.03	7	
ITALIAN W/ GRN PEP, FRZN, INVISIBLE CHEF	187	390	5.6	34.2					812	34	167	.64	3.4	607
1 serving		2.2							462		2.3	.32	47	
KNOCKWURST	100	278	2.2	23.2						8	154	.17	2.6	
3½ oz		14.1	0.0								2.1	.21		
MORTADELLA	50	157	0.3	12.5						6	119			
		10.2	0.0								1.5			
PEPPERONI, HORMEL	28	139	0.4	12.8					492					
1 oz		5.7												
POLISH	30	83	0.0	6.9	44	474	515	152		3	48	.05	0.8	0
1 slice		4.9	0.0		240	426	324	267			0.7	.06	0	0
POLSKA KIELBASA, SKINLESS, ECKRICH	28	95	1.0	8.5										
1 oz		3.5												
PORK & BEEF (CHOPPED TOGETHER)	60	252	0.0	24.7						4	73			0
1 piece		6.8	0.0								1.0		0	0
PORK, BROWN & SERVE, BROWNED	28	118	0.8	10.6										
1 oz		4.6	0.0											
PORK, CND	100	299	0.0	25.9	131	1104	1239	325	740	9	166	.20	3.0	0
2 patties		15.4	0.0		562	747	774	631	140	16	2.3	.24	0	0
PORK, COUNTRY STYLE	28	97	0.0	8.7						3	47	.06	0.9	
1 oz		4.2	0.0							4	0.6	.05		
PORK, LINKS/BULK, RAW	100	450	0.0	44.8	92	774	869	228	740	6	100	.43	2.3	0
3½ oz		10.8	0.0	4.4	394	524	543	442	140	9	1.6	.17	0	0
PORK, LINKS/BULK, CKD	20	94	0.0	8.8	32	258	292	76	192	2	35	.10	0.6	0
1 link		3.5	0.0	0.8	120	176	184	144	54	3	0.4	.05	0	0
PORK, McDONALDS	50	191	tr	17.2					487	13	57	.23	6.1	37
1 serving		8.9	tr						130	8	0.9	.13	1	36

	WT g	CAL / PRO	CHO / FIB	FAT / PUFA	TRY mg / PHE mg	LEU mg / ISO mg	LYS mg / VAL mg	MET mg / THR mg	Na mg / K mg	Ca mg / Mg mg	P mg / Fe mg	THI mg / RIB mg	NIA mg / ASC mg	A iu / D iu
SALAMI	30	130	0.0	11.0	61	514	577	152		4	78	.08	0.9	0
1 slice		7.2			262	348	360	194			1.1	.06	0	0
SCRAPPLE	57	209	26.0	9.0						4	65	.08	0.5	100
1 slice		4.7	tr								0.6	.03	0	0
SMOKEES, ECKRICH	45	160	1.5	13.5										
1 piece		6.0	0.0											
SMOKIE LINKS, OSCAR MAYER	43	133	0.6	12.0					390	3	39	.12	1.2	
1 piece		5.6		1.5					71	5	0.5	.19	9	
SOUSE	60	100	0.0	7.4						4	85			0
1 piece		7.9									1.2		0	0
THURINGER	28	86	0.4	6.9						3	60	.03	1.2	
1 oz		5.2	0.0								0.8	.07		
VIENNA, LINK	18	39	0.0	3.0						2	31	.02	0.6	0
1 avg		2.8									0.4	.02	0	0
VIENNA, CND	60	127	0.0	9.6	80	680	763	200	590	5	102	.02	1.6	
½ can, 2 oz		9.5	0.0		346	460	476	388			1.4	.09	0	0
Turkey														
ALL CLASSES, DARK MEAT, ROASTED	100	203	0.0	8.3					99	30	400	.04	4.2	
3½ oz		30.0	0.0						398	28	2.3	.23	0	
ALL CLASSES, FLESH ONLY, RAW	100	162	0.0	6.6		1836	2173	664	66	8	212		8.0	
3½ oz		24.0	0.0		960	1260	1187	1014	315		1.5	.14		
ALL CLASSES, FLESH ONLY, ROASTED	100	190	0.0	6.1					130	8	251	.05	7.7	
3½ oz		31.5	0.0						367	28	1.8	.18		
ALL CLASSES, FLESH & SKIN, ROASTED	100	223	0.0	9.6		2363	2800	857	60	30	400	.07	8.1	
3½ oz		31.9	0.0		1287	1623	1530	1307	490	28		.16		
ALL CLASSES, LIGHT MEAT, ROASTED	100	176	0.0	3.9					82	5	185	.05	11.1	
3½ oz		32.9	0.0						411	28	1.2	.14		
ALL CLASSES, SKIN ONLY, ROASTED	100	421	0.0	39.2								.03	4.6	
3½ oz		17.0	0.0									.04		
ALL CLASSES, TOTAL ED, RAW	100	218	0.0	14.7		1542	1825	558	40	23	320	.09	8.0	
3½ oz		20.1	0.0		806	1058	997	852	320	28	3.8	.14	0	
ALL CLASSES, TOTAL ED, ROASTED	100	263		16.4						28				
3½ oz		27.0												
BREAST, BAKED, WATSONS	100		0.2	6.7					16	10	285			
3½ oz		27.6	0.1						380	16	0.8			
CND	100	202	0.0	12.5		1611	1904	586		10		.02	4.7	130
3½ oz		20.9	0.0		837	1109	1046	900			1.4	.14		
FAT MATURE BIRD*, TOTAL ED, RAW	100	343	0.0	29.3										
3½ oz		18.4	0.0											
MED FAT BIRD†, DARK MEAT, RAW	100	127	0.0	4.3										
3½ oz		20.8	0.0											
MED FAT BIRD†, FLESH–SKIN, RAW	100	197	0.0	11.6										
3½ oz		21.6	0.0											
MED FAT BIRD†, LIGHT MEAT, RAW	100	115	0.0	1.2										
3½ oz		24.7	0.0											
MED FAT BIRD, TOTAL ED, RAW	100	227	0.0	15.8					66	23	320	.12	7.9	
3½ oz		19.9	0.0						367	28	3.8	.19		
TURKEY & HAM, FRZN, LAND O' LAKES	100	140	1.0	5.8					1000					
3½ oz		20.0												
TURKEY & HAM LOAF	100	247	14.3	14.7					326	49	148	.27	4.0	153
3½ oz		13.3	tr						263		2.1	.18	1	
TURKEY ROLL W/ BROTH, SWIFTS	100	123	1.2	2.3					690	1	230	.01	8.7	
3½ oz		22.9							280	24	1.0	.26		
TURKEY, SCALLOPED	200	253	25.0	5.6					1272	54	197	.09	6.5	27
1 serving		23.6	0.2						295		2.2	.20	1	
TURKEY TETRAZZINI, FRZN, STOUFFERS	200	282	18.6	16.6					740	90		.12	3.2	240
		13.6									1.2	0.26	0	
TURKEY W/GVY & DRESSING, FRZN, CAMPBELLS 1 pkg	214	260	16.0	12.0					855	32	242	.15	5.3	0
		24.0	0.0						325		1.5	.15	0	

*More than 32 weeks
†26-32 weeks

	WT	Proximate			Amino Acids				Minerals			Vitamins		
		CAL	CHO	FAT	TRY	LEU	LYS	MET	Na	Ca	P	THI	NIA	A
		PRO	FIB	PUFA	PHE	ISO	VAL	THR	K	Mg	Fe	RIB	ASC	D
	g	g	g	g	mg	mg	mg	mg	mg	mg	mg	mg	mg	iu
YOUNG BIRD,* DARK MEAT, RAW	100	121	0.0	4.3										
3½ oz		20.6	0.0											
YOUNG BIRD,* FLESH & SKIN, RAW	100	151	0.0	7.4										
3½ oz		19.8	0.0											
YOUNG BIRD,* LIGHT MEAT, RAW	100	108	0.0	4.3										
3½ oz		24.5	0.0											
YOUNG BIRD,* TOTAL ED, RAW	100	145	0.0	6.0		1654	1955	602						
3½ oz		21.4	0.0		859	1139	1074	924						
Veal														
ARM STK, LEAN & FAT, CKD	100	298	0.0	19.0	389	2169	2476	677	51	10	242	.12	6.1	0
3½ oz		29.6	0.0	0.6	1202	1562	1530	1283	503	20	3.8	.31	0	0
ARM STK, LEAN ONLY, CKD	90	180	0.0	4.8	422	2350	1676	734	46	8	252	.10	5.4	0
1 steak		32.0	0.0	0.2	1302	1692	1656	1390	452	20	4.0	.27	0	0
BLADE, LEAN & FAT, CKD	100	276	0.0	16.6	391	2177	2481	680	55	11	244	.13	6.5	0
3½ oz		29.7	0.0	0.5	1207	1568	1536	1288	539	22	3.8	.33	0	0
BLADE, LEAN ONLY, CKD	108	228	0.0	8.4	472	2632	3000	822	60	12	290	.14	7.0	0
1 steak		36.0	0.0	0.2	1460	1896	1856	1558	582	26	4.6	.35	0	0
BREAST, STEW MEAT, RAW	100	346	0.0	25.2	365	2038	2323	637	87	11	233	.11	7.2	0
3½ oz		27.9	0.0	0.8	1131	1468	1438	1205	491	19	3.5	.18	0	0
BREAST, STEWED W/ GRAVY	74	256	0.0	18.6	269	1502	1711	469	64	8	172	.08	5.3	0
4 pieces (¼ lb raw)		20.6	0.0	0.6	833	1082	1060	888	363	14	2.6	.13	0	0
CHUCK, FAT CLASS, TOTAL ED, RAW	100	198	0.0	13.0					90	11	191	.14	6.4	0
3½ oz		19.0	0.0						320	30	2.8	.25	0	0
CHUCK, MED FAT, TOTAL ED, RAW	100	173	0.0	10.0	255	1422	1620	444	48	11	199	.14	6.5	0
3½ oz		19.4	0.0		788	1024	1003	841	359	23	2.6	.26	0	140
CHUCK, MED FAT, TOTAL ED, BRAISED	100	235	0.0	12.8					80	12	151	.09	6.4	0
3½ oz		27.9	0.0						500		3.5	.29	0	0
CHUCK, THIN, TOTAL ED, RAW	100	139	0.0	6.0					90	12	206	.15	6.7	0
3½ oz		19.9	0.0						320	30	3.0	.26	0	
CUTLET, ROUND, LEAN & FAT, CKD	100	277	0.0	15.0	435	2429	2769	758	54	10	288	.12	6.4	0
3½ oz		33.2	0.0	0.4	1346	1751	1714	1438	527	23	4.2	.32	0	0
CUTLET, ROUND, LEAN ONLY, CKD	70	194	0.0	12.8	370	2065	2354	644	46	9	245	.10	5.4	0
		23.2	0.0	0.4	1144	1488	1457	1222	448	20	3.6	.27	0	0
CUTLET, BREADED	100	319	10.5	15.0	435	2429	2769	758	54	0	288	.12	6.4	0
3½ oz		34.2	0.0	0.4	1346	1751	1714	1438	527	23	4.2	.32	0	
CUTLET, BREADED, INVISIBLE CHEF	100	479	15.7	32.1					906	23	193	.07	2.9	
3½ oz		32.0							4		3.0	.12		
FLANK, FAT CLASS, TOTAL ED, RAW	100	387	0.0	36.0					90	8	126	.11	4.9	0
3½ oz		14.5	0.0						320	30	2.2	.19	0	0
FLANK, MED FAT, TOTAL ED, RAW	100	314	0.0	27.0					90	10	155	.12	5.5	0
3½ oz		16.5	0.0						320	30	2.5	.22	0	0
FLANK, MED FAT, TOTAL ED, STEWED	100	390	0.0	32.3					80	11	117	.05	4.2	0
3½ oz		23.2	0.0						500		3.0	.22	0	
FLANK, THIN CLASS, TOTAL ED, RAW	100	240	0.0	18.0					90	10	179	.13	6.1	0
3½ oz		18.1	0.0						320	30	2.7	.24	0	0
FORESHANK, FAT CLASS, TOTAL ED, RAW 3½ oz	100	173	0.0	10.0	255	1422	1620	444	90	11	199	.14	6.5	0
		19.4	0.0		788	1024	1003	841	320	30	2.9	.26	0	0
FORESHANK, MED FAT, TOTAL ED, RAW 3½ oz	100	156	0.0	8.0					90	11	203	.14	6.6	0
		19.7	0.0						320	30	3.0	.26	0	0
FORESHANK, MED FAT, TOTAL ED, STEWED 3½ oz	100	216	0.0	10.4					80	12	154	.05	5.0	0
		28.7	0.0						500		3.6	.26	0	0
FORESHANK, THIN CLASS, TOTAL ED, RAW 3½ oz	100	131	0.0	5.0					90	12	209	.15	6.7	0
		20.1	0.0						320	30	3.0	.27	0	0
LOIN, FAT CLASS, TOTAL ED, RAW	100	215	0.0	15.0					90	11	187	.14	6.2	0
3½ oz		18.6	0.0						320	30	2.8	.25	0	0
LOIN, MED FAT, TOTAL ED, RAW	100	181	5.2	1.7					51		94	.14	6.4	70
3½ oz		19.2	0.0						143	15	2.9	.26	0	0
LOIN, MED FAT, TOTAL ED, BROILED	100	234	0.0	13.4					80	11	225	.07	5.4	0
3½ oz		26.4	0.0						500		3.2	.25	0	
LOIN, THIN CLASS, TOTAL ED, RAW	100	156	0.0	8.0					90	11	203	.14	6.6	0
3½ oz		19.7	0.0						320	30	3.0	.26	0	0

*24 weeks and under

	WT	Proximate			Amino Acids				Minerals			Vitamins		
		CAL	CHO	FAT	TRY	LEU	LYS	MET	Na	Ca	P	THI	NIA	A
		PRO	FIB	PUFA	PHE	ISO	VAL	THR	K	Mg	Fe	RIB	ASC	D
	g	g	g	g	mg	mg	mg	mg	mg	mg	mg	mg	mg	iu
LOIN CHOP, LEAN & FAT, CKD	100	421	0.0	35.9	298	1660	1889	517	44	6	187	.14	4.7	0
3½ oz		22.7	0.0	1.1	920	1195	1169	981	314	16	2.9	.21	0	0
LOIN CHOP, LEAN & FAT, CKD	122	514	0.0	43.8	362	2022	2305	631	54	7	228	.17	5.8	0
1 med chop		27.6	0.0		1121	1451	1426	1197	384	20	3.5	.26	0	0
LOIN CHOP, LEAN ONLY, CKD	72	149	0.0	4.8	323	1801	2053	562	47	6	188	.15	5.1	0
1 med chop		24.6	0.0		998	1298	1270	1066	342	15	3.1	.23	0	0
PLATE, FAT CLASS, TOTAL ED, RAW	100	281	0.0	23.0		1435	1573	478	90	10	168	.13	5.8	0
3½ oz		17.3	0.0		778	852	886	812	320	30	2.6	.23	0	0
PLATE, MED FAT, TOTAL ED, RAW	100	231	0.0	17.0	240	1341	1528	419	90	11	182	.13	6.1	0
3½ oz		18.3	0.0		744	966	946	793	320		2.7	.24	0	0
PLATE, MED FAT, TOTAL ED, STWD	100	303	0.0	21.2					80	12	138	.05	4.6	0
3½ oz		26.1	0.0						500		3.3	.24	0	0
PLATE, THIN CLASS, TOTAL ED, RAW	100	190	0.0	12.0					90	11	193	.14	6.4	0
3½ oz		19.1	0.0						320	30	2.9	.25	0	0
RIB, FAT CLASS, TOTAL ED, RAW	100	248	0.0	19.0					90	10	178	.13	6.0	0
3½ oz		18.0	0.0						320	15	2.7	.24	0	0
RIB, MED FAT, TOTAL ED, RAW	100	207	0.0	14.0					90	11	190	.14	6.3	0
3½ oz		18.8	0.0						320	15	2.8	.25	0	0
RIB, MED FAT, ROASTED	100	269	0.0	16.9	362	2022	2305	631	80	12	248	.13	7.8	0
3½ oz		27.2	0.0		1121	1457	1426	1197	500	20	3.4	.31	0	0
RIB, THIN CLASS, TOTAL ED, RAW	100	164	0.0	9.0	256	1429	1629	446	90	11	200	.14	6.5	0
3½ oz		19.5	0.0		792	1030	1008	846	320	15	2.9	.26	0	0
RIB CHOP, LEAN & FAT, CKD	100	318	0.0	22.2	362	2022	2305	631	49	10	192	.17	5.8	0
3½ oz		27.6	0.0		1121	1457	1426	1197	466	17	3.5	.26	0	0
RIB CHOP, LEAN & FAT, CKD	83	264	0.0	18.4	300	1672	1906	522	41	8	159	.14	4.8	0
1 med chop		22.9	0.0		927	1205	1179	990	387	14	2.9	.22	0	0
RIB CHOP, LEAN ONLY, CKD	58	125	0.0	4.6	256	1429	1629	446	35	7	132	.12	4.1	0
1 med chop		19.5	0.0		792	1030	1008	846	329	11	2.5	.19	0	0
ROUND W/ RUMP, FAT CLASS, TOT ED, RAW 3½ oz	100	190	0.0	12.0					90	11	193	.14	6.4	0
		19.1	0.0						320	30	2.9	.25	0	0
ROUND W/ RUMP, MED FAT, TOT ED, RAW 3½ oz	100	164	0.0	9.0	256	1429	1629	446	124	11	200	.14	6.5	0
		19.5	0.0		792	1030	1008	846	245	11	2.9	.26	0	0
ROUND W/ RUMP, MED FAT, TOT ED, BROILED 3½ oz	100	216	0.0	11.1					80	11	231	.07	5.4	0
		27.1	0.0						500		3.2	.25	0	0
ROUND W/ RUMP, THIN CLASS, TOT ED, RAW 3½ oz	100	139	0.0	6.0					90	12	206	.15	6.7	0
		19.9	0.0						320	30	3.0	.26	0	0
RUMP, LEAN, MARB & FAT, CKD	100	232	0.0	12.3	371	2072	2362	647	68	8	230	.15	7.5	0
3½ oz		28.3	0.0		1148	1494	1462	1227	468	19	3.6	.18	0	0
RUMP, LEAN & MARB, ROASTED	97	169	0.0	4.8	386	2151	2450	671	73	8	240	.15	7.9	0
2 slices		29.5	0.0		1192	1548	1516	1272	492	20	3.7	.20	0	0
RUMP, LEAN ONLY, ROASTED	50	78	0.0	1.1	210	1167	1329	364	66	8	132	.19	8.9	0
2 slices		15.9	0.0		647	840	823	690	521	11	2.0	.22	0	0
SIRLOIN, LEAN, MARB & FAT, CKD	100	274	0.0	19.1	315	1758	2004	549	53	8	221	.15	7.1	0
3½ oz		23.9	0.0		974	1267	1240	1041	476	19	3.0	.21	0	0
SIRLOIN, LEAN & MARB, ROASTED	98	172	0.0	6.2	362	2022	2305	631	124	8	246	.17	8.2	0
2 slices		27.4	0.0		1120	1457	1426	1197	1114	22	3.5	.24	0	0
SIRLOIN, LEAN ONLY, ROASTED	59	90	0.0	1.9	226	1264	1441	394	38	5	152	.11	5.1	0
2 slices		17.1	0.0		700	911	891	798	342	13	2.2	.15	0	0
SIRLOIN STK, LEAN & FAT, CKD	88	266	0.0	17.4	334	1871	2133	583	39	8	199	.09	4.7	0
½ steak		25.3	0.0		1036	1348	1319	1107	384	15	3.2	.23	0	0
SIRLOIN STK, LEAN ONLY, CKD	68	139	0.0	4.1	312	1740	1984	544	42	8	180	.10	5.0	0
½ steak		23.7	0.0		964	1254	1228	1031	412	14	3.0	.25	0	0
STEWMEAT, MARB, RAW	100	346	0.0	25.2	367	2052	2338	641	87	11	233	.11	7.2	0
4 pieces		27.9	0.0		1138	1478	1447	1213	491	19	3.5	.18	0	0
VEAL & PEPPERS, SARA LEE	200	164	9.4	1.0					464	24	196	.16	6.0	1064
		29.8	0.4						614	28	3.2	.26	34	0
VEAL PARMIGIANA, FRZN, CAMPBELLS	214	295	17.0	13.9					1825	99	289	.30	0.4	618
1 serving		25.0	0.6						468		2.4	.38	12	0
VEAL SCALLOPINI	116	199	2.7	11.3					918	106	228	.05	10.1	94
1 serving		20.4	0.1						385		2.1	.29	tr	0

	WT	Proximate			Amino Acids				Minerals			Vitamins		
		CAL	CHO	FAT	TRY	LEU	LYS	MET	Na	Ca	P	THI	NIA	A
		PRO	FIB	PUFA	PHE	ISO	VAL	THR	K	Mg	Fe	RIB	ASC	D
Milk	g	g	g	g	mg	mg	mg	mg	mg	mg	mg	mg	mg	iu
MILK, COW, DRY														
BUTTERMILK (FROM SKIM MILK)	6	23	3.0	0.3	29	202	163	52	30	71	56	.02	tr	13
1 T		2.1	0.0	tr	99	125	138	93	96	7	tr	.10	tr	
SKIM, CALCIUM REDUCED	28	99	1.6	0.1	140	974	788	249	638	78	283	.05	0.2	2
¼ cup		9.9	0.0	tr	480	601	665	449	190	17		.46		
SKIM SOLIDS, INSTANT	60	215	31.0	0.4	300	2119	1669	535	315	777	603	.21	0.5	18
1 cup		21.4	0.0	tr	1027	1370	1477	1006	1034	86	0.4	1.06	4	
SKIM SOLIDS, REGULAR	70	252	36.6	0.6	350	2475	2008	635	372	916	11	.29	0.6	25
1 cup		25.1	0.0	tr	1222	1532	1694	1142	1256	66	0.2	1.09	5	
WHOLE, INSTANT	100	499	38.2	26.8	371	2578	2087	657	371	912	776	.29	0.7	922
1 cup		26.3	0.0	0.7	1271	1592	1762	1188	1330	85	0.5	1.21	9	
MILK, COW, FLUID														
BUTTERMILK (FROM SKIM MILK)	245	88	12.4	0.2	96	872	730	202	318	296	232	.10	0.2	tr
1 cup		8.8	0.0	tr	467	554	660	414	342	34	tr	.44	2	
BUTTERMILK (FROM WHOLE MILK)	244	92	9.5	2.4	89	802	672	196	212	293	220	.10	0.2	100
1 cup		8.1	0.0	tr	429	510	608	381	388	33		.41		
BUTTERMILK, BUTTERFLAKE, BORDEN	244	110	11.0	4.0										
1 cup		8.1												
CONDENSED, SWEET, CND	20	64	10.9	1.7	22	158	125	38	22	52	41	.02	tr	72
1 T		1.6	0.0	tr	77	102	110	74	63	5	tr	.08	tr	
CONDENSED, SWEET, CND	100	321	54.3	8.7	113	802	632	194	112	262	206	.08	0.2	360
⅓ cup		8.1	0.0	tr	389	518	559	373	314	25	0.1	.38	1	
EVAPORATED, CND	32	44	3.1	2.5	31	218	172	53	38	81	66	.01	0.1	105
2 T		2.2	0.0	tr	106	141	152	101	97	8	tr	.11	tr	28
EVAPORATED, CND	100	137	9.7	7.9	98	693	546	168	118	252	205	.04	0.2	320
½ cup		7.0	0.0	tr	336	448	483	322	303	25	0.1	.34	1	88
EVAPORATED (FROM SKIM MILK), CND 1 oz	30	23	3.4	tr	32	222	180	57	35	87	59	.01	tr	118
		2.3	0.0	tr	109	137	152	102	100	8	tr	.09	tr	26
EVAPORATED, FILLED, DAIRYMATE 1 T	15	19	1.5	1.0										
		1.1		0.4										
LOW-FAT (1% FAT)	244	102	11.7	2.7	112	786	637	203	122	300	234	.10	0.2	500
1 cup		8.0	0.0	0.1	388	486	537	364	381	34	0.1	.41	2	100
LOWFAT (1% FAT) W/ NONFAT MILK SOLIDS 1 cup	245	105	12.2	2.4	120	835	676	213	127	314	245	.10	0.2	500
		8.6	0.0	0.1	412	517	571	385	397	34	0.1	.42		100
LOWFAT (2% FAT)	244	122	11.7	4.7	115	795	644	205	122	298	232	.10	0.2	500
1 cup		8.1	0.0	0.2	393	490	544	366	374	34	0.1	.40	2	100
LOWFAT (2% FAT) W/ NONFAT MILK SOLIDS 1 cup	245	125	12.2	4.9	120	835	676	213	127	314	245	.10	0.2	500
		8.5	0.0	0.2	412	517	571	385	397	34	0.1	.42	2	100
LOWFAT (2% FAT), PROTEIN FORTIFIED 1 cup	246	137	13.5	4.9	137	952	771	244	145	352	276	.11	0.3	500
		9.7	0.0	0.2	469	588	650	439	447	40	0.2	.48	3	100
SKIM	246	89	11.9	0.4	120	846	669	212	128	303	251	.10	0.2	502
1 cup		8.0	0.0	tr	418	506	561	396	408	27	0.1	.34	2	100
SKIM W/ NONFAT MILK SOLIDS 1 cup	245	91	12.3	0.6	124	858	693	221	130	316	255	.10	0.2	500
		8.7	0.0	tr	422	529	586	394	419	37	0.1	.43	2	100
SKIM, PROTEIN FORTIFIED 1 cup	246	100	13.7	0.6	137	954	773	244	144	352	275	.11	0.3	500
		9.7	0.0	0.2	470	589	652	440	446	40	0.2	.48		
SKIM-LINE, BORDEN 1 cup	246	100	13.0	1.0										
		10.1												
WHOLE (3.25% FAT)	244	150	11.5	8.1	115	795	644	205	120	291	228	.09	0.2	307
1 cup		8.1	0.0	2.6	393	490	544	366	370	33	0.1	.39	2	100
WHOLE (3.3% FAT)	244	149	11.4	8.1	117	800	654	210	122	290	217	.07	0.2	307
1 cup		8.0	0.0	0.3	388	486	537	373	370	32	0.1	.41	2	100
WHOLE (3.5% FAT)	244	159	12.0	8.6	119	842	663	204	122	288	227	.07	0.2	340
1 cup		8.5	0.0	0.2	408	544	586	391	352	37	0.1	.42	2	100
WHOLE (3.7% FAT)	244	161	12.0	9.0	118	832	655	202	122	285	224	.07	0.2	370
1 cup		8.4	0.0	0.3	403	538	580	386	342	37	0.1	.42	2	100
WHOLE, LOW SODIUM	244	149	10.9	8.4	107	742	600	190	5	246	210	.05	0.1	317
1 cup		7.6	0.0	0.3	366	459	505	342	617	12		.26		
MILK, OTHER, DRY														
SOYBEAN MILK, POWDER	28	120	7.8	5.7	176	1053	925	187		77				
1 oz		11.7	tr		667	597	644	609						
SOYBEAN MILK, POWDER, SWEET	28	127	13.6	6.5					tr	32	80	.08	0.4	6
1 oz		5.7	0.1						256		1.4	.07	0	

	WT	Proximate			Amino Acids				Minerals			Vitamins		
		CAL	CHO	FAT	TRY	LEU	LYS	MET	Na	Ca	P	THI	NIA	A
		PRO	FIB	PUFA	PHE	ISO	VAL	THR	K	Mg	Fe	RIB	ASC	D
	g	g	g	g	mg	mg	mg	mg	mg	mg	mg	mg	mg	iu
MILK, OTHER, FLUID														
COCONUT	244	615	12.7	60.8					129	39	244	.07	2.0	
1 cup		7.8	0.0						464		3.9		5	0
COCONUT WATER	240	53	11.3	0.5					60	48	31		0.2	0
1 cup		0.7							353	67	0.7		5	
GOAT	244	163	11.2	9.8	94	663	741	156	83	315	259	.10	0.7	390
1 cup		7.8	0.0	0.4	289	203	328	515	439	41	0.2	.27	2	5
HUMAN	30	23	2.8	1.2	5	27	19	6	5	10	4	3	0.1	72
1 oz (2 T)		0.3	0.0	tr	13	16	18	13	15	1	tr	12	2	
HUMAN	100	77	9.5	4.0	18	98	70	22	16	33	14	.01	0.2	240
3½ oz		1.1	0.0	tr	47	59	67	48	51	4	0.1	.04	5	
INDIAN BUFFALO	244	237	12.6	16.8	129	893	683	237	127	412	285	.13	0.2	434
1 cup		9.2	0.0	0.4	395	495	534	444	434	76	0.3	.33	5	
REINDEER	100	234	4.1	19.6					157	254	198			
3½ oz		10.8	0.0						159		0.1			
SHEEP	245	265	13.1	17.2	1098			380	108	473	387	.16	1.0	360
1 cup		14.7	0.0		696		86	206	333	44	0.2	.87	10	
SOYBEAN	263	87	5.8	3.9	134	802	707	142	51	55	126	.21	0.5	105
1 cup		8.9	0.0		513	460	489	463	340	57	2.1	.08	0	
SOYBEAN, JOLLY JOAN	240	62	5.9	0.2	257	1342	1118	235	47	43	109	.16	1.0	13
1 cup		8.7	0.6		854	893	898	655	311	52	2.2	.07		
SOYBEAN, CONC, SWEET	30	38	3.7	2.2					13	9	18	.02	tr	
1 oz		1.4	tr						71		0.2	.01	0	
YOGURT (COW'S MILK)														
LOWFAT W/ NONFAT MILK SOLIDS	227	143	16.0	3.4	75	1201	1069	177	159	415	326	.09	0.2	150
1 cup		11.9	0.0	0.1	409	649	579	490	531	40	0.2	.49	2	
LOWFAT, VAN & COFFEE FLAVORED, W/ NONFAT MILK SOLIDS 1 cup	227	194	1128	2.8	63	1128	1003	330	149	389	306	.10	0.2	123
		11.2	0.0	0.1	610	610	926	460	498	37	0.2	.46	2	
LOWFAT, FRUIT FLAVORED W/ NONFAT MILK SOLIDS 1 cup	227	231	43.2	2.5	56	1000	889	292	133	345	271	.08	0.2	104
		9.9	0.3		541	541	821	407	442	33	0.2	.40	2	
SKIM W/ NONFAT MILK SOLIDS	227	127	17.4	0.4	73	1311	1166	383	174	452	355	.11	0.3	16
1 cup		13.0	0.0	tr	709	709	1076	534	579	43	0.2	.53	2	
WHOLE	227	141	10.4	7.7	45	795	706	232	107	275	216	.07	0.2	279
1 cup		7.9	0.0	0.2	429	429	651	320	352	27	0.1	.32	1	
Milk Beverages (Cow's Milk)														
CHOC MILK (1% FAT)	250	158	26.1	2.5	114	793	642	203	152	287	256	.10	0.3	500
1 cup		8.1	0.2	0.1	391	490	542	366	426	33	0.6	.42	2	100
CHOC MILK (2% FAT)	250	179	26.0	5.0	113	786	636	201	150	284	254	.09	0.3	500
1 cup		8.0	0.2	0.2	387	486	537	362	422	33	0.6	.41	2	100
CHOC MILK (WHOLE MILK)	250	208	25.9	8.5	112	776	629	199	149	280	251	.09	0.3	302
1 cup		7.9	0.2	0.3	383	479	530	358	417	33	0.6	.41	2	100
CHOC MILK W/ MALT (WHOLE MILK)	265	233	29.2	9.1	38	265	209	68	168	304	265	.14	0.7	326
1 cup		9.4	0.1	0.4	136	162	187	124	500	48	0.5	.43	2	100
CHOC MILK, W/ OVALTINE (SKIM MILK)	250	171	28.2	1.2					302	383	313	.79	9.3	2268
1 cup		9.9							605		2.8	1.21	29	182
CHOC MILK, W/ OVALTINE (WHOLE MILK) 1 cup	250	242	28.2	9.6					292	373	313	.76	9.3	2621
		9.6							605		2.8	1.21	29	188
COCOA HOT CHOC (WHOLE MILK)	250	218	25.8	9.1	128	891	722	228	123	298	270	.10	0.4	318
1 cup		9.1	0.2	0.3	439	551	609	411	480	56	0.8	.44	2	100
COCOA MIX W/ WATER, CARNATION	240	110	20.0	3.0	44	303	235	76	154	107	108	.03	0.2	23
1 cup		4.1			156	195	218	148	176	20	0.6	.15	2	
COCOA MIX W/ WATER, OVALTINE	240	144	26.4	3.6					187	120	136	.18	2.4	600
1 cup		1.2							417		2.2	.20	7	
EGGNOG (NONALCOHOLIC)	254	342	34.4	19.0	137	937	758	222	138	330	278	.09	0.3	294
1 cup		9.7	0.0	0.9	463	583	643	444	420	47	0.5	.48	4	57
FROSTED, BORDEN	200	214	24.5	8.4					140	222	174	tr	1.8	
		6.9							350		0.2	.14		
INSTANT BRKFST, VAN (WHOLE MILK), CARNATION 1 cup	276	280	35.1	8.0	219	1524	1157	394	242	407	386	.40	5.2	1250
		14.9		0.4	762	91	1008	687	711	115	4.6	.56	26	100
MALTED MILK	265	236	26.6	9.9	143	944	712	239	215	347	307	.20	1.3	376
1 cup		10.8	0.1	0.6	479	567	630	433	529	52	0.3	.54	2	100

	WT g	CAL / PRO g	CHO / FIB g	FAT / PUFA g	TRY / PHE mg	LEU / ISO mg	LYS / VAL mg	MET / THR mg	Na / K mg	Ca / Mg mg	P / Fe mg	THI / RIB mg	NIA / ASC mg	A / D iu
MILKSHAKE														
HOMEMADE	345	421	58.0	17.9						362	321	.10	0.5	687
1 cup		11.2	0.3								1.0	.55	4	3
CHOC, THICK	300	356	63.5	8.1	129	896	726	229	333	396	378	.14	0.4	258
1 avg		9.2	0.8	0.3	442	554	612	413	672	48	0.9	.67	0	
CHOC (3.5% FAT), BORDEN	300	390	60.0	12.0					333					
		12.0							672					
CHOC (6% FAT), BORDEN	300	450	63.0	18.0										
		9.0												
CHOC, DAIRY QUEEN	241	340	51.0	11.0						300		.06	0.4	400
1 small		10.0									1.8	.34	2	
CHOC, McDONALDS	289	324	51.7	8.4					329	338	292	.12	0.8	347
1 avg		10.7	0.3						656	51	1.0	.88	3	35
CHOC, WENDYS	250	390	55.0	15.0						250		.15	0.0	300
1 avg		8.0									0.7	.60	0	
CHOC MALT, DELMARK	240	229	28.3	9.1										
1 cup		9.0												
FROM SOFT-SERVE MACHINE	226	287	37.4	9.0						380	391			
1 avg		11.1												
STRAWBERRY, McDONALDS	293	346	57.5	8.5					257	340	299	.12	0.5	322
1 avg		10.3	0.3						545	35	0.2	.66	3	31
VANILLA, THICK	313	350	55.6	9.5	170	1184	958	303	299	457	361	.09	0.5	357
1 avg		12.1	0.2	0.4	583	731	809	545	572	37	0.3	.61	0	
VAN (3.5% FAT), BORDEN	300	360	18.0	12.0					285					
		12.0							549					
VAN (6% FAT), BORDEN	300	420	57.0	18.0										
		9.0												
VAN, BURGER CHEF	305	326	47.0	11.0					307	411	356	.11	0.3	10
1 avg		11.0	tr						518	34	0.2	.57	2	
VAN, BURGER KING	315	331	48.5	9.5					306	479	386	.01	0.3	9
1 avg		13.2	0.3						592	42	0.3	.05	tr	
VAN, McDONALDS	289	324	51.7	7.8					250	361	266	.12	0.6	347
1 avg		10.7	0.0						500	35	0.2	.66	3	35
VAN/STRAWBERRY, DELMARK	240	229	29.3	8.9										
1 cup		8.3												
ORANGE NOG, FROM MIX, (WHOLE MILK), DELMARK 1 cup	240	290	37.5	8.8					220	468	379	.30	3.9	965
		15.2							590		3.7	.65	12	54
Milk Mixes (For Cow's Milk)														
CHOC FLAVORED MIX, OVALTINE	21	79	3.3	1.0					168	79	79	.66	8.9	2216
1 serving		1.0							247		2.7	.75	27	177
CHOC FLAVORED POWDER, PDQ	16	236	14.2	0.3										
1 T		0.6												
CHOC MALT FLAVORED POWDER	21	83	17.8	0.9	9	73	39	17	49	13	37	.04	0.4	20
3 T		1.4	0.1	0.1	52	39	47	37	130	15	0.4	.04		
CHOC MILKSHAKE, MALTED, DRY MIX, DELMARK	20	82	17.8	0.9					41	8		.01	0.4	6
		0.6		tr					20		0.2	.02	tr	
CHOC POWDER, BOSCO	20	56	12.8	0.3					33				2.8	
1 T		0.6	0.2						869			.16		
CHOC POWDER, INSTANT, HERSHEY	8	28	6.4	0.2					14	2	13	tr	tr	2
1 T		0.4	tr	0.0					34		2.0	.01	tr	
COCOA, DRY, BAKERS	6	13	3.1	0.8					4	10	44	.01	0.1	1
1 T		1.4	0.3						180		0.5	.03	0	
COCOA, BRKFST, DRY, BAKERS	6	18	3.2	1.3					tr	8	39	.01	0.1	2
1 T		1.0	0.3						91		0.6	.03	0	
COCOA, DRY, HERSHEY	6	25	2.8	0.8					8	9	47	.01	0.1	8
1 T		1.6	0.3	0.0					105		1.0	tr	tr	
COCOA MIX	7	27	5.2	0.7					27	19	20	tr	tr	1
1 T		0.7	tr	0.0					42	8	tr	.03	tr	
COCOA MIX, INSTANT, CARNATION	28	109	19.8	3.0	44	300	232	75	153	106	107	.03	0.1	23
1 pkg		3.9	0.3	tr	154	192	216	146	174	19	0.6	.16	1	
COCOA MIX, INSTANT, HERSHEY	8	28	6.4	0.2					14	2	13	tr	tr	2
1 T		0.4	tr	0.0					34		2.4	.02	tr	
COCOA MIX, INSTANT, NESTLÉ QUICK 1 serving	28	109	21.7	1.0					141					
		4.0							347					

	WT	Proximate			Amino Acids				Minerals			Vitamins		
		CAL	CHO	FAT	TRY	LEU	LYS	MET	Na	Ca	P	THI	NIA	A
		PRO	FIB	PUFA	PHE	ISO	VAL	THR	K	Mg	Fe	RIB	ASC	D
	g	g	g	g	mg	mg	mg	mg	mg	mg	mg	mg	mg	iu
COCOA MIX, OVALTINE	28	119	21.7	3.0					154	99	112	.15	2.0	494
1 serving		1.0							343		1.8	.17	6	
COCOA MIX, LOW CAL, OVALTINE	13	51	8.2	0.7					143	102	86	.15	0.7	507
1 serving		2.4							324		4.2	.17	6	
COCOA POWDER, CAROB, CARACOA	7	27	6.3	tr					1	20				
1 T		0.3	0.4						6		0.1			
COCOA POWDER, HIGH FAT	7	21	3.4	1.7					tr	9	45	.01	0.2	2
1 T		1.2							107	31	0.7	.03	0	
COCOA POWDER, HIGH-MED FAT	7	19	3.6	1.3					tr	9	45	.01	0.2	1
1 T		1.2	0.3						107	30	0.7	.03	0	
COCOA POWDER, HIGH-MED FAT, PROC W/ ALKALI 1 T	7	18	3.4	1.3					50	9	45	.01	0.2	1
		1.2	0.3						46		0.7	.03	0	
COCOA POWDER, LOW-MED FAT	7	15	3.8	0.9					tr	11	48	.01	0.2	1
1 T		1.3	0.4	0.0					107	29	0.7	.03	0	
COCOA POWDER, LOW-MED FAT, PROC W/ ALKALI 1 T	7	15	3.5	0.9					50	11	48	.01	0.2	1
		1.3	0.4						46		0.7	.03	0	
COCOA POWDER, LOW FAT	7	13	4.1	0.6					tr	11	53	.01	0.2	1
1 T		1.4	0.4						107	29	0.7	.03	0	
COCOA POWDER W/ NONFAT DRY MILK 1 T	7	25	5.0	0.2					37	41	38	.01	tr	1
		1.3	tr						56		0.1	.05	tr	
EGGNOG MILK FLAVORING, PDQ	14	56	13.5	0.2										
1 t		tr												
INSTANT BREAKFAST MIX														
CHOC, CARNATION	36	130	23.0	1.0	740	273	504	167	110	100	150	.30	5.0	1000
1 pkg		7.0		0.1	354	356	407	282	380	80	4.5	.07	24	0
CHOC MALT, PILLSBURY	36	130	26.0	0.0					140					
1 pkg		6.0												
STRAWBERRY, PILLSBURY	36	130	27.0	0.0					125					
1 pkg		5.0												
VANILLA, CARNATION	35	130	24.0	0.0	100	685	493	184	120	100	150	.30	5.0	1000
1 pkg		7.0		0.0	347	363	422	293	360	80	4.5	.07	24	0
VANILLA, PILLSBURY	36	130	28.0	0.0					125					
1 pkg		5.0												
MALT FLAVORED DRY MIX, OVALTINE	21	79	16.8	1.0					67	79	79	0.7	8.9	2205
1 serving		2.0							168		2.7	0.8	27	178
MALT POWDER, CARNATION	5	22	3.8	0.5	6	35	16	8	23	15	18	.03	0.3	15
1 t		0.6	tr	tr	20	18	21	15	40	5	tr	.05	tr	
STRAWBERRY FLAVORING, PDQ	15	59	14.8	tr										
1 T		0.0												
STRAWBERRY POWDER, NESTLÉ QUICK 1 t	8	30	8.0	0.0										
		0.0												
TIGER MILK PLUS, POWDER	70	248	35.8	0.6					238	769	843	4.00	16.6	
1 cup		23.8							1039	72	14.2	4.20		
VAN/STRAWBERRY, DRY MIX, DELMARK	10	41	9.4	0.3					12	2		tr	tr	3
		tr		tr					3		tr	tr	tr	

	WT g	CAL / PRO	CHO / FIB	FAT / PUFA	TRY / PHE mg	LEU / ISO mg	LYS / VAL mg	MET / THR mg	Na / K mg	Ca / Mg mg	P / Fe mg	THI / RIB mg	NIA / ASC mg	A / D iu
Nuts and Nut Products														
ALMONDS, CHOCOLATE	28	142	16.9	7.9					10	42	60	.03	0.4	15
1 oz; 10 med dried seeds		2.4	0.8						75		0.6	.11	tr	0
ALMONDS, DRIED, UNBLANCHED	15	90	2.9	8.1	26	218	87	39	tr	38	71	.04	0.7	0
12 to 15 nuts		2.8		1.6	172	131	169	92	104		0.7	.10	tr	0
DRIED, UNBLANCHED	100	547	19.6	54.1	176	1454	582	259	3	254	475	.25	4.6	0
⅔ cup		18.6	2.7	10.8	1146	873	1124	610	690		4.4	.67	tr	0
DRIED, SALTED, UNBLANCHED	15	93	2.9	8.5	26	218	87	39	24	38	71	.04	0.7	0
12 to 15 nuts		2.8		1.7	172	131	169	92	106		0.7	.10	tr	0
ALMONDS, ROASTED & SALTED	28	176	5.5	16.2	49	407	163	73	55	66	141	.01	1.0	0
1 oz		5.2	0.7		321	244	315	171	216		1.3	.26	0	
ALMOND PASTE	226	1153	116	73	238	1963	786	350	477	343	641	.34	6.4	0
8 oz		25.1		14.6	1547	1178	1517	823	931		5.9	.91	tr	0
ALMONDS, UNSHELLED, EP	78	238	7.8	21.6	70	582	233	104	1.2	102	190	.10	1.8	0
1 cup		7.4		4.3	458	349	450	244	276		1.8	.27	tr	0
BEECHNUTS	28	171	5.7	14.1										
1 oz		5.4	1.0	5.7										
BRAZIL NUTS, SHELLED	15	97	1.7	9.9	28	169	66	141	tr	28	104	.13		tr
4 med nuts		2.2	0.3	2.8	93	89	123	63	100		0.5		0	0
SHELLED	100	646	11.0	65.9	187	1129	443	941	1	186	693	1.09	7.7	tr
⅓ cup; 25 med nuts		14.4	2.1	18.4	617	593	823	422	670			.12	10	0
NOT SHELLED, EP	100	323	5.5	33.0	94	564	222	470	1	93	347	.43		tr
8 to 9 nuts		7.2	1.1	9.2	308	296	412	211	335		1.7		0	0
BUTTERNUTS	15	96	1.3	9.2										
4 to 5 nuts		3.6									1.0			
CASHEWS, ROASTED	15	84	4.4	6.9	64	212	111	49	2	6	56	.06	0.3	15
6 to 8 nuts		2.6	0.2	0.4	132	170	222	103	70	40	0.6	.04		
ROASTED	50	280	14.6	22.9	215	705	370	164	8	19	186	.22	0.9	50
20 to 26 nuts		8.6	0.7	1.5	438	567	740	344	232	134	1.9	.13		
ROASTED	100	561	29.3	45.7	430	1410	740	327	15	38	373	.43	1.8	100
1 cup; 3½ oz		17.2	1.4	3.0	877	1135	1479	688	464	267	3.8	.25		
CHESTNUTS, FRESH	15	29	6.2	0.2					tr	4	13	.03	0.1	
2 large; 3 small		0.4							62		0.3	.03		
FRESH	100	191	41.5	1.5					2	29	87	.23	0.5	
½ cup scant		2.8	1.1						410		1.7	.22		
DRIED	100	377	78.6	4.1					4	57	170	.34	0.8	
1 cup scant		6.7	2.5								3.3	.39		
COCONUT, FRESH MEAT	15	54	2.1	5.2	5	40	23	11	3	3	15	.01	0.1	0
1 piece (1×1×⅜)		0.5		0.1	26	27	32	19	116		0.3	.01	0.1	0
FRESH, SHREDDED	97	349	13.6	33.7	32	261	147	69	16	20	95	.04	0.6	0
1 cup		3.3		0.3	169	175	206	125	747		1.9	.03	4	0
SHREDDED, DRIED	15	83	8.0	5.9	5	40	23	11	3	7	29	.01	0.1	0
2 T		0.5		0.1	26	27	32	19	116		0.5	.01	0.1	0
SHREDDED, DRIED	62	344	33.0	24.2	21	175	98	46	11	27	118	.03	0.4	0
1 cup		2.2		0.2	113	117	138	84	500		2.2	.02	3	0
FILBERTS OR HAZEL NUTS	15	97	3.0	9.5	27	118	53	18	tr	38	48	.07	0.8	16
10 to 12 nuts		1.6	2.3	3.5	68	107	118	52	71		0.5	.08	1	0
HICKORY NUTS	15	101	2.0	10.1								.08		
15 nuts, small		2.1	0.3	1.9							0.4		0	0
LITCHI NUTS, DRIED	15	45	10.5	0.1					tr	4				
6 nuts		0.5							165		0.3		5	0
MACADAMIA NUTS, ROASTED	15	109	1.5	11.7						8	36	.03	0.2	0
6 whole nuts		1.4									0.3	.02	0	0
MIXED NUTS, SHELLED*	15	94	2.7	8.9	71	228	119	53	2	14	67	.09	0.6	3
8 to 12 nuts		2.5			142	183	239	111	84		0.5	.02	tr	0
SHELLED*	50	313	9.0	29.6	236	759	396	176	7	47	223	.30	2.0	10
27 to 40 nuts		8.3			473	611	796	368	280		1.7	.07	tr	0
PEANUTS, RAW, W/ SKIN	100	543	21.3	44.0	321	1768	1038	256	2	66	393	.91	17.6	16
3½ oz		25.5	4.3	12.7	1470	1195	1447	782	720		3.0	.21	1	0
RAW, W/O SKIN	100	560	17.5	47.3	337	1853	1088	268		46	466	.86	18.8	5
3½ oz		26.7	1.7	13.7	1541	1253	1517	820			3.2	.22	1	0
ROASTED W/ SKIN	15	86	3.3	7.0	50	277	162	40	tr	11	60	.04	2.5	tr
1 T nuts		4.0	0.5	2.0	230	187	226	122	111		0.5	.04	tr	0

*Amino acid values used for mixed nuts are those of cashew nuts.

	WT	CAL / PRO	CHO / FIB	FAT / PUFA	TRY / PHE	LEU / ISO	LYS / VAL	MET / THR	Na / K	Ca / Mg	P / Fe	THI / RIB	NIA / ASC	A / D
	g	g	g	g	mg	mg	mg	mg	mg	mg	mg	mg	mg	iu
PEANUTS, ROASTED W/ SKIN	100	572	22.0	46.7	335	1845	1083	267	2	72	400	.25	16.8	tr
2½ oz		26.5	3.4	13.5	1534	1248	1510	816	740		3.4	.26	1	0
ROASTED, W/ SKIN, SALTED	100	566	18.1	46.9	364	2003	1176	290	460	36	415	.24	19.0	tr
3½ oz		28.8	1.5	13.6	1666	1355	1639	886	700		3.2	.14	0	0
ROASTED W/O SKIN, SALTED	15	85	2.7	7.0	55	300	176	44	69	5	62	.04	2.8	tr
1 T nuts		4.3	0.2	2.0	250	203	246	133	105		0.5	.02	tr	0
PEANUT BUTTER	15	86	3.2	7.2	50	272	160	39	18	11	59	.02	2.4	0
1 T		3.9		1.8	227	184	223	120	123		0.3	.02	0	0
PEANUT BUTTER	100	576	21.0	47.8	330	1816	1066	263	120	74	393	.12	16.2	0
6 to 7 T		26.1	2.0	11.9	1510	1228	1487	803	820		1.9	.13	0	0
PEANUT BUTTER	258	1486	54.2	123.3	851	4685	2750	679	310	191	1014	.31	41.8	0
1 cup		67.3	5.2	30.9	3896	3168	3836	2072	2116		4.9	.34	0	
PECANS, SHELLED, NO SALT	15	104	2.0	11.0	21	116	65	23	tr	11	49	.11	0.1	8
12 halves; 2 T chopped		1.4		2.3	85	83	79	58	63		0.4	.02	tr	
SHELLED, NO SALT	100	696	13.0	73.0	138	773	435	153	tr	74	324	.72	0.9	50
1 cup halves		9.4	2.2	15.3	564	553	525	389	420		2.4	.11	2	0
PIGNOLIA NUTS	15	84	1.7	7.3										
2 T		4.7												
PILI NUTS	15	90	1.4	9.4						23	9	.10	0.1	2
2 T		1.6	0.3								0.6	tr	3	
PINYON OR PINE NUTS	15	95	2.5	9.2						2	77	.11	1.5	2
2 T		2.3	0.4								0.7	.03	tr	
PISTACHIO NUTS	15	88	2.8	8.0										
30 nuts		2.9	0.3	1.6										
SOYBEAN NUTS	28	127	6.6	5.5	tr	2	2	tr	19	68	190	.08		
1 oz		13.3	1.0		2	1	1	1	336		1.4	.07		
WALNUTS, BLACK	15	94	2.8	8.7								.05		11
2 T; 8 to 10 halves		2.7		4.8							0.9		0	0
BLACK	100	628	15.1	59.6					3	tr		.22	0.7	302
3½ oz		20.7		32.8					460		6.0	.11		
WALNUTS, ENGLISH	8	49	1.2	4.8	14	98	35	24	tr	6	28	.04	0.1	2
1 T chopped		1.1	0.2	3.6	61	61	78	47	36		0.2	.01	tr	0
ENGLISH	15	98	2.3	9.7	26	184	66	46	tr	12	57	.07	0.2	5
8 to 15 halves		2.3	0.3	7.2	115	115	146	88	68		0.3	.02	tr	
ENGLISH	100	654	15.6	64.4	175	1228	441	306	2	83	380	.48	1.2	30
1 cup halves		15.0	2.1	47.5	767	767	974	589	450		2.1	.13	3	
Seeds														
PUMPKIN & SQUASH KERNELS	100	553	15.0	46.7	522	2291	1334	551		51	1144	.24	2.4	70
3½ oz		29.0	1.9		1624	1624	1566	870			11.2	.19		
SAFFLOWER SEED KERNELS*	100	615	12.4	59.5	306	1242	688	325						
3½ oz		19.1			1184	860	1108	649						
SESAME SEEDS, DECORTICATED†	100	582	17.6	53.4	309	1583	546	601		100	592	.18	5.4	
3½ oz		18.2	2.4		1383	892	837	673			2.4	.13	0	
SUNFLOWER SEED KERNELS	100	560	19.9	47.3	360	1824	912	456	30	120	837	1.96	5.4	50
3½ oz		24.0	3.8		1272	1320	1416	936	920		7.1	.23		

*Amino acid values are those for safflower seed meal.
†Amino acid values are those for the entire seed.

	WT	CAL	CHO	FAT	TRY	LEU	LYS	MET	Na	Ca	P	THI	NIA	A
		PRO	FIB	PUFA	PHE	ISO	VAL	THR	K	Mg	Fe	RIB	ASC	D
	g	g	g	g	mg	mg	mg	mg	mg	mg	mg	mg	mg	iu
APPLE, CELERY & WALNUT SALAD	154	137	15.9	8.3						32	48	.08	0.4	355
3 hp T, 2 leaves lettuce		1.9	1.1								0.8	.07	5	
CARROT & RAISIN SALAD	134	153	27.9	5.8						48	65	.08	0.5	4700
3 hp T, 2 leaves lettuce		1.9	0.7								1.5	.08	6	
CHICKEN SALAD	100	127	3.9	7.5						22	68	.03	2.3	199
½ cup		11.0	0.3								0.9	.09	3	
CHICKEN SALAD, CND, CARNATION	28	55	2.2	3.7					108	3	17	.01	0.4	30
SPREADABLES 1 oz		3.2		1.7					43		0.2	.02	tr	0
CHICKEN SALAD, CND, CONTINENTAL	100	239	19.4	12.9					600					
COFFEE CO ½ cup		11.4												
CHICKEN SALAD, CND, KELLOGGS	28	70	5.3	2.0					115	13	26	.08	0.5	0
1 oz		3.4	tr						69	53	0.5	.04	0	
CHICKEN W/ CELERY SALAD	146	185	5.7	10.9						32	100	.05	3.4	290
3 hp T, 2 leaves lettuce		16.1	0.5								1.3	.13	5	
COLESLAW W/O DRESSING	120	33	6.9	0.5					42	57	36	.06	0.4	1367
1 cup		1.6	1.0	tr					279	16	0.5	.06	47	0
COLESLAW W/ HOMEMADE FRENCH	120	155	6.1	14.8					157	50	30	.05	0.4	130
DRESSING 1 cup		1.3	0.8						236		0.5	.05	35	
COLESLAW W/ COMMERCIAL FRENCH	120	114	9.1	8.8					322	50	31	.05	0.4	130
DRESSING 1 cup		1.4	0.8						246		0.5	.05	35	
COLESLAW W/ MAYONNAISE	120	173	5.8	16.8					144	53	35	.06	0.4	190
1 cup		1.6	0.8						239		0.5	.06	35	
COLESLAW W/ MAYONNAISE-TYPE	120	119	8.5	9.5	12	58	67	13	149	52	34	.06	0.4	180
SALAD DRESSING 1 cup		1.4	0.8	1.9	30	56	43	41	230	14	0.5	.06	35	
CRAB SALAD	100	145	4.9	8.5						38	129	.06	1.3	97
		11.8								26	0.6	.06	2	
CRAB W/ CELERY SALAD	144	137	6.1	8.4						47	116	.06	1.6	290
3 hp T, 2 leaves lettuce		9.6	0.5								0.9	.08	5	
ENDIVE W/ BACON DRESSING	150	228	5.8	21.7						83	91	.19	1.0	3000
1 large serving		4.5	0.8								1.9	.15	11	
FRUIT SALAD, CND	193	155	30.1	5.9						23	31	.03	0.6	500
3 hp T, 2 leaves lettuce		1.3	0.8								0.8	.05	6	
FRESH*	195	174	21.2	11.1						45	44	.08	0.4	685
3 hp T, 2 leaves lettuce		1.9	0.7								0.8	.09	32	
GELATIN W/ FRUIT	188	139	21.6	5.7						23	25	.04	0.3	391
1 square, 2 leaves lettuce		2.1	0.5								0.5	.05	16	
GELATIN W/ CHOPPED VEGETABLES	164	115	15.1	5.7						24	27	.04	0.3	1977
1 square mold, 2 leaves lettuce		2.2	0.6								0.5	.06	8	
HAM SALAD, CND, CARNATION	28	51	2.2	3.6					174	2	31	.09	1.2	78
SPREADABLES 1 oz		2.6		1.1					56		0.5	.04	tr	0
LETTUCE W/ FRENCH DRESSING	130	133	6.9	10.8						22	25	.04	12	540
1 wedge, 8 small leaves		1.4	0.6								0.5	.08	8	
LETTUCE, TOM & MAYONNAISE	115	80	6.5	5.8						20	36	.06	0.5	1115
4 leaves lettuce, 3 slices tomato		1.5	0.7								0.8	.07	19	
LOBSTER SALAD	100	110	2.3	6.4					124	36	95	.09		
½ cup, 2 leaves lettuce		10.1		0.9					264		0.9	.08	18	
MACARONI, ONION & MAYONNAISE	190	335	48.5	11.8					1012	21	103	.03	0.7	40
SALAD 1 cup; large serving		7.7	tr						199		1.0	.02	2	
POTATO SALAD W/ CKD SALAD	250	248	40.8	7.0					1320	80	160	.20	2.8	350
DRESSING 1 cup		6.8							798		1.5	.18	28	
POTATO SALAD W/ MAYONNAISE &	250	363	33.5	23.0					1200	48	158	.18	2.3	450
FRENCH DRESSING 1 cup		7.5							740		2.0	.15	28	
POTATO SALAD W/ PARSLEY, CKD	100	99	16.3	2.8					528	32	64	.08	1.1	140
DRESSING ½ cup		2.7	.4						319		0.6	.07	11	
STUFFED PRUNE W/ LETTUCE	145	414	32.9	28.5						63	228	.11	7.9	800
4 prunes, peanut butter		13.5	0.8								2.3	.14	5	
TUNA SALAD†	100	170	3.5	10.5						20	142	.04	5.0	290
½ cup		14.6									1.3	.11	1	
TUNA SALAD, CND, CARNATION	28	53	2.1	3.6					93	3	23	.00	1.3	58
SPREADABLES 1 oz		30							41		0.2	.01	1	0

*Fresh orange, grapefruit and grapes, canned pineapple and peaches, with mayonnaise, whipped cream and fruit juice dressing

†W/celery, mayonnaise, pickle, onion, and egg

	WT	Proximate			Amino Acids				Minerals			Vitamins		
		CAL	CHO	FAT	TRY	LEU	LYS	MET	Na	Ca	P	THI	NIA	A
		PRO	FIB	PUFA	PHE	ISO	VAL	THR	K	Mg	Fe	RIB	ASC	D
	g	g	g	g	mg	mg	mg	mg	mg	mg	mg	mg	mg	iu
TUNA SALAD, CND, CONTINENTAL	100	223	21.6	11.0					640					
COFFEE CO ½ cup		9.5												
TUNA SALAD, CND, KELLOGGS	28	59	1.7	4.5					132	1	26	.08	1.3	0
1 oz		3.4	tr						66	53	0.2	.03	0	
WALDORF SALAD	100	127	17.3	7.0					72	17	30	.06	0.2	124
1 serving		0.9	0.8						185		0.6	.04	3	

	WT	CAL	CHO	FAT	TRY	LEU	LYS	MET	Na	Ca	P	THI	NIA	A
	g	PRO	FIB	PUFA	PHE	ISO	VAL	THR	K	Mg	Fe	RIB	ASC	D
	g	g	g	g	mg	mg	mg	mg	mg	mg	mg	mg	mg	iu
BACON, LETTUCE & TOM ON WHITE BREAD*	148	282	28.8	15.6						53	89	.16	1.6	870
		6.8	0.6								1.5	.14	13	
BACON, CHICKEN, TOM & LETTUCE ON 3 SLICES WHITE BREAD*	315	590	41.7	20.8						103	394	.38	10.2	1705
		35.6	1.2								4.3	.41	27	
CHICKEN SALAD ON WHITE BREAD*	110	245	26.6	8.6						50	101	.14	3.2	10
		14.3									1.5	.14	1	
CHICKEN W/ 3 T GVY ON WHITE BREAD	160	356	29.8	15.3						49	219	.18	6.5	0
		21.9	0.1								2.4	.21	0	
CHICKEN W/ LETTUCE ON WHITE BREAD*	131	303	26.6	14.4						52	165	.16	4.6	320
		15.8	0.3								1.8	.17	2	
CORNED BEEF ON RYE BREAD	130	296	24.0	10.8	292	2071	2004	599	1214	51	158	.10	3.5	17
		25.7	0.2		1114	1285	1410	1069	140	43	4.5	.24	0	
CREAM CHEESE & JELLY ON WHITE BREAD	119	368	50.4	16.0						60	74	.12	1.0	575
		6.6	0.1								1.1	.14	2	
EGG SALAD ON WHITE BREAD*	138	279	30.6	12.5						68	153	.16	1.0	580
		10.5	0.3								2.4		2	
HAM ON WHITE BREAD*	81	281	23.9	15.4						40	93	.28	2.3	165
		10.9	0.1								1.7	.14	0	
HAM SALAD ON WHITE BREAD*	114	321	30.4	16.9						45	102	.28	2.3	30
		11.3	0.4								2.0	.14	2	
HOT DOG	80	287	23.9	23.9										
		10.0												
LIVERWURST ON RYE BREAD*	91	251	26.8	12.3						38	143	.13	2.2	1745
		9.4	0.1								2.5	.38	0	
PNT BTR† ON WHITE BREAD	83	328	30.0	19.5						61	177	.10	5.4	165
		11.8	0.8								1.0	.80	0	
PNT BTR† & JELLY ON WHITE BREAD	100	374	50.0	15.1					475	72	169	.19	5.7	2
		12.3	0.6	3.2					257	61	2.3	.17	1	0
PNT BTR† & JELLY ON WHOLE WHEAT BREAD	114	385	55.0	15.2	165	910	442	152	418	72	220	.16	5.6	4
		12.6	1.3	11.0	710	584	678	392	344	71	2.2	.12	2	
ROAST BEEF ON WHITE BREAD*	83	328	19.5							61	177	.10	5.4	165
		11.8	0.8								1.0	.08	0	
ROAST BEEF W/ 3 T GVY ON WHITE BREAD	160	429	29.8	24.5						43	163	.17	4.9	0
		19.3	0.1								2.9	.21	0	
ROAST PORK ON WHITE BREAD*	116	288	26.6	14.7						46	123	.38	2.5	155
		11.4	0.3								1.9	.17	2	
ROAST PORK W/ 3 T GVY ON WHITE BREAD	180	503	29.8	30.9						46	239	.79	5.1	0
		23.3	0.1								3.5	.29	0	
SLOPPY JOE	200	473	6.8	33.5					540	19	301	.75	8.4	372
		36.4	0.2	1.5					756	37	5.0	.32	4	
SUBMARINE SANDWICHES (FROM SANDWICH SHOPS)														
HAM–SALAMI–PROVO–LETT–TOM–ONION–PICK–OIL–VINEGAR–SPICES 8"	280	582	53.1	28.0					1868	316	302			
		29.6	1.1						522	106	3.0			
HAM, SALAMI, CHEESE, LETT, TOM, OIL, VINEGAR & ONION 8"	326	639	65.7	23.4					3528	378	578			
		41.3	0.6						704	116	4.8			
ROAST BEEF, LETT, TOM, ONION, TOM, LETT, OIL & VINEGAR 6"	249	368	55.7	6.6					482	39	147			
		21.6	0.8						294	30	1.5			
ROAST BEEF, MAYON, TOM & LETT 8"	295	611	55.7	22.5					1442	62	307			
		46.4	0.8						590	120	5.0			
SPICED HAM–SALAMI–CHSE–ONION–TOM–LETT–OIL–VINEGAR 6"	228	449	61.2	16.6					1032	50	240			
		13.7	0.6						145	44	2.1			
TUNA, MAYON, LETT & TOM 8"	306	685	54.8	33.6					1760	77	280			
		41.2	0.7						532	123	4.0			
TUNA, MAYON, LETT, TOM, ONION, OIL & VINEGAR 6"	256	581	61.2	28.0					824	83	168			
		21.0	0.6						160	27	1.2			
TUNA, GRILLED*	100	277	38.0	30.5					857	89	255	.21	8.0	234
		25.1	0.4	2.2					214	39	3.4	.29	3	13
TUNA SALAD–WHITE BREAD*	105	278	25.8	14.2						48	135	.14	4.1	231
		11.0	0.1								1.2	.11	1	
TURKEY–WHITE BREAD*	156	402	28.0	18.4						62	370	.18	6.8	45
		29.2	0.1								5.8	.22	0	

*Approximately 5 gm butter or margarine was used. †Approximately 2 T peanut butter

Note: A serving is 1 sandwich.

	WT g	CAL / PRO g	CHO / FIB g	FAT / PUFA g	TRY / PHE mg	LEU / ISO mg	LYS / VAL mg	MET / THR mg	Na / K mg	Ca / Mg mg	P / Fe mg	THI / RIB mg	NIA / ASC mg	A / D iu
ASPARAGUS, CREAM OF, CND	200	140	18.8	5.6	14	320	176	52	1780	46	62	.20	1.8	342
		3.8	0.2		132	186	206	292	276		1.6	.18		
ASPARAGUS, CREAM OF, CND, COND 1 can	298	154	23.2	4.2	21	477	262	77	2444	65	92	.09	1.8	757
		6.0	1.8		197	277	307	435	411		1.8	.21		
ASPARAGUS, CREAM OF, CND, MADE W/ WATER 1 cup	240	84	11.3	3.4					996	28	36	.12	1.1	386
		2.3	0.1		79				166		1.0	.11		
ASPARAGUS, CREAM OF, CND, MADE W/ MILK 1 serving†	198	88	12.9	1.5	56	503	359	112	860	145	128	.07	0.7	252
		5.5	0.6		236	315	342	306			0.7	.25		
BEAN, HOMEMADE ¾ cup	190	195	18.6	11.2						52	120	.12	0.5	1364
		6.1	1.4								1.9	.07	0	0
BEAN, CND 1 serving	241	169	21.7	4.8					831	89	121	.07	0.5	877
		7.2	1.5						362		1.9	.05	1	
BEAN W/ BACON, CND, COND 1 can	326	435	55.7	13.7	202	913	2315	287	2445	125	321	.26	2.3	1793
		22.2	4.2	4.5	1108	1174	1017	1011	978		6.2	.16	4	
BEAN W/ BACON, CND, MADE W/ WATER 1 serving†	207	145	18.5	4.6	67	304	772	96	814	42	107	.09	0.8	597
		7.4	1.4	1.5	369	391	339	337	311		2.1	.05	1	
BEAN W/ HAM, HOMEMADE	200	160	36.5	5.5					228	85	295	.43	2.3	83
		17.3	2.3						588	8	4.3	.18	9	4
BEAN W/ HAM, CND 1 cup	240	190	23.3	6.4						666	165			
		9.9	1.4								2.1			
BEAN W/ PORK, CND, COND 1 can	326	421	48.6	16.3					2813	202	339	.36	2.6	1695
		19.9	4.0						1030		5.3	.20	7	
BEAN W/ PORK, CND, MADE W/ WATER 1 serving†	207	140	16.2	5.4					937	67	107	.10	0.8	538
		6.6	1.3						327		2.1	.06	2	
BEEF, CND, COND 1 can	312	293	24.3	6.9	184	1420	1894	337	2412	10	102	.03	3.7	3329
		28.4	1.2		671	555	1008	805			2.8	.19		
BEEF, CND, MADE W/ WATER 1 serving†	203	91	8.1	2.3	61	473	631	112	804	3	34	.01	1.2	1110
		9.5	0.4		224	185	336	268			0.9	.06		
BEEF BROTH, INSTANT, DRY 1 T	17	30	2.6	0.7					3611					
		3.2												
BEEF BROTH (BOUILLON, CONSOMME), CND, COND 1 can	298	68	6.3	0.0	6	417	453	80	2354	tr	73	tr	3.1	tr
		10.7	0.3		280	158	343	468	322		0.6	.06		
BEEF BROTH (BOUILLON, CONSOMME), CND, MADE W/ WATER 1 serving†	198	23	2.1	0.0	2	139	151	27	784	tr	24	tr	1.0	tr
		3.6	0.1		93	53	114	156	107		0.2	.02		
BEEF BARLEY, HOMEMADE 1 serving	113	97	18.1	0.7					367	13	79	.05	1.4	10
		5.6	0.2						108		0.9	.04	1	
BEEF CELERY, CND	200	212	29.0	7.4					296	106	106	.31	4.2	528
		7.4							380	tr	1.9	.36	32	42
BEEF, CHUNKY, CND, LOW-SODIUM 1 cup	185	167	14.8	7.4					65	11	94	.11	1.9	4053
		9.3	0.7						355		1.3	.15	7	
BEEF CUBE 1 cube	4	7	0.4	0.2	9	65	70	20	960	0	12	tr	tr	0
		0.8	0.1	tr	14	42	44	35	4	2	0.2	.07	0	0
BEEF FLAVORED, DEHYD, CUP-A-SOUP 1 pkg	8	20	3.8	0.1					959	1	2	.01		tr
		0.9							5					
BEEF FLAVORED NOODLE, DEHYD, CUP-A-SOUP 1 pkg	12	35	6.6	0.3					931	3	9	.03	0.3	442
		1.7	0.1						36	3	0.2	.02	7	
BEEF NOODLE, CND 1 serving	218	76	8.7	2.2					839	11	39	.07	1.3	305
		4.4	0.2						74		0.9	.07	0	
BEEF NOODLE, CND, COND 1 can	298	175	21.5	5.7	83	387	924	334	1937	tr	112	.18	3.6	155
		9.5	0.3	1.2	387	477	456	447	206		2.4	.21	0	
BEEF NOODLE, CND, MADE W/ WATER 1 serving†	198	58	7.2	1.9	28	129	308	111	645	tr	37	.06	1.2	52
		3.2	0.1	0.4	129	159	152	149	69		0.8	.07	0	
BEEF NOODLE, DEHYD 1 pkg	57	208	32.8	4.2					1350	10	141	.29	2.0	116
		8.9	0.5						131		1.0	.12	8	
BEEF NOODLE, DEHYD, MADE W/ WATER 1 serving‡	182	52	8.2	1.0					319	2	35	.07	0.5	29
		2.2	0.1						31		0.2	.03	2	
BEEF NOODLE VEG, DEHYD 1 serving	17	57	9.4	1.2					1003	9	33	.19	0.9	273
		2.4	0.2	tr					58		0.5	.07	1	
BEEF NOODLE W/ DUMPLINGS, CND, COND 1 cup	240	104	12.0	3.6					108	40	75	.04	0.5	846
		5.8	0.2						106		1.4	.08	3	

*Soups are ready to eat unless otherwise indicated. Many canned soups do not require dilution.

†For canned, condensed soups which have been diluted w/milk or water, a serving (unless otherwise indicated) is ⅓ of the can plus an equal volume of milk or water.

‡Four servings come from 1 package.

	WT	Proximate			Amino Acids				Minerals			Vitamins		
		CAL	CHO	FAT	TRY	LEU	LYS	MET	Na	Ca	P	THI	NIA	A
		PRO	FIB	PUFA	PHE	ISO	VAL	THR	K	Mg	Fe	RIB	ASC	D
	g	g	g	g	mg	mg	mg	mg	mg	mg	mg	mg	mg	iu
BEEF RICE	113	39	5.5	0.5					344	7	20	.03	1.1	278
1 serving		2.8		0.6					50		0.5	.02	1	
BEEF, VEG & BARLEY, CND, COND	298	173	20.6	4.8					2300	39	146			
1 can		12.2	0.8								2.1			
BEEF, VEG & BARLEY, CND, MADE W/ WATER 1 serving†	198	58	6.9	1.6					766	13	49			
		4.1	0.3								0.7			
BLACK BEAN, CND, COND	326	261	39.8	4.6	153	1304	962	143	2804	80	247	.20	1.3	841
1 can		15.3	2.6		782	769	857	1359			6.2	.23		
BLACK BEAN, CND, MADE W/ WATER	207	87	13.3	1.5	51	435	321	48	934	27	82	.07	0.4	280
1 serving*		5.1	0.9		260	256	285	453			2.1	.08		
BORSCHT, CND	250	80	19.3	0.0										
1 serving		0.6												
CAULIFLOWER, CREAM OF, HOMEMADE	113	136	6.3	10.7					517	98	92	.05	1.0	555
1 serving		4.1	0.2						157		0.4	.15	12	
CELERY, CREAM OF, CND, COND	298	193	22.9	9.5	30	268	507	209	2562	101	95	.03	2.4	516
1 can		3.9	1.8	5.9	179	206	149	143	269		2.4	.12	2	
CELERY, CREAM OF, CND, MADE W/ WATER 1 cup†	240	79	7.7	4.7	12	108	204	84	984	41	38	.01	1.0	209
		1.7	0.2	2.4	72	84	60	58	108		1.0	.05	0	
CELERY, CREAM OF, CND, MADE W/ MILK 1 serving†	198	100	12.7	3.3	59	433	441	156	905	135	129	.05	0.9	172
		4.8	0.6	2.1	230	292	290	209	238		0.9	.22	2	
CELERY, CREAM OF, HOMEMADE	155	157	10.9	10.5						171	138	.07	0.3	425
¾ cup		5.4	0.2								0.5	.23	1	1
CHEDDAR CHEESE, CND, MADE W/ WATER 1 serving†	200	136	8.5	8.2					780	134	106	.01	1.0	566
		4.0	0.1						120		0.5	.12		
CHICKEN, CND	200	212	29.0	7.4					296	106	106	.31	4.2	528
		7.4							380	tr	1.9	.36	32	42
CHICKEN BROTH, INSTANT, DRY 1 T	12	22	2.2	0.5					2431					
		2.2												
CHICKEN CONSOMME, CND, COND	245	44	3.7	0.2					1475	24.5	145			
1 cup		6.9	0.1								2.5			
CHICKEN CONSOMME, CND, MADE W/ WATER 1 cup	245	22	2.0	0.5					737	12	74			
		3.4	tr								1.2			
CHICKEN CHOWDER, FRZN	200	148	17.6	4.4					1160	34	88	2.00	2.6	2312
		9.8	0.8						366		1.4	.24	4	
CHICKEN, CHUNKY, CND, LOW-SODIUM	160	120	11.2	4.8					48	10	66	.10	2.6	693
1 cup		8.0	0.2						149		0.8	.18	2	
CHICKEN, CREAM OF, CND	234	117	7.0	9.4	51	281	538	197	983	35	35	.02	0.9	541
1 serving		2.3	0.1		246	281	220	234	80		0.7	.07	0	
CHICKEN, CREAM OF, CND, COND	298	235	17.9	14.9	66	358	685	250	2384	64	87	.03	1.8	1442
1 can		7.5	0.6	3.5	313	358	280	298	203		1.5	.12	0	
CHICKEN, CREAM OF, CND, MADE W/ WATER 1 cup	240	91	7.6	5.3	26	144	276	101	1176	26	36	.01	0.7	581
		3.4	0.1	1.4	127	144	113	120	82		0.7	.05	0	
CHICKEN, CREAM OF, CND, MADE W/ MILK 1 serving†	198	146	10.9	8.9	71	463	506	169	844	139	122	.05	0.7	640
		6.0	0.2	4.0	274	581	333	260	216	18	1.5	.21	2	
CHICKEN CUBES 1 cube	4	10	1.1	0.3										
		0.6												
CHICKEN FLAVORED, DEHYD, CUP-A-SOUP 1 pkg	8	25	3.4	0.7					841	1	3	.01	0.1	36
		1.1		0.1					9	1				
CHICKEN GUMBO, CND	240	85	13.2	2.1					783	37	168	.03	1.2	120
1 cup		3.3	0.2						115		1.4	.05	2	
CHICKEN GUMBO, CND, COND	298	138	21.8	3.0	60	346	474	95	2682	42	60	.06	4.2	533
1 can		6.0	0.9	0.8	197	188	244	200	179		1.8	.12	0	
CHICKEN GUMBO, CND, MADE W/ MILK 1 serving†	198	82	12.4	1.1	69	460	431	118	945	139	117	.06	1.5	177
		5.5	0.1		236	286	322	228			0.7	.06		
CHICKEN GUMBO, CND, MADE W/ WATER 1 cup	240	58	8.8	1.4					972	17	24	.02	1.7	216
		2.6	0.2	0.3					72		0.7	.05	5	
CHICKEN NOODLE, CND	234	70	7.0	2.3	61	351	842	204	866	12	33	.07	1.6	51
1 serving		4.7	0.1		351	374	257	304	47		0.7	.05	0	
CHICKEN NOODLE, CND, COND	298	152	20.6	3.9	77	447	1073	259	2265	17	84	.03	2.1	140
1 can		8.6	0.6	1.1	447	477	328	387	137		1.2	.06	0	

*Soups are ready to eat unless otherwise indicated. Many canned soups do not require dilution.

†For canned, condensed soups which have been diluted w/ milk or water, a serving (unless otherwise indicated) is ⅓ of the can plus an equal volume of milk or water.

Food	WT	Proximate CAL/PRO	CHO/FIB	FAT/PUFA	Amino Acids TRY/PHE	LEU/ISO	LYS/VAL	MET/THR	Minerals Na/K	Ca/Mg	P/Fe	Vitamins THI/RIB	NIA/ASC	A/D
	g	g	g	g	mg	mg	mg	mg	mg	mg	mg	mg	mg	iu
CHICKEN NOODLE, CND, MADE W/ WATER 1 serving†	198	51	6.9	1.3	26	149	357	86	754	6	28	.01	0.7	47
		2.9	0.2	0.4	149	159	109	129	40	8	0.4	.02	0	
CHICKEN NOODLE, DEHYD 1 serving	18	67	10.2	1.9					1047	24	4	.10	0.7	62
		2.2	0.1	0.2					19	5	0.3	.05	1	
CHICKEN NOODLE W/ DUMPLINGS 1 cup	240	87	8.7	3.4						40	52			
		5.2	0.2								0.7			
CHICKEN RICE, CND 1 serving	218	55	6.5	2.2					730	15	33	.02	0.9	610
		2.2	0.0						50		0.4	.02	0	
CHICKEN RICE, CND, COND 1 can	301	95	11.7	1.5	66	391	1054	283	2276	10	53	tr	1.8	400
		8.7	0.3	1.0	313	361	241	301	135		0.9	.06	0	
CHICKEN RICE, CND, MADE W/ WATER 1 serving†	199	32	3.9	0.5	22	130	351	94	758	3	18	tr	0.6	133
		2.9	0.1	0.3	104	120	80	100	46		0.3	.02	0	
CHICKEN RICE, DEHYD 1 pkg	45	187	23.5	5.5					1963	6	72	.03	2.6	16
		8.2	0.1						32		0.5	.30	0.5	
CHICKEN RICE, DEHYD, MADE W/ WATER 1 serving‡	178	47	5.9	1.4					461	2	18	.01	0.6	4
		2.0	tr						7		0.1	.08	tr	
CHICKEN RICE W/ MUSHROOMS, CND 1 cup	240	84	12.0	2.4					988	24	52	.05	1.9	99
		4.2	0.2						106		0.5	.10	3	
CHICKEN VEG, CND, COND 1 can	304	194	21.0	7.6	109	553	526	125	2493	44	105	.06	2.7	4378
		10.3	0.9		283	371	340	319	413		1.5	.09	1	
CHICKEN VEG, CND, MADE W/ WATER 1 serving†	200	65	7.0	2.5	36	184	175	42	830	15	35	.02	0.9	1458
		3.4	0.3		94	124	113	106	136		0.5	.03		
CHILI BEEF, CND 1 serving	225	169	22.5	6.8					866	47	133	.09	1.1	801
		6.8	1.1						801		2.7	.09	3.4	
CHILI BEEF, CND, COND 1 can	311	407	56.9	11.5					2675	68	365	.22	3.4	2065
		19.3	2.8								8.1	.19		
CHILI BEEF, CND, MADE W/ WATER 1 serving†	202	136	18.9	3.8					891	23	122	.07	1.1	688
		6.4	0.9								2.7	.06		
CLAM CHOWDER, MANHATTAN, CND 1 serving	218	76	10.9	2.2					872	22	35	.02	0.7	942
		2.2	0.4						164		1.3	.02	3	
CLAM CHOWDER, MANHATTAN, CND, COND 1 can	308	196	28.3	6.5	59	403	410	80	2218	76	104	.06	2.8	1284
		5.9	0.9	3.1	194	179	240	228	462	37	3.4	.06	10	
CLAM CHOWDER, MANHATTAN, CND, MADE W/ WATER 1 serving	201	65	9.4	2.2	20	134	137	27	739	25	35	.02	0.9	428
		2.0	0.3	2.1	65	60	80	76	151		1.1	.02	4	
CLAM CHOWDER, NEW ENGLAND, FRZN	200	124	15.8	3.8					1750	32	82	.06	1.0	2698
		6.6	0.8						518		1.2	.08	10	
CLAM CHOWDER, NEW ENGLAND, FRZN COND 1 can	291	314	25.9	18.6	119	873	661	183	2532	218	199	.09	1.2	146
		10.8	0.6		454	567	541	762	538		2.3	.20		
CLAM CHOWDER, NEW ENGLAND, FRZN MADE W/ WATER 1 serving†	196	173	13.5	10.1	89	635	492	147	894	191	159	.07	0.2	209
		7.1	0.2		321	790	420	415	329		0.9	.11		
CLAM CHOWDER, NEW ENGLAND, FRZN MADE W/ WATER 1 serving†	193	104	8.6	6.2					840	71	66	.03	0.4	48
		3.6	0.2						179		0.8	.07	3	
CORN CHOWDER, HOMEMADE 1 serving	116	105	11.0	5.1					379	53	61	.07	1.3	212
		3.4							118		0.4	.10	4	
FISH CHOWDER, ARTHUR TREACHERS 1 serving	170	112	11.2	5.4					835	61	87	.07	0.4	340
		4.6	0.9						228	12	0.1	.14	2	19
GUMBO, CREOLE, CND, COND 1 can	298	173	27.4	5.1					2078	80	89			
		4.2	1.0								1.9			
GUMBO, CREOLE, CND, MADE W/ WATER 1 serving†	198	58	9.1	1.7					693	27	30			
		1.4	0.3								0.6			
MINESTRONE, CND 1 serving	218	87	10.9	2.2					818	37	50	.04	0.9	1498
		4.4	0.7						279		8.7	.04	15	
MINESTRONE, CND, COND 1 can	298	248	27.4	8.9	101	584	465	92	2348	81	139	.18	2.7	5712
		14.6	2.1	2.7	334	402	420	319	760		2.1	.15	4	
MINESTRONE, CND, MADE W/ WATER 1 serving†	198	83	9.1	3.0	33	194	155	31	782	27	46	.06	0.9	1902
		4.9	0.9	0.9	111	134	140	106	253		0.7	.05	14	
MULLIGATAWNY, HOMEMADE 1 serving	125	64	6.0	3.5					421	8	20	.03	1.1	142
		2.2	tr						39		0.3	.01	tr	
MUSHROOM, CREAM OF, CND 1 serving	218	131	8.7	10.9					883	37	35	.02	0.7	0
		2.2	0.0						65		0.4	.07	0	

*Soups are ready to eat unless otherwise indicated. Many canned soups do not require dilution.

†For canned, condensed soups which have been diluted w/ milk or water, a serving (unless otherwise indicated) is ⅓ of the can plus an equal volume of milk or water.

‡Four servings come from 1 package.

	WT	CAL	CHO	FAT	TRY	LEU	LYS	MET	Na	Ca	P	THI	NIA	A
		PRO	FIB	PUFA	PHE	ISO	VAL	THR	K	Mg	Fe	RIB	ASC	D
	g	g	g	g	mg	mg	mg	mg	mg	mg	mg	mg	mg	iu
MUSHROOM, CREAM OF, CND, COND	298	338	25.6	24.1	39	232	536	250	2324	68	94	.03	1.8	235
1 can		4.5	0.9		229	226	179	161	179		1.2	.24	0	
MUSHROOM, CREAM OF, CND, MADE	198	149	13.6	8.1	62	421	450	169	827	146	128	.05	0.7	78
W/ MILK 1 serving†		5.0	0.3	3.7	246	298	300	215	208		0.5	.03		
MUSHROOM, CREAM OF, CND, MADE	240	134	10.1	9.4					955	41	50	.02	0.7	70
W/ WATER 1 cup		2.4	0.2	4.4					98		0.5	.12	0	
MUSHROOM, CREAM OF, DEHYD	43	133	25.3	0.7										
1 pkg		6.3												
MUSHROOM, CREAM OF, DEHYD, MADE W/ WATER 1 serving†	178	33	6.3	0.2										
		1.6												
MUSHROOM, CREAM OF, DEHYD,	19	86	9.7	4.5					530	35	99	.02	0.2	3
CUP-A-SOUP 1 serving		1.8		0.1					133	5	0.1	.10		
ONION, CND, COND	298	155	11.6	6.3	36	507	510	89	2608	70	68	tr	0.9	tr
1 can		13.1	1.2		280	232	343	587	283		1.2	.06	3	
ONION, CND, MADE W/ WATER	198	52	3.9	2.1	12	169	170	30	868	23	23	tr	0.3	tr
1 serving†		4.4	0.4		93	77	114	196	95		0.4	.02		
ONION, DEHYD	11	35	5.8	0.8					841	11	24	.06	0.2	1
1 serving		1.2	0.3	0.1					96	8	0.3	.02	2	
ONION, DEHYD, MADE W/ WATER	214	34	53	0.8						12	20	.01	1.1	7
1 serving†		1.6	0.3								0.2	.01	2	
ONION, FRENCH, HOMEMADE	113	38	2.8	2.9					744	8	5	.01	0.1	117
1 serving		0.8	0.1						16		tr	.01	1	
OYSTER STEW, HOMEMADE§	100	97	4.5	6.4					339	114	111	.06	0.9	340
1 serving		5.2	0.0						133		1.9	.08		
OYSTER STEW, HOMEMADE¶	100	86	4.7	5.3					203	117	109	.06	0.7	280
1 serving		4.9							138		1.4	.18		
OYSTER STEW, FRZN, COND	284	292	19.6	17.9	57	1610	1088	321	1931	371	329	.17	0.9	543
1 can		13.1	0.3	0.4	733	795	994	1338		3.1		.45		
OYSTER STEW, FRZN, MADE W/ MILK	193	133	11.6	6.1	68	881	635	193	696	247	207	.10	0.4	181
1 serving†		8.1	0.1	0.3	414	488	571	607	340	1.1	1.2	.23		
OYSTER STEW, FRZN, MADE W/	190	97	6.5	6.1					646	125	110	.06	0.4	190
WATER 1 serving†		2.1							194		1.1	.15		
PEA GREEN, CND	225	146	24.8	2.3					844	22	108	.09	1.1	234
1 serving		6.8	0.5						146		1.6	.07	2	
PEA GREEN, CND, COND	319	339	53.9	6.4	134	1021	2201	265	2424	78	307	.03	3.2	657
1 can		16.6	4.8		861	829	255	638	415		4.1	.16	4	
PEA GREEN, CND, MADE W/ MILK	200	116	18.5	1.6	44	334	718	86	828	24	96	.01	1.0	206
1 serving†		6.9	0.5		282	270	82	208	130		1.3	.05	6	
PEA GREEN, CND, MADE W/ WATER	204	113	17.9	2.1	45	340	733	88	808	26	10	.01	1.1	219
1 serving†		5.5	1.6		287	276	85	212	312		1.4	.05		
PEA GREEN, DEHYD	42	137	21.2	3.2					1166	35	82	.05	0.3	27
1 pkg		6.4	1.1	0.4					361	65	0.7	.04	8	
PEA GREEN, DEHYD, MADE W/ WATER	214	34	53	0.8					696	12	20	.01	0.1	7
1 serving†		1.6	0.3						257		0.2	.01	2	
PEA GREEN, DEHYD, CUP-A-SOUP	35	127	21.8	1.4					694	13	86	.33	1.2	31
1 serving		6.9	0.3	0.1					271	47	1.5	.12		
PEA GREEN W/ HAM, FRZN, COND	298	298	42.3	5.7	149	1511	1380	158	2235	74	304	.45	3.0	551
1 can		19.4	4.2		745	924	1055	1358	599		4.8	.18		
PEA GREEN W/ HAM, FRZN, MADE W/	198	99	14.1	1.9	50	504	460	53	745	25	101	.15	1.0	184
WATER 1 serving†		6.5	1.4		248	308	352	451	198		1.6	.06		
PEA SPLIT, CND	240	172	22.1	4.7					912	42	153	.12	1.3	348
1 serving		10.1	0.7	0.1					37	11	1.7	.20	2	33
PEA SPLIT, CND, COND	305	357	50.6	7.9					2342	76	372	.61	3.4	1099
1 can		21.4	1.2						671		3.4	.37	3	
PEA SPLIT, CND, MADE W/ WATER	200	119	16.9	2.6	60	500	420	60	781	25	124	.20	1.2	360
1 serving†		7.1	0.4	0.0	1600	380	380	260	220	12	1.1	.12	0	
PEPPER POT, CND, COND	312	261	23.7	10.0	125	1145	1289	284	3120	55	107	.06	2.8	1491
1 can		19.0	0.9		636	484	895	683			2.5	.06		

*Soups are ready to eat unless otherwise indicated. Many canned soups do not require dilution.
†For canned, condensed soups which have been diluted w/ milk or water, a serving (unless otherwise indicated) is ⅓ of the can plus an equal volume of milk or water.

†Four servings come from 1 package.
§1 part oysters to 2 parts milk.
¶1 part oysters to 3 parts milk.

	WT	Proximate			Amino Acids				Minerals			Vitamins		
		CAL	CHO	FAT	TRY	LEU	LYS	MET	Na	Ca	P	THI	NIA	A
		PRO	FIB	PUFA	PHE	ISO	VAL	THR	K	Mg	Fe	RIB	ASC	D
	g	g	g	g	mg	mg	mg	mg	mg	mg	mg	mg	mg	iu
PEPPER POT, CND, MADE W/ WATER	203	87	7.9	3.3	42	381	430	95	1039	18	36	.02	0.9	496
1 serving†		6.3	0.3		212	161	298	228			0.8	.02		
POTATO, CND	200	212	29.0	7.4					296	106	106	.31	4.2	528
1 serving		7.4							380	tr	1.9	.36	32	42
POTATO, CREAM OF, FRZN, COND	291	262	29.4	12.5	64	698	393	113	2852	140	148	.12	1.2	1004
1 can		7.9	0.9	1.0	268	329	419	527	538		2.0	.15		
POTATO, CREAM OF, FRZN (W/ MILK) 1 serving	200	158	15.0	8.2					974	166	146	.08	0.5	506
		6.2	0.3	0.4					336		0.8	.22		
POTATO, CREAM OF, FRZN, MADE W/ WATER 1 serving	196	87	9.8	4.2	21	233	131	38	951	47	49	.04	0.4	335
		2.6	0.3	0.3	89	110	140	176	182		0.7	.05		
SCOTCH BROTH, HOMEMADE	120	62	9.8	0.7					714	13	47	.03	0.9	729
1 serving		4.3	0.2						89		0.7	.03	1	
SCOTCH BROTH, CND, COND	305	227	26.2	7.3	128	781	1251	207	3050	31	138	.03	4.3	2162
1 can		14.0	1.2		448	412	656	497			2.4	.09		
SCOTCH BROTH, CND, MADE W/ WATER 1 serving†	200	76	8.7	2.4	43	260	417	69	1017	10	46	.01	1.4	721
		4.7	0.4		149	137	219	166			0.8	.03		
SHRIMP, CREAM OF, FRZN, COND	284	377	19.6	28.1	74	795	716	227	2442	90	113	.11	0.9	264
1 can		11.4	0.9		443	452	488	920	136		1.1	.14		
SHRIMP, CREAM OF, FRZN, MADE W/ MILK 1 serving†	193	162	11.6	9.5	74	609	511	162	866	153	38	.08	0.4	88
		7.3	0.3		318	374	403	468	191		0.5	.23		
SHRIMP, CREAM OF, FRZN, MADE W/ WATER 1 serving†	190	125	6.6	9.4					817	30	38	.03	0.3	89
		3.8	0.3						46		0.4	.05		
SOUP BASE, BEEF, GOOD SEASON 1 T	15	32	10.3	0.7					263	15	3	.01	0.0	2
		3.4	tr						8		tr	.00	tr	
SOUP BASE, BEEF, MAGGI 1 T	15	24	7.2	0.6					3930	30	65			
		2.8							62					
SOUP BASE, BEEF, UNIVERSAL FOODS 1 T	15	54	3.4	3.8					2364					
		1.7	tr											
SOUP BASE, CHICKEN, GOOD SEASONS 1 T	15	31	2.9	1.4					3591	26	3	tr	tr	2
		1.7	tr						10		0.1	tr	tr	
SOUP BASE, CHICKEN, MAGGI 1 T	15	31	3.2	0.8					3525	42	50			
		2.7							86					
SOUP BASE, ONION, MAGGI 1 T	15	47	8.4	0.7					1395	27	42			
		1.7							59					
TOMATO, CND	230	104	16.1	2.3	25	129	345	147	771	14	32	.09	1.6	589
1 serving		2.3	0.5		170	152	113	124	242		1.8	.05	33	
TOMATO, CND, COND	298	218	36.4	6.0	33	167	447	191	2414	30	80	.18	3.6	1976
1 can		4.8	1.5	2.2	221	197	146	161	560		2.7	.09		
TOMATO, CND, MADE W/ WATER 1 serving†	198	73	12.1	2.0	11	56	149	64	805	10	27	.06	1.2	659
		1.6	0.5	0.8	74	66	49	54	208	14	0.9	.03	20	
TOMATO, DEHYD 1 pkg	64	222	44	4.3										
		2.8	1.5											
TOMATO, DEHYD, MADE W/ WATER 1 serving‡	185	56	11	1.1										
		0.7	0.4											
TOMATO, DEHYD, CUP-A-SOUP 1 serving	23	80	17.4	0.5					676	21	27	.05	0.7	782
		1.4	0.2						228	11	0.8	.05	11	
TOMATO RICE, CND 1 serving	200	86	14.7	2.4					662	16	26	.06	0.9	916
		1.6	0.2						260		0.8	.04		
TOMATO RICE, CND, COND 1 can	305	199	38.1	3.0					1830	50	82	.21	2.7	1220
		4.9									2.4	.12		
TOMATO RICE, CND, MADE W/ WATER 1 serving†	200	66	12.7	1.0					610	17	27	.07	0.9	407
		1.6									0.8	.04		
TOM VEG, DEHYD 1 pkg	74	258	43.9	5.7					4197	28	82	.27	2.1	3189
		7.8	1.4	0.4					469	36	1.3	.11	32	
TOM VEG, DEHYD, MADE W/ WATER 1 serving†	215	64	11.0	1.4						7	20	.07	0.5	797
		2.0	0.4								0.3	.03	8	
TOM VEG W/ NOODLES, DEHYD 1 pkg	74	258	46.4	5.9					4541	34	83	.22	1.9	1776
		6.4	1.1						128		1.5	.14	19	
TOM VEG W/ NOODLES, DEHYD, MADE W/ WATER 1 serving‡	200	54	10.2	1.2					854	6	16	.04	0.4	400
		1.2	0.2						24		0.2	.02	4	

*Soups are ready to eat unless otherwise indicated. Many canned soups do not require dilution.

†For canned, condensed soups which have been diluted w/ milk or water, a serving (unless otherwise indicated) is ⅓ of the can plus an equal volume of milk or water.

‡Four servings come from 1 package.

	WT	Proximate			Amino Acids				Minerals			Vitamins		
		CAL	CHO	FAT	TRY	LEU	LYS	MET	Na	Ca	P	THI	NIA	A
		PRO	FIB	PUFA	PHE	ISO	VAL	THR	K	Mg	Fe	RIB	ASC	D
	g	g	g	g	mg	mg	mg	mg	mg	mg	mg	mg	mg	iu
TURKEY NOODLE, CND	240	92	11.5	2.6						33	54	.05	1.4	3602
1 serving		5.2	0.5						306		0.9	.12	5	
TURKEY NOODLE, CND, COND	298	175	22.4	5.7	95	638	483	134	2264	26	115	.18	3.3	358
1 can		8.6	0.3	1.5	316	545	414	341	179		1.8	.18		
TURKEY NOODLE, CND, MADE W/ WATER 1 serving†	198	58	7.5	1.9	32	213	161	45	755	9	38	.06	1.1	119
		2.9	0.1	0.5	105	182	138	114	59		0.6	.06	4	
TURKEY RICE, CND	240	94	12.2	2.9					1094	160	56	.04	1.7	106
1 cup		5.2	0.5						122		0.9	.09	3	
TURKEY VEG, CND	240	94	10.3	2.6					976	23	54	.06	1.4	696
1 cup		5.6	0.2						122		0.7	.10	3	
TURTLE, CND, COND	298	128	20.9	1.5					3147	48	83			
1 can		7.4	0.4								2.7			
TURTLE, CND, MADE W/ WATER 1 serving†	198	43	7.0	0.5					1049	16	28			
		2.5	0.1								0.9			
VEGETABLE, CND	218	87	15.3	2.2					709	15	37	.04	1.1	1134
1 serving		2.2	0.4						185		0.9	.04	2	
VEGETABLE, CND, COND	305	193	31.4	4.0	58	448	522	88	1708	44	105	.09	3.0	7588
1 can		7.9	1.8	2.1	296	165	336	259	598		2.7	.06		
VEGETABLE, CND, MADE W/ WATER 1 serving†	200	64	10.5	1.3	19	149	174	29	569	15	35	.03	1.0	2529
		2.6	0.6	0.7	99	55	112	86	195		0.9	.02		
VEG BEAN, CND, COND	305	250	40.9	3.7					2538	95	187	.21	2.4	686
1 can		13.4	2.4								1.5	.15		
VEG BEAN, CND, MADE W/ WATER 1 serving†	200	83	13.6	1.2					846	32	62	.07	0.8	229
		4.5	0.8								0.5	.05		
VEG BEEF, CND	218	101	10.5	2.6					918	36	85	.05	1.9	1690
1 serving		8.3	0.7						403		1.3	.07	3	
VEG BEEF, CND, COND	312	192	18.4	5.0	128	1039	1426	221	2459	14	109	.09	2.5	7020
1 can		18.4	1.6		562	434	752	574	468	37	1.6	.13	5	
VEG BEEF, CND, MADE W/ WATER 1 serving†	203	64	6.1	1.7	43	346	475	74	820	5	36	.03	0.8	2340
		6.1	0.5	0.0	187	145	251	191	152	20	0.5	.04	4	
VEG BEEF, FRZN, COND	291	200	19.2	6.7	102	1327	861	218	2305	63	184	.12	4.4	6539
1 can		15.7	1.5	3.6	582	611	687	1053	422		2.3	.20		
VEG BEEF, FRZN, MADE W/ WATER 1 serving†	196	67	6.4	2.2	34	442	287	73	768	21	61	.04	1.5	2180
		5.2	0.5		194	204	229	351	143		0.8	.07		
VEG BEEF BROTH, CND, COND	305	186	30.8	4.9					2080	55	92	.15	3.1	3172
1 can		5.5	1.3						519	37	1.5	.12	6	
VEG BEEF BROTH, CND, MADE W/ WATER 1 serving†	200	62	10.3	1.6					673	18	31	.04	1.0	2600
		1.8	0.4						196	20	0.6	.02		
VEG BROTH, INSTANT 1 T	15	34	15.6	0.5										
		2.8												
VEG, CREAM OF, CND, COND	305	250	23.8	14.6					2708	130	134	.09	0.9	2949
1 can		5.8	2.4								1.2	.21		
VEG, CREAM OF, CND, MADE W/ MILK 1 serving†	200	83	7.9	4.9					903	43	45	.03	0.3	983
		1.9	0.8								0.4	.07		
VEG W/ BEEF STOCK, CND	240	136	11.1	2.3						33	39			
1 cup		8.0	0.6								0.9			
VEG W/ GROUND BEEF, CND	200	70	6.1	2.6						25	45			
1 serving		5.3	0.3								0.8			
VEGETARIAN VEG, CND	218	55	13.5	1.5					1032	47	62	.08	0.2	2991
1 serving		2.0	0.9						213		1.3	.05	7	
VEGETARIAN VEG, CND, COND	305	190	31.7	4.6	40	275	610	223	1708	45	94	.09	2.1	7070
1 can		5.5	1.5	2.1	235	229	195	183	610	37	2.4	.09	5	
VEGETARIAN VEG, CND, MADE W/ WATER 1 serving	200	63	10.6	1.5	13	92	203	74	569	15	31	.03	0.7	2357
		1.8	0.5	1.2	78	76	65	61	200	20	0.8	.03	6	

*Soups are ready to eat unless otherwise indicated. Many canned soups do not require dilution.

†For canned, condensed soups which have been diluted w/ milk or water, a serving (unless otherwise indicated) is ⅓ of the can plus an equal volume of milk or water.

WT	Proximate			Amino Acids				Minerals			Vitamins		
	CAL	CHO	FAT	TRY	LEU	LYS	MET	Na	Ca	P	THI	NIA	A
	PRO	FIB	PUFA	PHE	ISO	VAL	THR	K	Mg	Fe	RIB	ASC	D
g	g	g	g	mg	mg	mg	mg	mg	mg	mg	mg	mg	iu
ACCENT (MONOSODIUM GLUTAMATE) 5	14	0.0	0.0					62					
1 t	2.3												
ALLSPICE 2	5	1.4	0.2					2	13	2	tr	0.1	11
1 t	0.1	0.4	tr					21	3	0.1	tr	1	
ANISE SEED 2	7	1.0	0.3					tr	13	9			
1 t	0.4	0.3	0.1					29		0.7			
BASIL 1	3	0.6	tr	2	11	6	2	tr	21	5	tr	0.1	94
1 t	0.1	0.2		7	6	7	6	34	4	0.4	tr	1	
BAY LEAVES 1	3	0.7	0.1					tr	8	1	tr	tr	62
1 t	0.1	0.3	tr					5	1	0.4	tr	1	
CARAWAY SEEDS 2	7	1.0	0.3					tr	14	11	.01	0.1	7
1 t	0.4	0.3	0.1					27	5	0.3	.01	tr	
CARDAMON SEED 2	6	1.4	0.1					tr	8	4	tr	tr	0
1 t	0.2	0.2	tr					22	5	0.3	tr	tr	
CELERY SEED 2	8	0.8	0.5					3	35	11	.01		1
1 t	0.4	0.2	0.1					28	9	0.9	.01	tr	
CHERVIL, RAW* 50	28	5.7	0.4										
¼ cup	1.7											4	
CHOCOLATE, CHOCO-BAKE, NESTLÉ 28	168	11.9	13.8					3					
1 oz	1.4							286					
CHOC, BAKING, HERSHEY 28	185	6.7	15.5					1	20	112	.02	0.2	6
1 oz	4.0	0.7	1.0					221		2.0	.13	tr	
CHOC, BITTER OR BAKING 28	141	8.1	14.8					1	22	108	.01	0.4	17
1 oz	3.0	0.7						232	82	1.9	.07	0	0
CHOC NABS, UNSWEET BAKERS 28	134	8.4	14.1					3	22	108	.01	0.4	17
1 oz	3.1	0.8						255		1.8	.07	0	
CINNAMON 2	5	1.6	0.1					1	25	1	tr	tr	5
1 t	0.1	0.5	tr					10	1	0.8	tr	1	
CLOVES, GROUND 2	6	1.2	0.4					5	tr	2	tr	tr	11
1 t	0.1	0.2	0.0					22	5	0.2	.01	2	
CORIANDER LEAF, DRIED 1	3	0.5	0.1					2	12	5	.01	0.1	
1 t	0.2	0.1						45	7	0.4	.02	6	
CORIANDER SEED 2	6	1.1	0.4					1	14	8	.01	tr	0
1 t	0.2	0.6	tr					25	6	0.3	.01	tr	
CUMIN SEED 2	8	0.9	0.5					3	19	10	.01	0.1	25
1 t	0.4	0.2						36	7	1.3	.01	tr	
CURRY POWDER 2	7	1.2	0.3					1	10	7	.01	0.1	20
1 t	0.3	0.3						31		0.6	.01	tr	
DILL SEED 2	6	1.1	0.3		19	21	3	tr	30	6	.01	0.1	1
1 t	0.3	0.4	tr	13	15	22	12	24	5	0.3	.01	tr	
DILL WEED, DRIED 1	3	0.6	tr					2	18	5	tr	tr	
1 t	0.2	0.1						33	5	0.4	tr		
FENNEL SEED 2	7	1.0	0.3	5	20	15	6	2	24	10	.01	0.1	3
1 t	0.3	0.3	tr	13	14	18	12	34	8	0.4	.01	tr	
FENU GREEK SEED 4	13	2.3	0.3					3	7	12	.01	0.1	
1 t	0.9	0.4						31	8	1.3	.02	tr	
GARLIC CLOVE, RAW 3	4	0.9	tr					1	1	6	.01	tr	tr
1 clove	0.2							16		tr	tr	tr	
GARLIC POWDER 3	10	2.2	tr	7	31	17	10	1	2	13	.01	tr	0
1 t	0.5	0.1		15	19	21	14	33	2	0.1	tr	tr	
GINGER, GROUND 2	7	1.4	0.1	1	8	6	1	1	2	3	tr	0.1	3
1 t	0.2	0.1	tr	5	5	8	4	27	3	0.2	tr	tr	
GINGER ROOT, FRESH 28	14	2.7	0.3					2	6	10	.01	0.2	3
1 oz	0.4	0.3						74		0.6	.01	1	
GRAVYMASTER SEASONING & 10	2	0.4	tr					20	tr	tr	tr	0.0	0
BROWNING SCE	0.1	tr						1		tr	.08	1	
MACE, GROUND 2	10	1.0	0.6					2	5	2	.01	tr	16
1 t	0.1	0.1	0.1					9	3	0.3	.01		
MARJORAM, DRIED 1	3	0.6	0.1					1	20	3	tr	tr	81
1 t	0.1	0.2						15	4	0.8	tr	1	
MEATLOAF MIX, DELMARK 10	34	6.4	0.3					513	14		.02	0.3	7
	1.1	tr						25		0.3	.03	tr	

*An aromatic annual old world herb that is cultivated for its finely divided, often curled leaves, which are used especially in soups and salads.

	WT g	CAL / PRO g	CHO / FIB g	FAT / PUFA g	TRY / PHE mg	LEU / ISO mg	LYS / VAL mg	MET / THR mg	Na / K mg	Ca / Mg mg	P / Fe mg	THI / RIB mg	NIA / ASC mg	A D iu
MEATLOAF MIX W/ TVP,* DELMARK	10	33	5.1	0.3					390	13		.01	0.6	4
		2.5		tr					23		0.6	.02	tr	
MUSTARD, DRIED	2	12	0.4	0.9					tr	6	16	.01	0.2	4
1 t		0.6	tr						14	9	0.2	.01	tr	
NUTMEG	2	11	1.0	0.7					tr	4	4	.01	tr	2
1 t		0.1	0.1	tr					7	4	0.1	tr	tr	
OREGANO	2	6	1.3	0.2					tr	32	4	.01	0.1	138
1 t		0.2	0.3	0.1					33	5	0.9	tr	tr	
PARSLEY, DRIED	1	3	0.5	tr					5	15	4	tr	0.1	233
1 t		0.2	0.1						38	2	1.0	.01	1	
PEPPER, BLACK, GROUND	2	5	1.3	0.1					1	8	4	tr	tr	4
1 t		0.2	0.3						25	4	0.6	tr	tr	
PEPPER, RED	2	6	1.1	0.4					1	3	6	.01	0.2	832
1 t		0.2	0.5	0.2					40	3	0.1	.02	2	
PEPPER, WHITE	2	6	1.4	tr					tr	5	4	tr	tr	0
1 t		0.2	0.1						2	2	0.3	tr		
POPPY SEED	3	16	0.7	1.3	8	45	33	14	1	43	25	.02	.03	0
1 t		0.5	0.2	0.9	26	27	39	27	21	10	0.3	.01	tr	
POULTRY SEASONING	2	6	1.3	0.2					1	20	3	.01	0.1	53
1 t		0.2	0.2						14	5	0.7	tr	tr	
PUMPKIN PIE SPICE	2	7	1.4	0.3					1	14	2	tr	tr	5
1 t		0.1	0.3						13	3	0.4	tr		
ROSEMARY, GROUND	1	3	0.6	0.2					1	13	1	.01	tr	31
1 t		tr	0.2						10	2	0.3		1	
ROSEMARY LEAVES	1	4	0.7	0.2					tr	15	1	.01	tr	2
1 t		tr	0.2						10		tr	tr	1	
SAFFRON	1	3	0.7	0.1					2	1	3			
1 t		0.1	tr						17		0.1			
SAGE	1	3	0.6	0.1					tr	17	1	.01	0.1	59
1 t		0.1	0.2	tr					11	4	0.3	tr	tr	
SALT, TABLE†	5	tr	tr	0.0					1955	5	2	.00	0.0	0
1 t		0.0	0.0	0.0					tr	tr	tr	.00	0	0
SALT SUBSTITUTES														
ADOLPHS	5								1					
1 t									2378					
ADOLPHS, SEASONED	5	5	1.2	0.1					1	2	2			
1 t		0.1							1697		0.1			
DIAMOND	5	tr	tr	0.0					1	19	9	.00	0.0	0
1 t		tr		0.0					2575	1	0.0	.01	0	
FEATHERWEIGHT	5								tr					
1 t									2					
FEATHERWEIGHT, SEASONED	5								1					
1 t									2					
MORTON	5	tr	0.1	0.0					1	13	23	.00	0.0	0
1 t		0.0		0.0					2515	0		.00	0	
MORTON LITE SALT	5	tr	tr	0.0					975	5	32	.00	0.0	0
1 t		0.0		0.0					1300	4		.00	0	0
SAVORY	1	3	0.7	0.1					tr	21	1	tr	tr	51
1 t		0.1	0.2						11	4	0.4	tr	tr	
SLOPPY JOE MIX, DELMARK	10	19	1.1	1.1					54					
		1.2							19					
TACO MIX, DELMARK	10	23	0.6	1.4					68					
		2.0							37					
TARRAGON	1	3	0.5	0.1					1	11	3	tr	0.1	42
1 t		0.2	0.1						30	4	0.3	.01	tr	
THYME	1	3	0.6	0.1	2	4	2	3	1	19	2	.01	tr	38
1 t		0.1	0.2	tr	5	5	5	3	8	2	1.2	tr	tr	
TUMERIC	2	7	1.3	0.2					1	4	5	tr	0.1	0
1 t		0.2	0.1						50	4	0.8	.01	1	
VANILLA, DOUBLE STRENGTH	5	15	1.5	0.0					0	0		.00	0.0	0
1 t		0.0							0		0.0	.00	0	

*TVP = texturized vegetable protein.   †Iodized table salt contains about 385 mcg iodine/5 gm salt.

	WT	Proximate			Amino Acids				Minerals			Vitamins		
		CAL	CHO	FAT	TRY	LEU	LYS	MET	Na	Ca	P	THI	NIA	A
		PRO	FIB	PUFA	PHE	ISO	VAL	THR	K	Mg	Fe	RIB	ASC	D
	g	g	g	g	mg	mg	mg	mg	mg	mg	mg	mg	mg	iu
BUTTER														
	5	36	0.0	4.0	1	4	3	1	41	1	1	tr	tr	153
1 t		0.0	0.0	0.2	2	3	3	2	1	tr	tr	tr	0	
	15	108	tr	12.2	2	13	10	3	124	4	4	tr	tr	459
1 T		0.1	0.0	0.5	6	8	9	6	4	tr	tr	.01	0	
SWEET	5	36	tr	4.1	tr	3	2	1	tr	1	1			165
1 t		tr	0.0	0.1	2	2	2	1	1	tr	0.0		0	
SWEET	15	107	0.6	12.2	1	9	7	2	1	3	2			495
1 T		0.9	0.0	0.3	4	6	6	4	4	tr	0.0		0	
WHIPPED	3.8	27	tr	3.1	tr	3	3	1	31	1	1	tr	tr	116
1 t		tr	0.0	0.1	2	2	2	1	1	tr	tr	tr	0	
WHIPPED	14	81	tr	9.3	tr	9	9	3	93	3	3	tr	tr	348
1 T		tr	0.0		6	6	6	3	3	tr	tr	tr	0	
CREAM														
COCONUT	14	48	1.2	4.5					1	2	18	tr	0.1	0
1 T		0.6		0.3					45		0.3	tr	tr	
HALF & HALF	15	20	0.6	1.7	6	43	35	11	6	16	14	.01	tr	65
1 T		0.4	0.0	0.1	21	27	30	20	19	2	tr	.02	tr	
LIGHT (COFFEE/TABLE)	15	29	0.6	2.9	6	40	32	10	6	14	12	.01	tr	108
1 T		0.4	0.0	0.1	20	25	27	18	18	1	tr	.02	tr	
MED (25% FAT)	15	37	0.5	3.8	5	36	29	9	6	14	11	tr	tr	141
1 T		0.4	0.0	0.1	18	22	25	17	17	1	tr	.02	tr	
SOUR, CULTURED	12	26	0.5	2.5	4	32	25	8	6	14	10	tr	tr	95
1 T		0.4	0.0	0.1	15	20	22	15	17	1	tr	.02	tr	1
SOUR, HALF & HALF, CULTURED	15	20	0.6	1.8					6	16	14	.01	tr	68
1 T		0.4	0.0	0.1					19	2	tr	.02		
SOUR, IMITATION (NONDAIRY)	15	31	1.0	2.9					15	tr	7	.00	0	0
1 T		0.4	0.0	tr					24			.00	0	
WHIPPING, HEAVY, FLUID	15	52	0.4	5.6	4	30	24	8	6	10	9	tr	tr	220
1 T		0.3	0.0	0.2	15	19	21	14	11	1	tr	.02	tr	15
WHIPPING, LIGHT, FLUID	15	44	0.4	4.6	5	32	26	8	5	10	9	tr	tr	169
1 T		0.3	0.0	0.1	16	20	22	15	15	1	tr	.02	tr	8
CREAM SUBSTITUTES														
COFFEE-MATE CARNATION	2	11	1.1	0.7					4	tr	6			
1 t		0.1							14					
LIQUID, FRZN W/ HYDROGENATED OIL* & SOY PROTEIN ½ fl oz	15	20	1.7	1.5	2	13	10	2	12	1	10	.00	0.0	13
		0.2	0.0	tr	8	8	8	7	29	tr	tr	.00	0	
LIQUID, FRZN W/ LAURIC ACID OIL† & SODIUM CASEINATE ½ fl oz	15	20	1.7	1.5	2	15	12	5	12	1	10	.00	0.0	13
		0.2	0.0	tr	8	9	11	6	29	tr	tr	.00	0	
POWDERED	2	11	1.1	0.7	1	9	8	3	4	tr	8	.00	0.0	4
1 t		0.1	0.0	PUFA	5	6	7	4	16	tr	tr	.00	0	
PET	2	10	1.0	1.0										
1 t		0.0												
PREAM	3	7	1.7											
1 t		0.2												
POLYPERX	5	7	0.6	0.5					1	0	2			
1 t		tr	0.0	0.2					5	0	0.0			
POLYRICH	14	22	2.1	1.4					3	tr	5			
½ oz		tr		0.7					11	tr	tr			
POLYRICH	112	172	17.0	11.4					25	2	43			
4 oz		0.3		5.3					85	1	0.2			
DIPS														
BEAN, FRITO-LAY	28	33	2.9	1.4					164	10	36	.02	tr	17
1 oz		1.6	0.9						87	6	0.4	.01	0	
ENCHILADA, FRITO-LAY	28	37	3.8	1.5					95	17	3	.01	0.3	113
1 oz		1.9	0.2						100	15	1.0	.04	0	
FRENCH ONION, SEALTEST	15	24	1.1	2.0					84	18	14	tr	0.1	78
1 T		0.5							26	2	tr	.02	0	1

*Usually soybean, cottonseed, safflower or blends thereof.   †Lauric oils include modified coconut oil, hydrogenated coconut oil and/or palm kernel oil.

	WT	Proximate			Amino Acids				Minerals			Vitamins		
		CAL	CHO	FAT	TRY	LEU	LYS	MET	Na	Ca	P	THI	NIA	A
		PRO	FIB	PUFA	PHE	ISO	VAL	THR	K	Mg	Fe	RIB	ASC	D
	g	g	g	g	mg	mg	mg	mg	mg	mg	mg	mg	mg	iu
SOUR CREAM, BORDEN	15	28	1.1	2.7						21	18			
1 T		0.4										.03		
GRAVY														
BROWN, HOMEMADE	72	164	8.0	14.0	14	94	29	14	720	0	8	.04	0.0	0
4 T		1.2	0.0	6.0	65	58	50	36	76	1	0.4	.03	0	
BROWN KNORR-SWISS	60	25	2.8	1.0					441	1	19			
4 T		0.8	tr						70		0.2			
BROWN MUSHROOM, DAWN	60	18	3.6	0.6					313	4	9	.01	0.1	8
FRESH 4 T		0.6	tr						21		0.2	.04	3	
CHICKEN	28	64	3.1	5.4					56	0	3	.01	tr	43
1 oz		0.5	0.0	0.3					2	1	0.1	.18	tr	
HAMBURGER	28	46	0.4	3.6					28	2	25	.01	0.7	33
1 oz		3.2	tr	1.6					57	3	0.4	.03	tr	0
ROAST BEEF AU JUS, KNORR-SWISS	60	6	0.1	0.4					250		7			
4 T		0.4	0.0						20		0.1			
SALAD DRESSING														
BLUE/ROQUEFORT	14	71	1.0	7.3					153	11	10	tr	0.9	29
1 T		0.7	tr						5	1	tr	.01	tr	
BLUE/ROQUEFORT, LOW-CAL	14	11	0.6	0.8					155	9	7		tr	24
1 T		0.4	tr						5		tr	.01	tr	
BLUE CHEESE W/ YOGURT, HENRI	14	24	3.1	1.1					188	12	10	tr	tr	17
1 T		0.4	tr	0.5					21		tr	.02	tr	
CAESAR, SEVEN SEAS	15	70	1.0	7.0										
1 T		0.0												
COLESLAW, LOW-CAL, MILANI	15	31	0.0	3.4					163					
1 T		0.1												
FRENCH	14	57	2.4	5.4					192	2	2			
1 T		0.1	tr						11	1	0.1			
FRENCH, HOMEMADE	14	88	0.5	9.8					92	1	tr			
1 T		tr							4		tr			
FRENCH, LOW-CAL	14	22	0.2	2.4					110	2	2			
1 T		0.1	tr						11	1	tr			
FRENCH, CREAMY, GOOD SEASONS	14	70	2.3	6.9					125	1	1			3
1 T		0.1	1.0						3		0.0			
FRENCH W/ YOGURT, HENRI	14	21	3.4	0.8					108	4	4	tr	0.1	39
1 T		0.2	tr	0.5					23		0.1	.01	tr	
GREEN GODDESS	14	68	1.2	7.0					150	2	1			
1 T		0.1		4.0					9					
GREEN GODDESS, LOW-CAL	14	27	2.2	2.0					57					
1 T		0.1												
HOMEMADE, CKD	15	25	2.3	1.5					109	13	14	.01	tr	74
1 T		0.7	0.0						17		0.1	.02		
ITALIAN	14	77	1.0	8.4					293	2	1	tr	tr	tr
1 T		tr	tr	4.8					2	1	tr	tr	0	
ITALIAN, LOW-CAL	14	7	0.4	0.7					110	tr	1			
1 T		tr							2		tr			
ITALIAN, CREAMY, HENRI	14	52	2.7	4.5					105	0				
1 T		0.1	tr											
ITALIAN W/ YOGURT, HENRI	14	20	2.9	0.8					178	6	6	tr	tr	17
1 T		0.2	tr						20		tr	.01	tr	
MAYONNAISE TYPE	14	61	2.0	5.9					82	2	4	tr	tr	31
1 T		0.1	tr						1	tr	tr	tr	tr	
MAYONNAISE TYPE, LOW-CAL	14	19	0.7	1.8					17	3	4	tr	tr	31
1 T		0.2	0.1						1		tr	tr		
MIRACLE WHIP, KRAFT	14	68	1.8	6.9					89	2	3			
1 T		0.1		4.1					2					
OIL & VINEGAR, KRAFT	15	69	0.6	7.5					244	2	1			
1 T		tr		4.7					4					
RANCH STYLE PREPARED W/MAYON	15	54	0.6	5.7					97					
HIDDEN VALLEY 1 T		0.4												
RUSSIAN	15	74	1.6	7.6					130	3	6	.01	0.1	104
1 T		0.2	0.1	1.8					24		0.1	.01	1	

	WT	Proximate			Amino Acids				Minerals			Vitamins		
		CAL	CHO	FAT	TRY	LEU	LYS	MET	Na	Ca	P	THI	NIA	A
		PRO	FIB	PUFA	PHE	ISO	VAL	THR	K	Mg	Fe	RIB	ASC	D
	g	g	g	g	mg	mg	mg	mg	mg	mg	mg	mg	mg	iu
RUSSIAN, LOW-CAL	14	27	2.7	1.9					85					
1 T		0.7												
SMOKY BITS, HENRI	14	53	tr	4.0					77	0				
1 T		0.2												
SPIN BLEND, HELLMANNS	16	56	2.8	5.3					108			.00	0.0	
1 T		0.1	0.0	2.8								.00		
SOYAMAISE, CELLU	14	100	0.0	11.0					6	2	6			
1 T		0.0							26					
SWEET 'N SOUR, KRAFT	15	29	6.9	0.3					68	1	1			
1 T		0.2							14					
TAS-TEE, HENRI	14	53	3.3	4.3					83	0				
1 T		0.1	0.5											
THOUSAND ISLAND	14	70	2.2	7.0					98	2	2	tr	tr	45
1 T		0.1	tr						16		0.1	tr	tr	
THOUSAND ISLAND, LOW-CAL	14	25	2.2	1.9					98	2	2	tr	tr	45
1 T		0.1	tr						16		0.1	tr	tr	
THOUSAND ISLAND W/ YOGURT, HENRI 1 T	14	23	3.8	0.8					125	5	6	tr	0.1	22
		0.2	tr						26		0.1	.01	tr	
SAND SPREAD W/ CHOPPED PICKLE	20	76	3.2	7.2					125	3	4	tr	0.3	56
1 T		0.1	0.1	3.9					18	2	0.1	.01	1	8
SAND SPREAD W/CHOP PICKLE, LOW-CAL 1 T	20	22	1.6	1.8					125	3	4	tr		56
		0.2	0.1						18		0.1	.01	1	
SPREADS														
MARGARINE*	5	36	tr	4.1	1	4	3	1	49	1	1	.00	0.0	165
1 t		tr	0.0	0.7	2	3	3	2	1	tr	0.0	.00	0	
	15	108	tr	12.2	2	11	9	3	148	3	2	.00	0.0	495
1 T		0.1	0.0	2.1	6	8	8	6	3	tr	0.0	.00	0	
unsalted*	5	36	tr	4.1	1	4	3	1	1	1	1	.00	0.0	165
1 t		tr	0.0	0.7	2	3	3	2	1	tr	0.0	.00	0	
	15	108	tr	12.2	2	11	9	3	2	3	2	.00	0.0	495
1 T		0.1	0.0	2.1	6	8	8	6	3	tr	0.0	.00	0	
whip*	4	29	0.0	3.2						1				132
1 t		tr									0.0		0	
	12	86	0.0	9.6						2				395
1 T		0.2									0.0		0	
low-cal*	5	16	0.0	1.7					49	tr	-			165
1 t		0.0							1					
	14	45	0.0	4.8					136	1				462
1 T		0.0							2					
MAYONNAISE	14	101	0.3	11.2					84	3	4	tr	tr	39
1 T		0.2	tr						5	tr	tr	.01		
MAYONNAISE, IMITATION, LADY LEE 1 T	14	50	0.0	5.0					100					
		0.0		1.0										

*See Table V for information on specific brands; PUFA are quite variable.

	WT	CAL	CHO	FAT	TRY	LEU	LYS	MET	Na	Ca	P	THI	NIA	A
	g	**PRO** g	**FIB** g	**PUFA** g	**PHE** mg	**ISO** mg	**VAL** mg	**THR** mg	**K** mg	**Mg** mg	**Fe** mg	**RIB** mg	**ASC** mg	**D** iu
Cereals, ready-to-serve														
BARLEY	14	54	10.5	0.6					1	94	107	.40	2.0	0
6 T		1.6	0.2		252				44		14.2	.30		
HIGH PROTEIN	14	53	6.2	0.9					1	94	107	.40	2.0	
6 T		5.0	0.3		227				212		14.2	.30		
MIXED	14	54	10.3	0.6					1	94	99	.40	2.0	
6 T		1.7	0.2		57				40		14.2	.30		
MIXED W/HONEY	18	70	12.6	0.9					18	191		.36	5.8	0
6 T		2.4	tr						67		10.6	.43		0
OATMEAL	14	56	9.3	1.1					1	94	105	.40	2.0	0
6 T		2.1	0.2		105				51		14.2	.30		
OATMEAL W/ HONEY	18	72	12.7	1.4					18	191		.36	5.8	0
6 T		2.3	0.1						55		10.8	.43		0
RICE	14	54	10.8	0.7					1	94	118	.40	2.0	0
6 T		1.0	0.1		37				45		14.2	.30		
RICE W/ HONEY	18	68	14.4	0.6					17	191		.36	5.8	0
6 T		1.5	tr						13		10.6	.43		0
WHEAT W/ HONEY	18	67	13.9	0.2					18	238		.36	5.8	0
6 T		2.0	tr						15		10.6	.43		0
Cereals w/Strained Fruit														
HIGH PROTEIN W/ APPLESAUCE &	134	125	23.5	0.7					162	40	111	.25	4.0	
BANANAS 1 jar		6.2	0.4						243		10.0	.60	26	
MIXED W/APPLESCE & BANANAS	134	111	24.0	0.9					134	5	74	.25	4.0	2
1 jar		1.6	0.4						33		5.0	.60	5	
OATMEAL W/ APPLESCE & BANANAS	134	99	20.2	1.2					196	8	99	.25	4.0	27
1 jar		1.9	0.5						44		5.0	.60	5	
RICE W/ APPLESCE & BANANAS	134	93	21.0	0.8					163	11	16	.25	4.0	2
1 jar		0.5	0.3						13		0.24	.60	26	
RICE W/ STRAWBERRIES	18	70	14.1	0.9					14	119	130	.51	2.5	
6 T dry		1.3	0.2						64		18.0	.38		
Baked Goods														
COOKIES, ANIMAL SHAPED	7	29	4.2	1.0					28	8	12	.08	0.8	
1 cookie		0.8	0.1		35				24		0.1	.13		
COOKIES, ARROWROOT	10	46	7.4	1.4					41	3	13	.05	0.6	
		0.8	tr						14		tr	.04	1	
PRETZEL	5	19	3.9	0.1					30	2	7	.01	tr	
1 avg		0.7							3		0.1			
TEETHING BISCUIT	11	43	81	0.6					60	53	49	.05	0.2	
1 avg		1.3	0.1						34		0.2	.04	0	
Strained Juices														
APPLE	131	65	15.9	0.1					3	4	7	tr	0.1	
1 can†		0.2	0.1						96		0.8	tr	52	
APPLE-CHERRY	131	59	14.1	0.1					2	6	8	.01	0.1	
1 can†		0.3	0.1						104		1.2	.01	52	
APPLE-GRAPE	131	85	21.0						4	7	10	.05	0.1	
1 can†		0.3	0.1						148		1.5	.05	52	
MIXED FRUIT	131	77	18.2	0.3					3	8	9	.03	0.1	9
1 can†		0.3	0.1						131		0.5	.01	52	
ORANGE	131	65	14.4	0.5					3	8	12	.05	0.2	39
1 can†		0.7	0.1						213		0.3	.02	52	
ORANGE-APPLE	131	71	16.2	0.5					3	7	11	.03	0.1	31
1 can†		0.4	0.1						170		0.6	.01	52	
ORANGE-APPLE-BANANA	131	86	20.1	0.3					4	10	20	.03	0.2	
1 can†		0.6	0.1						210		1.1	.02		
ORANGE-APRICOT	131	79	18.6	0.2					2	10	16	.07	0.2	196
1 can†		0.8	0.2						271		0.5	.03	52	
ORANGE-PINEAPPLE	131	78	18.4	0.2					2	9	13	.08	0.2	15
1 can†		0.7	0.2						186		0.3	.01	52	

*In addition to being used in infant meals, strained/junior & toddler foods may also be used for older children and adults on soft diets.

†One can of baby juice = 4.2 fl oz.

	WT	Proximate			Amino Acids				Minerals			Vitamins		
		CAL	CHO	FAT	TRY	LEU	LYS	MET	Na	Ca	P	THI	NIA	A
		PRO	FIB	PUFA	PHE	ISO	VAL	THR	K	Mg	Fe	RIB	ASC	D
	g	g	g	g	mg	mg	mg	mg	mg	mg	mg	mg	mg	iu
PINEAPPLE-GRAPEFRUIT	131	76	18.2	0.1					3	9	5	.03	0.1	3
1 can†		0.4	0.1		10				89		0.5	.01	52	
PRUNE-ORANGE	131	99	23.5	0.2					5	14	21	.05	0.4	22
1 can†		0.7	0.1						214		0.6	.01	52	
Strained Fruits														
APPLES & APRICOTS	134	99	24.8	0.0					3	19		.01	0.2	429
1 jar		0.0	0.5						146		0.4	.05	16	
APPLE BLUEBERRY	134	91	22.0	0.1					4	5	15	.03	0.2	
1 jar		0.3	0.8						106		0.3	.03	16	
APPLE RASPBERRY	134	92	22.1	0.1					4	5	13	.03	0.2	
1 jar		0.3	0.1						119		0.3	.03	16	
APPLESAUCE	134	109	26.5	0.2					2	4	8	.01	tr	21
1 jar		0.2	0.9		13				84		0.5	.01	40	
APPLESCE & APRICOTS	134	116	28.2	0.2					2	6	10	.01	0.2	328
1 jar		0.4	1.0		19				113		0.4	.01	17	
APPLESCE & CHERRIES	134	90	21.8	0.0					3	12		.03	0.0	54
1 jar		0.0	0.4						120		0.5	.07	16	
APPLESCE & PINEAPPLE	134	106	25.6	0.3					2	5	8	.04	0.1	19
1 jar		0.3	0.8						99		0.4	.01	17	
APPLESCE & RASPBERRIES	134	96	23.9	0.0					3	0		.01	0.0	0
1 jar		0.0	0.4						94		0.3	.07	16	
APRICOTS W/ TAPIOCA	134	107	26.0	0.1					42	9	13	tr	0.3	912
1 jar		0.5	0.4						143		0.4	.01	17	
BANANAS	134	110	26.3	0.3					8	13	9	.01	0.3	255
1 jar		0.5	0.1		12				142		0.3	.03	38	
BANANAS & PINEAPPLE	134	136	24.7	0.0					39	5	7	.03	0.2	58
1 jar		0.4							95		0.3	.03	16	
BANANAS & PINEAPPLE W/ TAPIOCA	134	84	20.9	0.0					31	19		.01	0.2	40
1 jar		0.0	0.1						110		0.4	.05	16	
BANANAS W/ TAPIOCA	134	117	28.5	0.1					39	5	12	.02	0.3	72
1 jar		0.6	0.3		33				156		0.5	.02	17	
PEACHES	134	107	25.8	0.1					3	4	15	.01	0.8	281
1 jar		0.8	1.1		33				168		0.5	.02	17	
PEARS	134	92	22.3	0.1					3	11	12	.01	0.2	
1 jar		0.5	2.0		19				117		0.3	.02	17	
PEARS & PINEAPPLE	134	95	22.9	0.2					3	12	13	.03	0.2	20
1 jar		0.5	1.7		7				129		0.5	.03	17	
PLUMS W/ TAPIOCA	134	132	32.2	0.2					6	8	10	.01	0.3	141
1 jar		0.4	0.5						94		0.3	.01	3	
PRUNES W/ TAPIOCA	134	118	28.3	0.2					26	19	23	.01	0.5	280
1 jar		0.8	1.0		17				205		1.02	.01	6	
Junior Fruits														
APPLES & CRANBERRIES	220	191	46.6	0.2					13	20	4	.01	0.0	0
1 jar		0.4	0.9		5				44		0.7	.01	3	
APPLESAUCE	220	179	43.1	0.5					4	7	12	.02	0.1	4
1 jar		0.4	1.5		8				135		0.4	.04	66	
APPLESCE & APRICOTS	220	192	45.7	0.7					3	9	14	.03	0.3	537
1 jar		0.7	1.3		11				180		0.7	.04	20	
APPLESCE & PINEAPPLE	220	162	39.4	0.3					4	9	13	.05	0.1	33
1 jar		0.5	2.3						153		0.6	.03	20	
APRICOTS W/ TAPIOCA	220	178	43.4	0.1					70	14	19	.01	0.5	641
1 jar		0.8	0.7		7				243		0.6	.02	20	
BANANAS & PINEAPPLE	220	167	40.5	0.2					176	9	11	.02	0.2	0
1 jar		0.7	0.7		9				119		0.2	.02	16	
BANANAS, PINEAPPLE & TAPIOCA	220	181	44.0	0.3					214	10	11	.02	0.2	67
1 jar		0.6	0.6						153		0.4	.04	20	
PEACHES	220	181	42.4	0.7					12	8	21	.02	1.4	332
1 jar		1.3	2.0		24				266		0.4	.07	20	

*In addition to being used in infant meals, strained/junior & toddler foods may also be used for older children and adults on soft diets.

† One can of baby juice = 4.2 fl. oz.

	WT	Proximate			Amino Acids				Minerals			Vitamins		
		CAL / PRO	CHO / FIB	FAT / PUFA	TRY / PHE	LEU / ISO	LYS / VAL	MET / THR	Na / K	Ca / Mg	P / Fe	THI / RIB	NIA / ASC	A / D
	g	g	g	g	mg	mg	mg	mg	mg	mg	mg	mg	mg	iu
PEARS	220	152	36.4	0.4					4	18	20	.02	0.4	7
1 jar		0.7	3.0		13				199		0.5	.04	20	
PEARS & PINEAPPLE	220	156	36.8	0.6					4	20	17	.05	0.3	21
1 jar		0.8	2.7		12				206		0.5	.04	20	
PLUMS W/ TAPIOCA	220	218	53.0	0.4					12	13	15	.02	0.5	171
1 jar		0.7	0.8		31				160		0.9	.04	3	
PRUNES W/ TAPIOCA	220	201	47.9	0.5					44	32	34	.03	0.9	418
1 jar		1.3	1.5		31				348		1.15	.14	7	
Strained Dinners														
BEEF & EGG NOODLES	128	61	9.2	1.1					163	9	37	.07	1.0	860
1 jar		3.6	0.4		93				73		0.8	.05	2	
BEEF W/ VEGETABLES	128	92	6.0	3.3					338	18	106	.16	2.9	768
1 jar		9.4	0.3								1.4	.25	3	
CEREAL, EGG YOLKS & BACON	128	92	8.9	5.0					172	29	54	.04	0.2	7
1 jar		2.8	0.1		128				46		0.6	.09	1	
CHICKEN A LA KING	128	125	6.4	7.2					307	210	148	.02	2.6	0
1 jar		8.8	0.1								0.4	.08		
CHICKEN NOODLE	128	60	10.0	.9					147	30	48	.06	0.6	658
1 jar		2.9	0.4		120				78		0.6	.05	2	
CHICKEN W/ RICE	128	119	8.1	6.0					295	202	150	.01	2.5	0
1 jar		8.2	0.3								0.1	.09		
CHICKEN W/ VEGETABLES	128	152	7.7	9.3					319	191	169	.26	3.8	691
1 jar		9.2	0.4								1.3	.31	2	
CORN GRITS W/ EGG YOLK	128	86	10.6	3.6					157	37	63	.05	0.4	156
1 jar		2.4	0.1						88		0.4	.07	tr	
HAM W/ VEGETABLES	128	152	11.4	8.3					307	23	108	.24	2.2	102
1 jar		7.7	0.3								0.9	.17		
HIGH MEAT BEEF	128	125	8.0	7.3					38	27		.03	1.7	1331
1 jar		7.3	0.4						195		0.9	.10	5	
HIGH MEAT BEEF W/ VEG	128	104	7.6	4.7					194	8	66	.01	1.8	583
1 jar		7.8	0.4		416				133		1.2	.08	3	
HIGH MEAT CHICKEN	128	92	9.1	4.1					18	51		.00	0.9	1216
1 jar		5.0	0.3						60		0.7	.06	2	
HIGH MEAT CHICKEN W/ VEG	128	110	7.5	5.3					172	47	73	.01	1.2	569
1 jar		8.0	0.2		403				74		2.1	.07	3	
HIGH MEAT COTTAGE CHEESE W/ VEG 1 jar	134	174	22.1	6.0					195	86	116	.07	0.1	22
		8.0	1.2						125		0.4	.20	3	
HIGH MEAT HAM	128	131	9.1	7.7					29	24		.13	1.6	294
1 jar		7.0	0.1						140		0.7	.12	3	
HIGH MEAT HAM W/ VEG	128	101	8.2	3.8					224	8	85	14	2.6	50
1 jar		8.4	0.3		320				173		0.9	.06	3	
HIGH MEAT TURKEY	128	108	9.1	5.4					35	54		.01	2.2	973
1 jar		6.3	0.1						120		1.0	.13	3	
HIGH MEAT TURKEY W/ VEG	128	96	7.6	4.1					195	17	54	.01	1.4	28
1 jar		7.1	0.2		294				84		1.0	.08	3	
HIGH MEAT VEAL	128	101	8.1	4.6					32	24		.03	2.0	922
1 jar		7.3	0.3						165		0.8	.10	4	
HIGH MEAT VEAL W/ VEG	128	82	8.3	1.8					161	8	77	.03	2.1	57
1 jar		8.1	0.3		333				162		1.3	.08	3	
MACARONI & CHEESE	128	82	10.4	2.8					247	69	115	.06	0.6	35
1 jar		3.5							44		0.5	.09	2	
MACARONI, TOM, BEEF & BACON	128	78	11.3	2.4					137	22	43	.07	0.8	448
1 jar		2.7	0.6		141				101		0.7	.06	3	
MACARONI, TOM & BEEF	128	74	10.8	2.0					18	35		.05	1.0	307
1 jar		2.9	0.3						120		0.7	.08	3	
MACARONI, TOM, MEAT & CEREAL	128	86	12.3	2.6					488	27	45	.18	1.3	640
1 jar		3.3	0.4						99		0.6	.15	1	
SPLIT PEAS, VEG & HAM/BACON	128	102	14.3	2.7					378	37	101	.10	0.6	768
1 jar		5.1							143		0.9	.06	1	

*In addition to being used in infant meals, strained/junior & toddler foods may also be used for older children and adults on soft diets.

	WT	Proximate			Amino Acids				Minerals			Vitamins		
		CAL	CHO	FAT	TRY	LEU	LYS	MET	Na	Ca	P	THI	NIA	A
		PRO	FIB	PUFA	PHE	ISO	VAL	THR	K	Mg	Fe	RIB	ASC	D
	g	g	g	g	mg	mg	mg	mg	mg	mg	mg	mg	mg	iu
TUNA, CREAM OF	128	93	11.5	3.1					242	51	69	.01	1.6	627
1 jar		4.7	0.5								0.1	.04		
TURKEY RICE	128	59	9.5	1.2					129	16	25	.01	0.3	635
1 jar		2.6	0.2						52		0.4	.03	2	
TURKEY & VEGETABLES	128	122	8.2	6.3					417	59	81	.02	1.8	243
1 jar		7.9	0.3						156		0.4	.10	3	
VEAL & VEGETABLES	128	92	5.6	3.2					438	19	104	.19	3.4	947
1 jar		10.0	0.4						122		1.3	.25	3	
VEGETABLES & BACON	128	94	12.5	3.9					183	13	39	.05	0.7	2773
1 jar		2.1	0.5		76				145		0.4	.03	3	
VEG, BACON & CEREAL	128	87	11.1	3.7					361	22	36	.09	0.8	2816
1 jar		2.2	0.5						166		0.8	.06	1	
VEGETABLES & BEEF	128	67	8.4	2.6					173	8	28	.03	0.6	1285
1 jar		2.5	0.2		154				77		0.5	.03	2	
VEG & BEEF W/ CEREAL	128	72	9.7	2.0					393	22	50	.04	1.2	3584
1 jar		3.5	0.5						183		1.0	.05	1	
VEGETABLES & CHICKEN	128	53	8.8	0.8					137	17	32	.03	0.2	997
1 jar		2.6	0.2						33		0.4	.03	3	
VEG & CHICKEN W/ CEREAL	128	67	9.9	1.8					393	42	42	.04	0.6	1280
1 jar		2.7	0.3						70		0.5	.05		
VEG, HAM W/ BACON	128	70	9.5	2.6					298	7	29	.08	0.5	491
1 jar		2.1	0.3						50		0.6	.03	2	
VEG & HAM W/ CEREAL	128	81.9	10.6	2.8					461	32	54	.10	0.6	1280
1 jar		3.6	0.4						115		0.4	.06	4	
VEGETABLES & LAMB	128	64	9.5	1.7					184	9	29	.02	0.5	1824
1 jar		2.6	0.4		111				98		0.4	.03	2	
VEG & LAMB W/ CEREAL	128	74	9.9	2.6					344	29	47	.04	0.9	2816
1 jar		2.8	0.4						189		0.9	.06	1	
VEG & LIVER W/ BACON	128	77	8.2	3.6					262	9	51	.03	1.5	1344
1 jar		2.9	0.2						100		2.73	.32	4	
VEG & LIVER W/ BACON & CEREAL	128	73	9.6	2.4					364	14	54	.04	1.7	5888
1 jar		3.1	0.4						168		3.3	.42	3	
VEG & LIVER W/ CEREAL	128	60	10.0	0.5					302	22	73	.05	2.0	6016
1 jar		4.0	0.4						207		3.5	.47	4	
VEGETABLES & TURKEY	128	57	10.7	0.5					157	11	24	.01	0.4	640
1 jar		2.3	0.3						57		0.5	.02	3	
VEG & TURKEY W/ CEREAL	128	56	9.2	1.0					393	28	33	.01	0.5	512
1 jar		2.7	0.3						59		0.4	.04	1	
Junior Dinners														
BEEF & EGG NOODLES	212	107	16.3	2.1					222	13	64	.05	1.4	607
1 jar		5.8	.4		176				137		1.3	.10	5	
CEREAL, EGG YOLKS & BACON	212	158	14.9	8.8					284	56	97	.07	0.3	16
1 jar		4.7	0.2		254				75		.9	.16	3	
CHICKEN NOODLE	212	93	16.6	1.0					320	24	51	.05	0.9	1019
1 jar		4.3	0.4		163				55		.9	.06	4	
HIGH MEAT BEEF W/ VEG	212	174	9.1	9.1					606	36	155	.17	3.4	2311
1 jar		14.0	0.6		466				212		2.1	.19	3	
HIGH MEAT CHICKEN W/ VEG	212	201	8.3	12.1					606	144	214	.42	5.9	1113
1 jar		15.0	0.6		530				403		1.3	.45	2	
HIGH MEAT HAM W/ VEG	212	233	19.7	11.2					536	59	212	.38	4.0	1071
1 jar		13.4	0.6		488				375		0.6	.34	3	
HIGH MEAT TURKEY W/ VEG	212	191	11.2	10.2					534	163	106	.02	1.5	1200
1 jar		13.1	0.4						121		1.5	.15	3	
HIGH MEAT VEAL W/ VEG	212	172	15.5	5.9					479	11	144	.04	3.3	778
1 jar		13.6	0.6						271		1.1	.15	4	
MACARONI, TOM, BEEF & BACON	212	132	21.3	3.0					265	33	61	.09	1.5	648
1 jar		5.0	0.6						173		1.0	.11	4	
SPAGHETTI, TOM SCE & BEEF	121	145	26.4	1.6					483	33	78	.19	2.4	718
1 jar		6.2	0.8		201				249		1.3	.18	6	

*In addition to being used in infant meals, strained/junior & toddler foods may also be used for older children and adults on soft diets.

	WT	CAL	CHO	FAT	TRY	LEU	LYS	MET	Na	Ca	P	THI	NIA	A
		PRO	FIB	PUFA	PHE	ISO	VAL	THR	K	Mg	Fe	RIB	ASC	D
	g	g	g	g	mg	mg	mg	mg	mg	mg	mg	mg	mg	iu
SPLIT PEAS W/BACON	212	179	26.0	5.4					267	41	104	11	0.9	1195
1 jar		6.6	0.6						186		1.1	.09	4	
TURKEY RICE	212	92	15.4	1.5					316	20	36	.01	0.7	1260
1 jar		4.3	0.5						77		.8	.06	4	
VEGETABLES & BACON	212	137	18.2	5.2					362	46	82	.09	0.8	1767
1 jar		4.3	0.5		170				187		1.0	.07	5	
VEGETABLES & BEEF	212	108	14.3	3.3					310	15	68	.08	1.5	1154
1 jar		5.2	0.4		180				179		1.1	.08	4	
VEGETABLES & CHICKEN	212	105	20.7	0.6					227	21	57	.03	0.9	1537
1 jar		4.1	0.5						130		1.7	.05	4	
VEGETABLES & HAM	212	131	15.5	5.7					577	55	80	.15	1.1	360
1 jar		4.7	0.4		189				165		0.6	.08		
VEGETABLES & HAM W/BACON	212	122	18.0	4.0					445	12	55	.09	0.8	551
1 jar		3.6	0.5		189				91		0.8	.05	4	
VEGETABLES & LAMB	212	106	15.7	2.9					273	13	49	.03	0.9	1284
1 jar		4.2	0.5						182		0.9	.05	4	
VEGETABLES & LIVER W/BACON	212	104	17.2	2.0					498	19	81	.06	2.2	4287
1 jar		4.2	0.7						213		3.5	.46	5	
VEGETABLES & TURKEY	212	89	17.3	0.6					224	15	41	.03	0.5	714
1 jar		3.6	0.4						53		1.1	.04	4	
Toddler Dinners														
BEEF STEW	170	97	11.0	1.9					836	15	107	.02	2.2	2805
1 jar		8.7	0.5						207		1.2	.10	5	
CHICKEN STEW	170	145	10.4	7.1					716	61	105	.05	2.0	1717
1 jar		9.2	0.5						109		1.2	.12	3	
SPAGHETTI & MEATBALLS	170	129	20.0	1.4					780	37	114	.09	2.6	119
1 jar		8.8	0.7						235		1.5	.14	7	
Strained meats, yolks, cottage cheese														
BEEF	99	90		4.1					180	7	106	.01	2.5	
1 jar		13.4			557				193		1.8	.14	3	
BEEF W/O ADDED SALT	14	14		0.7					9	1	19	tr	0.4	
1 T		1.9							29	2	0.2	.02	tr	
BEEF W/BEEF BROTH	99	93	0.6	4.1					365	31	156	.02	2.1	0
1 jar		14.4	0.1		510				288		1.6	.15	3	
BEEF HEART	99	85	0.7	3.7					149	6	108	.04	3.1	
1 jar		12.3	0.1						157		2.1	.38	3	
CHICKEN	99	131	0.1	8.5					168	37	125	.02	2.9	76
1 jar		13.5	0.0		672				118		1.3	.18	3	
CHICKEN W/O ADDED SALT	14	22	tr	1.5					5	7	20	tr	0.5	
1 T		2.1							19	2	0.1	.01	tr	
CHICKEN W/CHICKEN BROTH	99	115	0.6	6.4					288	58	100	.01	3.0	430
1 jar		14.3	0.1		490				123		0.8	.13	3	
COTTAGE CHEESE W/PINEAPPLE	99	131	19.0	3.0					153	71	91	.04	0.1	111
1 jar		6.6	0.9						97		0.2	.16	1	
EGG YOLKS	94	187	1.9	16.7					165	77	274	.17	tr	614
1 jar		9.2			376				64		2.9	.37	3	
EGG YOLKS W/HAM OR BACON	94	182	0.3	16.1					312	55	235	.28	0.6	270
1 jar		9.3	0.0						86		2.3	.34	2	
HAM	99	113	0.7	6.2					200	5	120	.22	2.5	46
1 jar		13.7							204		.9	.20	4	
HAM W/HAM BROTH	99	95	0.6	5.0					41	0		.17	2.5	0
1 jar		12.8	0.1						199		0.7	.16	5	
LAMB	99	96	0.8	4.0					164	6	116	.08	3.1	99
1 jar		15.1	0.1		757				191		1.3	.17	2	
LIVER	99	92	2.0	3.2					160	4	201	.10	8.3	25398
1 jar		14.0	0.1		693				203		4.3	2.26	19	
LIVER & BACON	99	12.3	1.3	6.6					302	6	157	.05	7.8	22,000
1 jar		13.7	0.1						192		4.2	1.99	7	

*In addition to being used in infant meals, strained/junior & toddler
foods may also be used for older children and adults on soft diets.

	WT g	Proximate CAL / PRO	CHO / FIB	FAT / PUFA	Amino Acids TRY / PHE	LEU / ISO	LYS / VAL	MET / THR	Minerals Na / K	Ca / Mg	P / Fe	Vitamins THI / RIB	NIA / ASC	A / D
LIVER W/LIVER BROTH	99	87	0.0	1.8					168	5	215	.07	5.3	26,880
1 jar		17.8	0.0		630				202		4.6	2.07	19	
PORK	99	109	0.5	6.0					218	6	118	.21	2.6	44
1 jar		13.7	0.1		752				224		1.0	.21	3	
PORK W/PORK BROTH	99	94		4.0					200	14	60	.25	2.7	
1 jar		15.6			520				195		1.5	.08		
TURKEY	99	129	0.3	8.1					180	20	108	.02	2.9	593
1 jar		13.7							125		1.6	.26	3	
TURKEY W/TURKEY BROTH	99	108	0.6	6.3					48	32		.02	3.1	360
1 jar		12.8	0.1						141		1.1	.22		
VEAL	99	89	0.1	4.0					177	6	107	.02	3.1	
1 jar		13.4	0.3		475				213		1.0	.17	4	
VEAL W/O ADDED SALT	14	13	tr	0.5					8	1	21	.03	0.6	
1 T		2.0							31	2	0.1	.02	7	
VEAL W/VEAL BROTH	99	91	0.6	3.6					326	12	70	.01	2.9	
1 jar		14.3	0.1		480				143		1.3	.14	4	
Junior meats, cottage cheese														
BEEF	99	95	0.0	4.0					154	8	110	.01	3.4	117
1 jar		14.9	0.0		557				190		1.9	.17	4	
BEEF W/BEEF BROTH	99	100	0.6	4.4					240	15	130	.02	2.3	0
1 jar		14.6	0.1		520				296		1.6	.13	4	
CHICKEN	99	133	0.5	8.2					206	27	121	.01	3.7	122
1 jar		14.4	0.0		672				122		1.6	.21	3	
CHICKEN STICKS	70	134	0.7	9.6					300	51	84	.01	1.4	
1 jar (7 sticks)		11.1	0.1						74		1.3	.17	2	
CHICKEN W/CHICKEN BROTH	99	104	0.0	5.8					223	49	112	.01	3.5	
1 jar		13.3	0.0		500				145		0.7	.10		
COTTAGE CHEESE W/BANANAS	220	169	35.9	0.7					31	37	62	.03	0.2	0
1 jar		4.8	0.2		242				92		3.1	.11	3	
HAM	99	116	0.6	6.0					204	6	127	.16	2.8	31
1 jar		14.8	0.0		693				219		1.3	.17	3	
LAMB	99	96	0.0	3.9					201	6	111	.02	3.3	19
1 jar		15.3	0.0		757				180		1.6	.21	0.2	
LAMB W/LAMB BROTH	99	98	0.0	4.7					263	10	117	.02	3.1	0
1 jar		13.8	0.0		480				208		1.6	.15	2	
MEAT STICKS	70	115	0.8	7.5					326	25	72	.10	1.0	
1 jar (7 sticks)		11.1	0.2						80		1.4	.15	2	
PORK	99	113	0.0	6.1					241	6	118	.19	2.8	
1 jar		14.5	0.0						191		1.0	.19	4	
TURKEY	99	105	0.1	4.7					166	26	114	.02	3.5	
1 jar		15.5	0.0						139		1.6	.21	3	
VEAL	99	99	0.0	4.2					166	6	119	.01	3.8	50
1 jar		15.4	0.0		723				202		1.2	.19	4	
VEAL W/VEAL BROTH	99	94		3.9					270	9	109	.02	3.3	
1 jar		13.8			540				210		1.1	.12		
Strained Soups														
CHICKEN SOUP	128	83	1.3	2.2					358	42	64	.06	.6	785
1 jar		2.6	1.2		91						0.6	.09		
CREAM OF CHICKEN SOUP	128	73	12.1	1.3					177	41	47	.02	0.4	528
1 jar		3.2	0.3						81		0.4	.05	2	
LIVER SOUP	128	50	4.9	1.3					268	18	84	.10	2.1	9779
1 jar		5.2	0.4								1.7	.47	2	
TOMATO SOUP	128	69	17.3	0.1					376	31	67	.06	0.9	1280
1 jar		2.4	0.3						384		0.5	.15	4	
VEGETABLE SOUP	128	54	10.6	0.4					252	33	49	.19	1.3	7296
1 jar		2.0	0.4								0.4	.08	2	
Junior Soups														
CHICKEN SOUP	212	112	17.6	3.4					555	81	42	.02	0.6	1802
1 jar		3.4	0.4		129				148		0.8	.06		

*In addition to being used in infant meals, strained/junior & toddler foods may also be used for older children and adults on soft diets.

	WT g	CAL	CHO	FAT	TRY	LEU	LYS	MET	Na	Ca	P	THI	NIA	A
		PRO	FIB	PUFA	PHE	ISO	VAL	THR	K	Mg	Fe	RIB	ASC	D
	g	g	g	g	mg	mg	mg	mg	mg	mg	mg	mg	mg	iu
CREAM OF CHICKEN SOUP	212	116	19.1	2.1					293	70	82	.03	0.7	746
1 jar		5.2	0.5						148		.9	.13	4	
Strained Vegetables														
BEETS	128	48	10.1	0.1					205	17	28	.02	0.1	tr
1 jar		1.7	0.9		16				302		0.6	.02	2	
CARROTS	128	37	8.0	0.2					156	27	26	.03	0.4	15147
1 jar		0.9	1.0		27				239		0.3	.03	6	
CORN, CREAMED	128	82	17.8	0.3					129	40	42	.02	0.3	26
1 jar		1.9	0.3		45				88		0.3	.07	3	
GARDEN VEGETABLES	128	41	6.7	0.3					142	45	54	.09	0.8	5319
1 jar		2.9	1.2						183		1.2	.06	9	
GREEN BEANS	128	36	7.1	0.1					146	48	33	.05	0.4	416
1 jar		1.7	1.4		27				175		1.1	.07	7	
MIXED VEGETABLES	128	49	10.4	0.1					218	17	33	.05	0.3	5184
1 jar		1.6	0.5						118		0.4	.02	4	
PEAS	128	56	8.5	0.4					129	19	73	.14	1.4	409
1 jar		4.5	1.6		188				120		1.6	.07	8	
PEAS, CREAMED	128	101	15.5	2.7					297	20	78	.20	1.7	428
1 jar		4.2	1.3		125						1.2	.10	15	
POTATOES	128	84	9.5	4.5					419	8	52	.05	1.2	
1 jar		1.4	0.2								.11	.01	9	
SPINACH, CREAMED	128	53	7.6	1.0					150	123	68	.05	0.3	3300
1 jar		3.5	0.7		161				228		1.1	.11	11	
SQUASH	128	35	7.1	0.2					129	31	24	.03	0.5	1692
1 jar		1.1	1.1		36				194		0.4	.03	10	
SWEET POTATOES	128	92	20.9	0.1					135	26	44	.06	0.5	6745
1 jar		1.8	0.8		81				314		0.5	.03	14	
Junior Vegetables														
CARROTS	212	63	13.4	0.4					259	46	45	.05	0.8	27690
1 jar		1.5	1.5		32				432		0.7	.07	11	
CARROTS & PEAS	212	79	13.3	0.7					213	42	80	.16	1.9	13256
1 jar		4.9	2.8						293		1.8	.11	10	
CORN, CREAMED	212	133	28.8	0.5					214	66	81	.02	0.6	44
1 jar		3.2	0.6						145		0.5	.15	4	
GREEN BEANS W/ BACON, CREAMED	212	141	18.1	5.9					652	64	68	.06	0.6	224
1 jar		3.9	0.9						188		1.0	.10	7	
MIXED VEGETABLES	212	85	17.3	0.3					286	28	69	.08	1.4	8362
1 jar		3.2	1.1		148				302		1.2	.05	7	
PEAS, CREAMED	212	153	24.2	3.6					42	34	100	.19	1.7	117
1 jar		5.7	0.6		233				248		1.7	.13	6	
SPINACH, CREAMED	212	95	11.0	3.0					320	182	106	.06	0.5	4836
1 jar		5.7	1.1		267				471		1.7	.21	20	
SQUASH	212	57	11.4	0.5					214	52	41	.04	0.9	2626
1 jar		1.8	2.0						360		0.7	.08	20	
SWEET POTATOES	212	153	34.6	0.2					222	35	71	.08	0.9	11758
1 jar		3.1	1.3						483		1.2	.07	27	
Strained Desserts														
APPLE BETTY	128	96	22.9	0.0					9	20		.01	0.0	0
1 jar		0.0	0.3						50		0.3	.05	3.3	
APPLE PIE	128	127	28.5	1.3					5	51	14	.01	0.1	0
1 jar		0.4	0.8						38		0.1	.01	3	
BLUEBERRY BUCKLE	128	105	25.9	0.1					57	4	4	tr	tr	8
1 jar		0.2	0.3						23		0.4	.01	2.0	
BUTTERSCOTCH PUDDING	128	129	23.1	3.1					157	62	67	.01	tr	5
1 jar		2.1	0.2						79		0.4	.25	2	
CARAMEL CREAM PUDDING	128	125	22.1	0.6					81	61	51	.01	tr	54
1 jar		1.7	0.1						33		0.8	.06	3	

*In addition to being used in infant meals, strained/junior & toddler foods may also be used for older children and adults on soft diets.

FOOD	WT g	CAL	PRO	CHO	FIB	FAT	PUFA	PHE mg	Na	K	Ca	Mg	P	Fe	THI	RIB	NIA	ASC	A iu	D
CHERRY-VANILLA PUDDING (1 jar)	134	115	0.4	25.4	0.4	1.3			69	56	9		9	0.5	tr	.06	tr	3	193	
CHOCOLATE CUSTARD PUDDING (1 jar)	128	122	2.4	23.3	1.1	2.1			129	103	74		80	0.8	.02	.18	tr	3	289	
COTTAGE CHEESE W/PINEAPPLE JCE (1 jar)	128	116	4.0	21.6	0.1	1.5			175	27	28		35	0.1	.02	.03	tr	17	52	
CUSTARD PUDDING (1 jar)	128	123	2.2	21.9		2.9		96	100	79	56		59	0.3	.03	.09	0.1	tr	128	
DUTCH APPLE DESSERT (1 jar)	134	124	0.1	27.9	0.6	1.3			36	73	5		5	0.3	tr	.01	tr	10	11	
FRUIT DESSERT (1 jar)	134	130	0.4	31.9	0.1	0.1		5	92	111	8		64	0.1	.03	.01	0.1	4	322	
FRUIT DESSERT W/TAPIOCA (1 jar)	134	120	0.4	28.7	0.3	0.4		9	51	122	10		10	0.5	.02	.01	0.1	4	203	
ORANGE PUDDING (1 jar)	134	133	1.5	29.0	0.4	1.2		56	129	89	39		41	0.6	.06	.10	0.1	7	89	
PEACH COBBLER (1 jar)	134	116	0.5	27.8	0.7	0.3			23	92	5		11	0.3	tr	.01	0.4	10	131	
PEACH MELBA (1 jar)	134	151	0.4	36.4	0.3	0.4			51	17	8		43	0.9	.03	.03	0.4	10	255	
PINEAPPLE DESSERT (1 jar)	128	95	0.0	23.8	0.1	0.0			24	55	36			0.0	.03	.05	0.0	15	0	
RASPBERRY COBBLER (1 jar)	128	102	0.3	24.9	0.3	0.1			62	35	6		5	0.3	tr	tr	tr	1	15	
VANILLA CUSTARD PUDDING (1 jar)	128	112	2.2	22.3	1.1	1.6		141	129	82	71		70	0.6	.01	.17	tr	3	292	
Junior Desserts																				
APPLE PIE (1 jar)	220	220	0.7	49.7	0.4	2.2			15	75	9		11	0.4	.02	.02	0.2	1	0	
BANANA PUDDING (1 jar)	220	210	2.1	46.9	1.3	1.6			273	114	18		52	1.1	.04	.11	0.2	3	53	
BLUEBERRY BUCKLE (1 jar)	220	185	0.5	45.2	0.5	0.2			101	24	9		8	0.6	.01	.01	0.1	3	6	
BUTTERSCOTCH PUDDING (1 jar)	212	195	3.6	38.9	0.8	2.8			254	110	105		112	0.7	.02	.41	tr	5	4	
CHERRY-VANILLA PUDDING (1 jar)	220	190	0.8	43.0	0.3	1.6			110	78	16		16	0.5	.02	.08	0.1	4	284	
CHOCOLATE CUSTARD PUDDING (1 jar)	220	208	4.5	40.1	1.0	3.3			221	153	125		135	0.9	.03	.31	0.1	4	479	
COTTAGE CHEESE W/ PINEAPPLE (1 jar)	220	199	6.8	37.2	0.2	2.6			301	46	48		59	0.2	.04	.06	tr	20	90	
CUSTARD PUDDING (1 jar)	220	213	3.7	36.3		5.9		174	370	145	106		101	0.9	.04	.15	0.2		0	
DUTCH APPLE DESSERT (1 jar)	220	204	0.3	47.0	0.8	1.6			61	74	9		9	0.5	.01	.01	tr	20	18	
FRUIT W/ TAPIOCA (1 jar)	220	202	0.6	49.2	0.5	0.3			91	197	16		18	0.9	.03	.02	0.3	6	274	
PEACH COBBLER (1 jar)	220	192	0.9	46.4	0.7	0.3			38	134	9		19	0.4	.01	.03	0.6	20	141	
RASPBERRY COBBLER (1 jar)	220	180	0.6	44.0	0.4	0.2			101	45	12		9	0.8	tr	.01	0.1	4	7	
VANILLA CUSTARD PUDDING (1 jar)	212	202	4.0	38.5	1.2	3.5		233	229	139	114		114	0.6	.03	.25	tr	3	483	

*In addition to being used in infant meals, strained/junior & toddler foods may also be used for older children and adults on soft diets.

	WT	Proximate			Amino Acids				Minerals			Vitamins		
		CAL	CHO	FAT	TRY	LEU	LYS	MET	Na	Ca	P	THI	NIA	A
		PRO	FIB	PUFA	PHE	ISO	VAL	THR	K	Mg	Fe	RIB	ASC	D
	g	g	g	g	mg	mg	mg	mg	mg	mg	mg	mg	mg	iu
APPLE BUTTER	20	37	9.1	0.2					tr	3	4	tr	tr	0
1 T		0.1	0.2						50		0.1	tr	tr	0
CANDIED FRUITS & VEGETABLES														
APRICOTS	30	101	26.0	0.1						4	6	tr	0.2	370
1 avg		0.2	0.2								0.3	.01	tr	
CHERRIES	15	51	13.0	tr										30
3 large		0.1	0.1										0	
CHERRIES, MARASCHINO	16	19	4.7	tr										
2 pieces		tr	0.1											
CITRON	28	89	22.7	0.1					81	24	7			
1 oz		0.1	0.4						34		0.2			
FIGS	30	90	22.1	0.1						20	12	tr	0.1	30
1 piece		1.1	1.3								0.2	tr	tr	0
GINGER ROOT	28	95	24.4	0.1										
1 oz		0.1	0.2											
GRAPEFRUIT PEEL	28	89	22.6	0.1					5					
1 oz		0.1	0.6						1					
LEMON PEEL	28	89	22.6	0.1					5					
1 oz		0.1	0.6						1					
ORANGE PEEL	28	89	22.6	0.1					5					
1 oz		0.1	0.6						1					
PEARS	28	85	21.3	0.2										
1 oz		0.4												
PINEAPPLE	38	120	30.4	0.2						12	9	.05	0.1	40
1 slice		0.3	0.3								0.2	.01	9	0
SWEET POTATO	100	168	34.3	3.3	22	74	61	23	42	37	43	.06	0.4	6300
1 piece		1.3	0.6		72	62	98	61	190		0.9	.04	10	
YAMS	100	176	37.9	2.5					75	26	26	.03	0.7	4503
1 piece		1.2	0.6						156		0.8	.03	8	
CHUTNEY, APPLE	100	206	52.3	0.1					170	27	34			
5 T		0.8							217	18	1.0			
CHUTNEY, TOMATO	100	155	38.8						130	26	36			
5 T		1.1							278	18	0.9			0
FRUIT FILLING														
APPLE	100	107	29.1	1.9						9				
1 serving		0.0									3.5		1	
APRICOT	100	117	29.1	1.9						38			0.3	1546
1 serving		0.0									2.2		17	
BLUEBERRY	100	125	30.4	1.9						6		.01	0.1	0
1 serving		0.2									0.4	.03	4	
CHERRY	100	107	28.2	1.0						32		.01	0.1	226
1 serving		0.0									2.0	.02	1	
MINCE	100	194	43.7	4.9						61			0.2	
1 serving		0.0									9.0	.02	2	
PEACH	100	107	24.3	1.9										690
1 serving		0.0									3.1		4	
RAISIN	100	117	30.1	1.0						54		.20		
1 serving		0.0									4.0		1	
STRAWBERRY	100	117	30.1	1.0						38			0.1	
1 serving		0.0									2.2		6	
GUAVA BUTTER	20	39	10.0	0.0										
1 T		0.0	0.0										16	0
GUM, WRIGLEY	3	9	2.2	0.0						28				
1 piece		0.0												
GUM, SUGARLESS	3	8	1.8	0.0					1					
1 piece		0.0												
HONEY	20	61	16.5	0.0					1	1	1	.00	0.1	0
1 T		0.1	tr	0.0					10	1	0.1	.01	tr	0
HONEY	100	306	78.0	0.0					5	20	16	.01	0.2	tr
5 T		0.2	0.1	0.0					51	3	0.8	.07	4	0
HONEY	338	1034	264	0.0					17	68	54	.03	0.7	0
1 cup		0.7	0.3	0.0					172	10	2.7	.24	13	0

	WT g	CAL / PRO g	CHO / FIB g	FAT / PUFA g	TRY / PHE mg	LEU / ISO mg	LYS / VAL mg	MET / THR mg	Na / K mg	Ca / Mg mg	P / Fe mg	THI / RIB mg	NIA / ASC mg	A / D iu
ICING														
CARAMEL	10	36	7.6	0.7					8	10	6	tr	tr	28
1 serving		0.1	0.0						5		0.2	.01	tr	
CHOCOLATE	10	38	6.7	1.4					6	6	11	tr	tr	21
1 serving		0.3	tr						20		0.1	.01	tr	
CHOC, DOUBLE DARK, PILLSBURY	10	42	8.2	1.5					24					
1 serving		0.0												
CHOCOLATE FUDGE	10	38	6.7	1.4					16	2	7	tr	tr	27
1 serving		0.2	0.1						6	26	0.1	tr	0	
COCONUT	10	36	7.5	0.8					12	1	3	tr	tr	0
1 serving		0.2	0.1						17		0.1	tr	0	
COCONUT ALMOND, PILLSBURY	10	48	6.4	2.4					16					
1 serving		0.4												
COCONUT PECAN, PILLSBURY	10	46	8.0	1.7					21					
1 serving		0.0												
CREAMY FUDGE	10	34	7.5	0.7					23	4	9	tr	tr	
1 serving		0.3	0.1						10		0.1	.01		
LEMON, PILLSBURY	10	42	7.4	1.3										
1 serving		0.0												
MILK CHOCOLATE	10	47	8.1	1.6					1					
1 serving		0.0												
STRAWBERRY, PILLSBURY	10	44	7.9	1.5										
1 serving		0.0												
VANILLA, PILLSBURY	10	42	7.6	1.6					8					
1 serving		0.0												
WHITE, UNCKD	10	38	8.2	0.7					5	2	tr	tr	tr	27
1 serving		0.1	0.0						2		tr	tr	tr	
WHITE, BOILED	10	32	8.0	0.0					14	tr	tr	tr	tr	0
		0.1	0.0						2		tr	tr	0	
WHITE, FLUFFY, PILLSBURY	10	39	9.4	0.0					47					
1 serving		0.0												
JAM														
ALL VARIETIES	20	55	14.2	0.1						2	2	tr	tr	2
1 T		0.1	0.1								0.1	tr	1	0
LOW-CAL	20	29	7.4	tr					16	1	1	.00	tr	0
1 T		0.1							31	1	tr	tr	tr	
GRAPE	20	59	15.0	tr						2	2	tr	tr	6
1 T		0.1									0.3	.01	tr	
PLUM	20	59	15.0	tr						2	2	tr	0.1	64
1 T		0.1									0.3	tr	2	0
JELLY														
ALL VARIETIES	20	55	14.1	tr					3	4	1	tr	tr	2
1 T		tr	0.0	0.0					15	1	0.3	.01	1	0
KRAFT	20	52	13.0	tr					tr	2	2			
1 T		tr	0.0						12					
LOW-CAL, SMUCKERS	20	26	6.6	0.0					tr					
1 T		0.0												
BLACKBERRY, SMUCKERS	20	51	13.2	0.0					4	4	2	.00	0.1	13
1 T		0.1							12	2	0.1	.01	tr	
BOYSENBERRY, SMUCKERS	20	52	13.4	0.0					3	4	2	.00	.01	13
1 T		0.0							12	1	0.1	.01	tr	
CHERRY, SMUCKERS	20	52	13.2	tr					3	3	3	tr	tr	0
1 T		0.1								1	0.2	.01	tr	
CHERRY, LOW-CAL, SMUCKERS	20	27	7.0	0.0					23	1	2	tr	tr	0
1 T		0.1							35	1	0.1	.01	tr	
CURRANT, SMUCKERS	20	52	13.4	0.0					8	4	2	.00	0.0	9
1 T		1.2							19	1	0.2	tr	3	
GRAPE	20	55	14.1	tr					3	4	1	tr	tr	2
1 T		tr	0.0						15	1	0.3	.01	1	
GRAPE, LOW-CAL, FEATHERWEIGHT	20	16	4.0	0.0										
1 T		0.0												
GRAPE, LOW-CAL, SMUCKERS	20	7	1.6	0.0					tr	1	1	.00	0.0	0
1 T		0.1							13	1	tr	.00	0	

	WT g	CAL / PRO	CHO / FIB	FAT / PUFA	TRY / PHE mg	LEU / ISO mg	LYS / VAL mg	MET / THR mg	Na / K mg	Ca / Mg mg	P / Fe mg	THI / RIB mg	NIA / ASC mg	A / D iu
GUAVA	20	39	10.0	0.0										
1 T		0.0	0.0										16	0
GUAVA, SMUCKERS	20	52	13.4	tr					3	3	3	.00	0.1	13
1 T		0.1							13	1	0.2	tr	3	
QUINCE, SMUCKERS	20	51	13.2	tr					2	2	2	tr	tr	3
1 T		0.1							16		0.2	tr	tr	
STRAWBERRY, SMUCKERS	20	51	13.2	0.0					2	3	2	.00	tr	5
1 T		.01							13	1	0.2	.01	tr	
MARMALADE														
BITTER ORANGE, SMUCKERS	20	54	14.0	0.0					8	3	2	tr	tr	9
1 T		0.1							9	0	tr	.00	tr	
CITRUS	20	51	14.0	tr					3	7	2	tr	tr	0
1 T		0.1	0.1						7	1	0.1	tr	1	
ORANGE	20	56	14.0	0.1					4	4	3	tr	tr	0
1 T		0.2							9		0.1	tr	2	0
PAPAYA	20	57	14.6	0.0										
1 T		0.0											7	0
SWEET ORANGE	20	54	14.0	0.0					4	9	2	tr	tr	21
1 T		0.1							13	0	tr	tr	1	
MOLASSES, CANE														
BARBADOS	20	54	14.0								10	.01		
1 T		0.0										.04		0
BARBADOS	100	271	70.0								50	.06		
5 T		0.0										.20		0
BARBADOS	328	889	230								164	.20		
1 cup		0.0										.66		0
FIRST EXTRACTION, LIGHT	20	50	13.0						16	33	9	.01	tr	
1 T		0.0							300		.9	.01		0
FIRST EXTRACTION, LIGHT	100	252	65.0						80	165	45	.07	.2	
5 T		0.0							1500		4.3	.06		0
FIRST EXTRACTION, LIGHT	328	826	213						262	541	148	.23	.7	
1 cup		0.0							4920		14.1	.20		0
SECOND EXTRACTION, MEDIUM	20	46	12.0							58	14			
1 T		0.0									1.2			0
SECOND EXTRACTION, MEDIUM	100	232	60.0							290	69			
5 T		0.0									6.0			0
SECOND EXTRACTION, MEDIUM	328	761	197							951	226			
1 cup		0.0									19.7			0
THIRD EXTRACTION, BLACKSTRAP	20	43	11.0							116	17.0	.06	.4	
1 T		0.0									2.3	.05		0
THIRD EXTRACTION, BLACKSTRAP	100	213	55.0							579	85	.28	2.1	
5 T		0.0									11.3	.25		0
THIRD EXTRACTION, BLACKSTRAP	328	699	180							1899	279	.92	6.9	
1 cup		0.0									37.1	.82		0
UNSULPHERED, GRANAMAS	20	50	14.4	0.0					11	20		.00	0.0	0
1 T		0.0							132		0.4	.00	0	
PRESERVES														
APRICOT, SMUCKERS	20	51	13.0	tr					2	2	2	.00	tr	42
1 T		0.1							19	1	0.2	tr	tr	
APRICOT/BLACKBERRY, LOW-CAL, FEATHERWEIGHT 1 T	20	21	5.0	0.0										
		0.0												
APRICOT-PINEAPPLE, SMUCKERS	20	51	13.0	tr					3	3	2	.02	tr	40
1 T		0.3							22	1	0.2	tr	tr	
BLACKBERRY, SMUCKERS	20	55	13.8	tr					2	3	2	.00	0.1	13
1 T		0.3							12	1	0.1	.01	tr	
BOYSENBERRY, SMUCKERS	20	54	13.8	tr					6	4	3	.00	0.1	13
1 T		0.2							12	1	0.1	.01	tr	
PEACH, SMUCKERS	20	51	13.2	0.0					4	2	2	.00	0.1	0
1 T		0.1							19	1	0.2	tr	tr	
RELISH, CRANBERRY-ORANGE	15	27	6.8	0.1					tr	3	1	.01	tr	11
1 T		0.1							11		0.1	tr	3	

	WT g	Proximate CAL	CHO	FAT	Amino Acids TRY	LEU	LYS	MET	Minerals Na	Ca	P	Vitamins THI	NIA	A
		PRO g	FIB g	PUFA g	PHE mg	ISO mg	VAL mg	THR mg	K mg	Mg mg	Fe mg	RIB mg	ASC mg	D iu
STRAWBERRY, SMUCKERS	20	53	13.6	0.0					2	3	2	.00	tr	5
1 T		0.1							14	1	0.2	.01	3	
STRAWBERRY-PINEAPPLE, SMUCKERS	20	54	14.0	0.0					3	3	2	tr	tr	5
1 T		1.0							14	1	0.2	tr	tr	
TOMATO, SMUCKERS	20	53	13.6	tr					19	2	3	tr	0.1	0
1 T		0.2							28	2	0.2	tr	tr	
SUGAR														
BROWN, CRUDE	14	14	12.7	.07						7	6	tr	.04	tr
1 T (weight varies)		0.1									0.6	.02	.3	0
BROWN, CRUDE	100	356	90.6	.5						51	44	.02	.3	tr
7 T (weight varies)		0.4	0.1								4.2	.11	2	0
BROWN, DARK	14	52	13.4	0					3.4	11	5	.00	0	0
1 T (weight varies)			0.0						32	9	0.4	.00		0
BROWN, DARK	100	370	95.5	0					24	76	37	.01	0.2	0
½ cup, scant			0.0						230	62	2.6	.03		0
BROWN, DARK	220	814	210	0					53	167	81	.02	0.4	0
1 cup			0.0						506	124	5.7	.06		0
CORN, NOT REFINED	100	348	90									.00	0	0
3½ ounces			0.0									.00	0	0
DEXTROSE, ANHYDROUS	100	385	99.5	0								.00	0	0
3½ ounces		0.0	0.0									.00	0	0
DEXTROSE, CRYSTALLINE	100	348	90.0	0.0								.00	0	0
3½ ounces		0.0	0.0									.00	0	0
DEXTROSE D-GLUCOSE POWDER	12	45	11.0	0.0										
1 T		0.0												
MAPLE	15	52	13.5							27	2			
1 piece (1¼ × 1 × ½)			0.0								0.5			0
MAPLE	100	348	90							180	10			
4 pieces (2 × 1 × ½)											3.5			0
POWDERED	6	23	6.0	0								.00	0	0
1 t rounded		0.0	0.0									.00		0
POWDERED	11	42	10.9	0								.00	0	0
1 T (weight varies)		0.0	0.0									.00		0
POWDERED	100	385	99.5	0								.00	0	0
9 to 12 T (varies)		0.0	0.0									.00		0
POWDERED	176	678	175	0								.00	0	0
1 cup (weight varies)		0.0	0.0									.00		0
WHITE, DOMINO	6	24	6.0	0						tr		.00	0	0
1 piece (1⅛ × ¾ × ⅜)		0.0	0.0									.00		0
WHITE, GRANULAR	4	16	4.0	0						tr		.00	0	0
1 t, level		0.0	0.0									.00		0
WHITE, GRANULAR	8	32	8.0	0						tr		.00	0	0
1 t, rounded		0.0	0.0									.00		0
WHITE, GRANULAR	12	46	11.9	0						tr		.00	0	0
1 T		0.0	0.0									.00		0
WHITE, GRANULAR	100	385	99.5	0					tr	5	1	.00	0	0
½ cup		0.0	0.0						1		0.1	.00		0
WHITE, GRANULAR	200	770	199.0	0					1	10	2	.00	0	0
1 cup		0.0	0.0						1		0.2	.00		0
SUGARCANE JUICE	20	16	4.1	tr						3	2	tr	tr	0
1 T		0.1									0.1	tr	tr	
SUGAR SUBSTITUTES														
ADOLPHS	5	18	4.4	0.0					21	17				
1 t		0.0	0.0	0.0										
DIAMOND	4	14	3.6	0.0					1	24				
1 t		0.0	0.0	0.0										
SUGARLIKE, FEATHERWEIGHT	14	53	13.3	0.0					0					
1 t		0.0	0.0	0.0					0					
SUGARTWIN, BROWN	1	4	0.9	0.0					6	10				
1 t		0.0	0.0	0.0					1					
SUGARTWIN, WHITE, GRANULATED	1	4	0.9						6	10				
1 T									1					

	WT	Proximate			Amino Acids				Minerals			Vitamins		
		CAL	CHO	FAT	TRY	LEU	LYS	MET	Na	Ca	P	THI	NIA	A
		PRO	FIB	PUFA	PHE	ISO	VAL	THR	K	Mg	Fe	RIB	ASC	D
	g	g	g	g	mg	mg	mg	mg	mg	mg	mg	mg	mg	iu
SUPEROSE 1 t	4	14							4					
SWEET 'N LOW 1 t	5	18	4.5						20					
									225					
SYRUPS														
CANE, LIGHT 1 T	20	53	13.6	0.0						12	6	.03	tr	0
		0.0	0.0						85		0.7	.01	0	0
CANE, LIGHT 5 T	100	263	68.0	0.0						60	29	.13	0.1	0
		0.0	0.0						425		3.6	.06	0	0
CORN, AVG 1 T	20	57	14.8	0.0						9	3	.00	tr	0
		0.0	0.0								0.8	tr	0	0
CORN, AVG 5 T	100	287	74.0	0.0						46	16	.00	0.1	0
		0.0	0.0								4.1	.01	0	0
CORN, AVG 1 cup	328	939	242	0.0						151	52	0	0.3	0
		0.0	0.0								13.4	.03	0	0
CORN, DARK, KARO 1 T	20	59	14.7	0.0					27					
		0.0	0.0											
MAPLE 1 T	20	50	12.8	0.0					3	33	3			0
		0.0	0.0						26	2	0.2		0	0
MAPLE 5 T	100	252	65.0	0.0					10	104	8			0
		0.0	0.0						176	10	1.2		0	0
MAPLE, IMITATION 1 T	20	53	13.0	0.3					3	2	1	.00	0.0	14
		0.0	0.0						1		0.2	.00	0	
MAPLE, IMITATION 5 T	100	266	64.8	1.6					65	10	4	.00	0.0	70
		0.0	0.0						3		0.9	.00	0	
MAPLE, LOW-CAL, FEATHERWEIGHT 1 T	15	9	2.3	0.0										
		0.0												
MAPLE-HONEY-FLAVORED, LOG CABIN 1 T	20	53	13.7	0.0					3		1	.00		0
			0.0						1		0.2			
SORGHUM 1 T	20	52	13.4						4	30	5	.02		
									120		2.4	.02		0
STRAWBERRY, LOW-CAL, FEATHERWEIGHT 1 T	15	14	3.0	0.0										
		0.0												
TABLE BLEND (MAINLY CORN) 1 T	20	57	14.8	0.0					13	9	3	.00	tr	0
		0.0	0.0						1	tr	0.8	tr	0	0
TABLE BLEND (MAINLY CORN) 5 T	100	286	74.0	0.0					68	46	16	.00	0.1	0
		0.0							4	2	4.1	.01	0	0
TABLE BLEND (MAINLY CORN) 1 cup	328	938	243	0					223	151	52	.00	0.3	0
		0	0						13	7	13.4	.03	0	0
TABLE BLEND (CANE & MAPLE) 1 T	20	50	13.0	0.0					tr	3	tr	.00	0.0	0
		0.0	0.0						5			.00	0	
TREACLE, BLACK* 1 T	20	53	13.4	0					19	99	6			
		0.2							297		1.8			0
5 T	100	265	67.2	0					96	495	31			
		1.2	0.0						1470	144	9.2			0

*Treacle is a word of many meanings but generally refers to the residual molasses left from sugar making. It is high in mineral and color, and relatively low in sugar. Values given here are from McCance and Widdowson.

	WT	Proximate			Amino Acids				Minerals			Vitamins		
		CAL	CHO	FAT	TRY	LEU	LYS	MET	Na	Ca	P	THI	NIA	A
		PRO	FIB	PUFA	PHE	ISO	VAL	THR	K	Mg	Fe	RIB	ASC	D
	g	g	g	g	mg	mg	mg	mg	mg	mg	mg	mg	mg	iu
ALFALFA SPROUTS, RAW	56	20	2.0	0.0						0	0	.03	0.4	400
2 oz		2.0									0.7	.03		
ALFALFA SPROUTS, CKD	100									28		.12	0.8	
3½ g		5.1	1.7								1.4	.20	11	
ARTICHOKE, FRENCH/GLOBE, RAW*	200	88	21.2	0.4					86	102	176	.16	2.0	320
1 large bud, AP		5.8	4.8						860		2.6	.10	24	
CKD, EP	100	44	9.9	0.2					30	51	69	.07	0.7	150
base & soft ends of leaves		2.8	2.4						301		1.1	.04	8	
ARTICHOKE HEARTS, FRZN	100	26	6.4	0.4					47	17	53	.06	0.9	150
3½ oz		2.6	0.9						248	27	0.3	.16	7	
ASPARAGUS, RAW	100	26	5.0	0.2	30	110	118	38	2	22	62	.18	1.5	900
5 to 6 spears		2.5	.7		78	90	120	76	278	20	1.0	.20	33	
CKD	100	20	3.6	0.2	26	97	103	33	1	21	50	.16	1.4	900
⅔ cup cut pieces		2.2	0.7		68	79	106	66	183		0.6	.18	26	
CND, GREEN	115	21	3.3	0.3	26	97	103	33	271	21	49	.07	0.9	586
6 med spears		2.2	0.6		68	79	106	66	191		2.0	.10	17	
CND, WHITE	115	21	3.7	0.3	22	79	85	27	271	17	38	.06	0.8	58
6 med spears		1.8	0.6		56	65	87	54	161		1.0	.07	17	
FRZN	100	23	3.6	0.2	40	145	155	50	2	23	66	.16	1.2	850
6 spears		3.3	0.8		102	119	158	100	239	14	1.3	.14	25	
ASPARAGUS IN BUTTER SCE, FRZN	180	81	5.4	5.4					747	36	83	.16	3.6	1487
1 cup		3.6	2.5						270		0.7	.25	31	
AVOCADO, CALIFORNIA	120	185	6.5	18.4	17		89	14	10	9	32	.08	1.4	501
½ Avocado		2.4	1.9						574	42	1.3	.10	7	
AVOCADO, FLORIDA	180	196	13.5	16.8					6	15	64	.17	2.4	440
½ Avocado		2.0							924		0.9	.31	21	
BAMBOO SHOOTS, RAW	133	36	6.9	0.4						17	78	.20	0.8	27
1 cup		3.5	0.9						709		0.7	.09	5	
BAMBOO SHOOTS, CND	133	21	3.5	0.1					5	14	20			
1 cup		1.9	0.9						128	6	0.4			
BEANS, BLACK, BROWN, BAYO,	100	339	61.2	1.5					25	135	420	.55	2.2	30
RAW ½ cup		22.3	4.4						1038	171	7.9	.20		
BEANS, COMMON WHITE, RAW	100	340	61.3	1.6	201	1918	1650	223	19	144	425	.65	2.4	0
½ cup		22.3	4.3		1227	1271	1360	959	1196	170	7.8	.22		
CKD	100	118	21.2	0.6	70	671	577	78	7	50	148	.14	0.7	0
½ cup		7.8	1.5		429	445	476	335	416		2.7	.07	0	
W/PORK & TOMATO, CND	125	152	23.8	3.2	68	654	562	76	579	68	115	.10	0.8	162
½ cup		7.6	1.8		418	433	464	327	263	50	2.2	.04	2	
W/PORK & MOLASSES, CND	125	188	26.4	5.9	70	670	578	78	475	79	142	.08	0.6	
½ cup		7.8	2.1		429	444	476	336		58	2.9	.05		
W/O PORK, CND	125	150	28.8	0.6	71	679	585	79	422	85	151	.09	0.8	75
½ cup		7.9	1.8		434	450	482	340	335	73	2.5	.05	2	
BEANS, RED KIDNEY, RAW	100	343	61.9	1.5	202	1935	1665	225	10	110	406	.51	2.3	20
½ cup		22.5	4.2		1238	1260	1350	968	984	163	6.9	.20		
CKD	100	118	21.4	0.5	70	671	577	78	3	38	140	.11	0.7	tr
⅖ cup		7.8	1.5		429	437	468	335	340		2.4	.06		
CND	100	90	16.4	0.4	51	490	422	57	3	29	109	.05	0.6	tr
⅖ cup		5.7	0.9		314	319	342	245	264		1.8	.04		
BEANS, PINTO CALICO, RAW	100	349	63.7	1.2	206	1969	1695	229	10	135	457	.84	2.2	
½ cup		22.9	4.3		1260	1305	1397	985	984		6.4	.21		
BEANS, FRIJOLES†	193	232	29.9	6.6					1102	234	291			
1 serving		13.0	1.9						693	116	2.0			
BEANS, GREEN, RAW	100	32	7.1	0.2	27	110	99	28	7	56	44	.08	0.5	600
1 cup, 1 in pieces		1.9	1.0		46	86	91	72	243	32	0.8	.110	19	
CKD IN SMALL AMT WATER	125	31	6.8	0.2	28	116	104	30	5	62	46	.09	0.6	675
1 cup		2.0	1.2		48	90	96	76	189		0.8	.11	15	
CKD IN LARGE AMT WATER	125	31	6.8	0.2	28	116	104	30	5	62	46	.08	0.4	675
1 cup		2.0	1.2		48	90	96	76	189		0.8	.10	12	
BEANS, GREEN, CND	110	20	4.6	0.1	15	64	57	16	270	37	23	.03	0.3	320
½ cup		1.1	0.7		26	50	53	42	104	15	1.3	.04	4	

*Carbohydrate may be inulin, which is of doubtful availability. During storage, inulin is converted to sugars. Caloric values range from 9 per 100 grams for freshly harvested to 47 for stored product.

†Frijoles are whipped pinto beans and shredded mild cheddar cheese covered w/ red sauce.

	WT	Proximate			Amino Acids				Minerals			Vitamins		
	WT	CAL	CHO	FAT	TRY	LEU	LYS	MET	Na	Ca	P	THI	NIA	A
		PRO	FIB	PUFA	PHE	ISO	VAL	THR	K	Mg	Fe	RIB	ASC	D
	g	g	g	g	mg	mg	mg	mg	mg	mg	mg	mg	mg	iu
CND, SOLIDS & LIQUID	239	43	10.0	0.2	34	139	125	36	565	81	50	.07	0.7	695
1 cup		2.4	1.4		58	108	115	91	227		2.9	.09	10	
CND, DRAINED SOLIDS	125	30	6.5	0.3	25	104	94	27	295	56	31	.04	0.4	588
1 cup		1.8	1.2		43	81	86	68	109		1.9	.06	5	
FRZN, CUT	100	25	5.7	0.1	22	93	83	24	1	40	32	.07	0.4	580
3½ ounces		1.6	1.0		38	72	77	61	152	21	0.7	.09	5	
FRZN, FRENCH STYLE, CKD	100	26	6.0	.1	22	93	83	24	2	38	30	.06	.3	530
3½ ounces		1.6	1.1		38	72	77	61	136	20	.9	.08	7	
BEANS, GREEN, ALMANDINE, FRZN	71	46	3.8	2.8					396	28	20	.03	0.2	263
1 serving		1.2	0.6						76		0.7	.04	5	
BEANS, GREEN IN BUTTER SCE, FRZN	100	40	4.0	2.0					420	25	25	.03	0.2	469
1 serving		1.0	0.8						100		0.7	.07	7	
BEANS, GREEN, IN MUSHROOM SCE,	100	40	6.0	0.5					364	70	55	.04	0.3	270
FRZN 1 serving		2.0	0.5						150		0.5	.14	1	
BEANS, W/ONIONS & BACON BITS,	100	40	4.2	1.6					432	26	253	.03	0.2	270
FRZN 1 serving		1.5	1.1						110		0.6	.11	4	
BEANS, GRN W/PEARL ONIONS, FRZN	100	38	7.3	0.1					4	38	36	.07	0.3	380
1 serving		2.0	0.9						159	20	0.6	.08	15	
BEANS, W/SLICED MUSHROOMS, FRZN	100	37	7.2	0.2					169	38	31	.06	0.7	460
1 serving		1.7	0.1						185	21	0.6	.10	10	
BEANS, GRN, W/TOASTED ALMONDS,	100	63	9.1	1.9					394	45	39	.06	0.6	440
FRZN 1 serving		3.1	1.2						196	20	0.6	.07	10	
BEANS, LIMA, GREEN, RAW	100	123	22.1	0.5	76	697	563	134	2	142	52	.24	1.4	290
4 rounded T		8.4	1.8		496	487	529	395	650	67	2.8	.12	29	
CKD	100	111	19.8	0.5	68	631	509	122	1	47	121	.18	1.3	280
⅝ cup		7.6	1.8		448	441	479	357	422		2.5	.10	17	
CND, DRAINED SOLIDS	115	110	21.0	0.3	56	515	415	99	271	32	81	.03	.6	218
½ cup		6.2	2.1		366	360	391	291	255		2.8	.06	7	
FRZN, BABY LIMAS, CKD	100	99	19.1	0.1	54	498	402	96	101	20	90	.07	1.0	230
⅝ cup		6.0	1.6		354	348	378	282	426	48	1.7	.05	17	
FRZN, FORDHOOK, CKD	100	118	22.3	0.2	67	614	496	118	129	35	126	.09	1.2	220
⅝ cup		7.4	1.9		437	429	466	348	394	48	2.6	.05	12	
BEANS, LIMA, DRY SEEDS,	100	345	64.0	1.6	184	1693	1367	326	4	72	385	.48	1.9	tr
RAW ½ cup dry		20.4	4.3		1204	1183	1285	959	1529	180	7.8	.17		
CKD, BAKED	115	159	29.4	0.7	85	780	630	150	2	33	177	.15	.8	
⅝ cup ckd		9.4	2.0		555	545	592	442	704		3.6	.07		
BEANS, MUNG, DRY, RAW	100	340	60.3	1.3	169	2202	1646	266	6	118	340	.38	2.6	80
3½ oz		24.2	4.4		1162	1355	1428	750	1028		7.7	.21		
SPROUTED SEEDS, UNCKD	100	35	6.6	0.2	27	346	258	42	5	19	64	.13	.8	20
3½ oz		3.8	0.7		182	213	224	118	223		1.3	.13	19	
BOILED DRAINED	100	28	5.2	0.2	22	291	218	35	4	17	48	.09	.7	20
3½ oz		3.2	0.7		154	179	189	99	156		0.9	.10	6	
BEANS, SOYA—see SOYBEANS														
BEANS, WAX, YELLOW, RAW	100	27	6.0	0.2	24	99	88	26	7	56	43	.08	0.5	250
1 cup, 1 in. pieces		1.7	1.0		41	76	82	65	243	32	0.8	.11	20	
CKD	100	22	4.6	0.2	20	81	73	21	3	50	37	.07	.5	230
1 cup		1.4	1.0		34	63	67	53	151		0.6	.09	13	
CND, SOLIDS & LIQUID	100	19	4.2	0.2	14	58	52	15	236	34	21	.03	0.3	60
⅔ cup		1.0	0.6		24	45	48	38	95		1.2	.04	5	
CND, DRAINED SOLIDS	100	24	5.2	0.3	20	81	73	21	236	45	25	.03	0.3	100
½ cup		1.4	0.9		34	63	67	53	95		1.5	.05	5	
FRZN, CKD	100	27	6.2	0.1	24	99	88	26	1	35	31	.07	0.4	100
3½ oz		1.7	1.1		41	76	82	65	164	21	0.7	.08	6	
BEANS, WINGED (GOA)–RAW	100	22	4.8	tr						57	37	.22	.7	330
2 cups (12 beans)		1.8									0.5	.01		
CKD, DRAINED	46	10	2.2	0						26	17	.10	.3	152
1 cup		0.8									0.2	.04	0	
BEAN SPROUTS, MUNG, RAW	100	35	6.6	.2	27	346	258	42	5	19	64	.13	0.8	20
1⅔ cup		3.8	0.7		182	213	224	118	223		1.3	.13	19	
CKD, DRAINED	100	28	5.2	.2	22	291	218	35	4	17	48	.90	0.7	20
1 cup		3.2	0.7		154	179	189	99	156		0.9	.10	6	

	WT	Proximate			Amino Acids				Minerals			Vitamins		
		CAL	CHO	FAT	TRY	LEU	LYS	MET	Na	Ca	P	THI	NIA	A
		PRO	FIB	PUFA	PHE	ISO	VAL	THR	K	Mg	Fe	RIB	ASC	D
	g	g	g	g	mg	mg	mg	mg	mg	mg	mg	mg	mg	iu
BEAN SPROUTS, SOY, RAW	100	46	5.3	1.4		267	211	43		48	67	.23	0.8	80
1 cup		6.2	0.8		186	223	223	161			1.0	.20	13	
CKD, DRAINED	100	38	3.7	1.4		228	180	37		43	50	.16	0.7	80
¾ cup		5.3	0.8		159	191		138			0.7	.15	4	
BEETS, RED, RAW	100	43	9.9	0.1	14	54	86	6	60	16	33	.03	0.4	20
2 beets (2 in. dia)		1.6	0.8		27	51	50	34	335	27	0.7	.05	10	
CKD, DRAINED	83	27	6.0	0.1	8	31	49	4	36	12	19	.03	0.3	17
½ cup, diced		0.9	0.7		15	29	28	19	172		0.4	.03	5	
CND, SOLIDS & LIQUID	100	34	7.9	0.1	8	31	49	4	236	14	17	.01	0.1	10
3½ ounces		0.9	0.5		15	29	28	19	167	15	0.6	.02	3	
CND, DRAINED SOLIDS	83	31	7.3	0.1	7	28	44	4	196	16	18	.01	0.1	17
½ cup, diced		0.8	0.7		14	29	25	19	138	12	0.6	.02	3	
BEETS, RED, CND, PICKLED	240	161	39.8	0.2					701	41	43	.05	0.5	24
1 cup		2.2	1.2						384	34	1.4	.10	5	
BEET GREENS, RAW	100	24	4.6	0.3	26	141	119	37	130	119	40	.10	0.4	6100
½ cup when cooked		2.2	1.3		128	92	110	84	570	106	3.3	.22	30	
CKD, DRAINED	100	18	3.3	0.2	20	109	92	29	76	99	25	.07	0.3	5100
½ cup		1.7	1.1		99	71	85	65	332		1.9	.15	15	
BLACKEYED PEAS, FRZN	100	130	22.1	9.0					387					
3½ oz		0.4												
BROCCOLI, RAW	100	32	5.9	0.3	40	176	158	54	15	103	78	.10	0.9	*2500
1 stalk (5½ in long)		3.6	1.5		130	137	184	133	382	24	1.1	.23	113	
CKD	100	26	4.5	0.3	34	152	136	46	10	88	62	.09	0.8	2500
1 large stalk; ⅔ cup		3.1	1.5		112	118	158	115	267		0.8	.20	90	
FRZN, CKD	100	26	4.6	0.3	32	142	128	44	15	54	56	.06	0.5	2600
3½ ounces		2.9	1.1		104	110	148	107	212	21	0.7	.12	57	
BROCCOLI IN CHEESE SCE, FRZN	100	60	6.0	3.0					350	106	95	.05	0.3	1050
		4.0	0.8						195		0.3	.22	35	
BROCCOLI W/ HOLLANDAISE SCE, FRZN	100	107	3.1	9.1					122	25	55	.04	0.3	1050
		3.0	0.9						192	10	0.6	.08	42	
BROCCOLI NORMANDY	100	25	5.0	0.2					19	35	42	.05	0.5	4333
3½ oz		1.9	1.0						214	13	0.6	.07	35	
BRUSSELS SPROUTS, RAW	100	45	8.3	0.4	49	216	221	49	14	36	80	.10	0.9	550
9 medium		4.9	1.6		162	206	216	172	390	29	1.5	.16	102	
CKD, DRAINED	100	36	6.4	0.4	42	185	189	42	10	32	72	.08	0.8	520
⅔ cup; 6 to 7		4.2	1.6		139	176	185	147	273		1.1	.14	87	
FRZN, CKD	100	33	6.5	0.2	32	141	144	32	14	21	61	.08	0.6	570
3½ oz		3.2	1.2		106	134	141	112	295	21	0.8	.10	81	
BRUSSELS SPROUTS AU GRATIN, FRZN	100	70	7.5	2.5					465	80	200	.05	0.3	150
3½ oz		4.5	0.9						245		0.6	.10	33	
BURDOCK, CKD	100	123	33.4	0.1						64	39			
½ cup		1.1												
CABBAGE, CHINESE, RAW	100	14	3.0	0.1					23	43	40	.05	0.6	150
2¼ cups, shredded		1.2	0.6						253	14	0.6	.04	25	
CKD	82	8	1.3	0						26	32	.07	0.8	40
½ cup		1.2									0.2	.07	26	
CABBAGE, COMMON, RAW	100	24	5.4	0.2	10	52	61	12	20	49	29	.05	0.3	130
1 cup shredded		1.3	0.8		27	51	39	36	233	13	0.4	.05	†47	
CKD IN SMALL AMT WATER	100	20	4.3	0.2	9	44	52	10	14	44	20	.04	0.3	130
⅗ cup		1.1	0.8		23	43	33	31	163		0.3	.04	33	
CKD IN LARGE AMT WATER	100	18	4.0	0.2	8	40	47	9	13	42	17	.02	0.1	120
⅗ cup		1.0	0.8		21	31	30	28	151		0.3	.02	24	
DEHYDRATED	100	308	73.7	1.7	99	496	583	112	190	405	287	.45	3.0	1300
3½ oz		12.4	10.3		260	484	372	347	2207		3.9	.40	211	
CABBAGE, RED, RAW	100	31	6.9	0.2	16	80	94	18	26	42	35	.09	0.4	40
1 cup shredded		2.0	1.0		42	78	60	56	268		0.8	.06	61	
CABBAGE, SAVOY, RAW	100	24	4.6	0.2					22	67	54	.05	0.3	200
2 cups shredded		2.4	0.8						269		0.9	.08	55	
CABBAGE SPOON, RAW	100	16	2.9	0.2					26	165	44	.05	0.8	3100
3½ oz		1.6	0.6						306		0.8	.10	25	
CKD	100	14	2.4	0.2					18	148	33	.04	0.7	3100
3½ oz		1.4	0.6						214		0.6	.08	15	

*Vitamin A value for leaves is 16,000 iu per 100 grams; flower clusters, 3000 iu; stalks, 400 iu.

†Ascorbic acid values average 51 mg per 100 grams for freshly harvested, 42 mg for stored cabbage.

	WT	Proximate			Amino Acids				Minerals			Vitamins		
		CAL	CHO	FAT	TRY	LEU	LYS	MET	Na	Ca	P	THI	NIA	A
		PRO	FIB	PUFA	PHE	ISO	VAL	THR	K	Mg	Fe	RIB	ASC	D
	g	g	g	g	mg	mg	mg	mg	mg	mg	mg	mg	mg	iu
CABBAGE, SWAMP, CKD	100	21	3.9	0.2						55	32	.05	0.5	5200
1 cup		2.2	0.9						88		1.5	.08	16	
CARROTS, RAW	100	42	9.7	0.2	9	59	48	9	47	37	36	.06	0.6	11000
1 large; 2 small		1.1	1.0		38	42	51	40	341	23	0.7	.05	8	
CKD, DRAINED	100	31	7.1	0.2	7	48	39	7	33	33	31	.05	0.5	10500
⅔ cup		0.9	1.0		31	34	41	32	222		0.6	.05	6	
CND, SOLIDS & LIQUID	100	28	6.5	0.2	5	32	26	5	236	25	20	.02	0.4	10000
⅖ cup		0.6	0.6	0.6	21	23	28	22	120		0.7	.02	2	
CND, DRAINED SOLIDS	100	30	6.7	0.3	6	43	34	6	236	30	22	.02	0.4	15000
⅔ cup		0.8	0.8		28	30	37	29	120		0.7	.03	2	
DEHYDRATED	100	341	81.1	1.3	53	356	284	53	268	256	234	.31	3.0	100000
3½ oz		6.6	9.3		231	251	304	238	1944		6.0	.30	15	
CARROTS, HONEY GLAZED, FRZN	85	63	11.5	1.6					196	18	18	.03	0.5	8959
1 serving		0.5	1.2						114		0.3	.06	6	
CARROTS W/BROWN SUGAR GLAZE,	100	82	15.6	2.3					537	24	17	.03	0.5	18840
FRZN		0.8	0.7						220	12	0.2	.03	6	
CAULIFLOWER, RAW	100	27	5.2	0.2	35	181	151	54	13	25	56	.11	0.7	60
1 cup, flower pieces		2.7	1.0		84	116	162	113	295	24	1.1	.10	78	
CKD, DRAINED	100	22	4.1	0.2	30	154	129	46	9	21	42	.09	0.6	60
⅞ cup		2.3	1.0		71	99	138	97	206		0.7	.08	55	
FRZN, CKD	100	18	3.3	0.2	25	127	106	38	10	17	38	.04	0.4	30
⅞ cup		1.9	0.8		59	82	114	80	207	13	0.5	.05	41	
CAULIFLOWER, HUNGARIAN, FRZN	100	75	7.0	4.0					560	70	155	.05	0.2	180
3½ oz		3.0	1.1						208		0.4		22	
CAULIFLOWER W/CHEESE SCE, FRZN	100	65	5.6	3.0					315	90	85	.04	0.2	672
3½ oz		3.7	1.0						180		0.4	.11	28	
CELERIAC ROOT, RAW	100	40	8.5	0.3					100	43	115	.05	0.7	
4 to 6 celery roots		1.8	1.3						300		0.6	.06	8	
CELERY, RAW	20	3	0.8	tr	2		3	3	25	8	6	.01	0.1	48
1 small inner stalk (5 in)		0.2	0.1						68	4	0.1	.01	2	
RAW	50	8	2.0	0.1	4		6	5	63	20	14	.02	0.2	120
3 inner stalks; 1 outer		0.4	0.3						170	11	0.2	.02	5	
RAW	100	17	3.9	0.1	8		12	10	126	39	28	.03	0.3	†240
1 cup diced		0.9	0.6						341	22	0.3	.03	9	
CKD	100	14	3.1	0.1	8		12	10	88	31	22	.02	0.3	230
⅘ cup diced		0.8	0.6						239		0.2	.03	6	
CHARD, RAW	100	25	4.6	0.3	24	130	94	7	147	88	39	.06	0.5	6500
⅗ cup when ckd		2.4	0.8		79	103	94	98	550	65	3.2	.17	32	
CKD	100	18	3.3	0.2	18	97	70	5	86	73	24	.04	0.4	5400
⅗ cup		1.8	0.7		97	77	70	74	321		1.8	.110	16	
CHAYOTTE, RAW	100	28	7.1	0.1	8		38	1	5	13	26	.03	0.4	20
½ med squash		0.6	0.7						102		0.5	.03	19	
CHERVIL, RAW	100	57	11.5											
3½ oz		3.4											9	
CHICKPEAS (GARBANZO BEANS)	100	360	61.0	4.8	164	1517	1415	266	26	150	331	.31	2.0	50
½ cup dried		20.5	5.0		1004	1189	1004	738	797		6.9	.15	tr	
CHICKPEAS (GARBANZO BEANS), CND	100	179	30.3	2.4						75	165	.15	1.0	25
3½ oz		10.2								54	3.0	.07	0	
CHICORY OR WITLOOF, RAW	25	4	0.8	tr	3		6	2	2	4	5			tr
10 small inner leaves		0.2							46	3	0.1			
CHICORY GREENS, RAW	100	20	3.8	0.3	28		58	18		86	40	.06	0.5	4000
30 to 40 inner leaves		1.8	.8						420	13	0.9	.10	22	
CHIVES, RAW	10	3	0.6	tr						7	4	.01	0.1	580
1 T chopped		0.2	0.1						25	3	0.2	.01	6	
COLLARDS, RAW	100	40	7.2	0.7	50	202	187	43	43	203	63	.20	1.7	6500
½ cup ckd		3.6	0.9		115	112	180	104	401	57	1.0	.31	92	
CKD	100	29	4.9	0.6	38	151	140	32	25	152	39	.14	1.2	5400
½ cup		2.7	0.8		86	84	135	78	234		0.6	.20	46	
FRZN, CKD	100	30	5.6	0.4	41	162	151	35	16	176	51	.06	0.6	6800
½ cup		2.9	1.0		93	90	145	84	236	35	1.0	.14	33	
CORNSALAD†, RAW	100	21	3.6	0.4										
		2.0	0.8							13				

*Vitamin A values range from 270 iu per 100 grams for green varieties to 140 iu for yellow varieties.

†Cornsalad—European herb widely cultivated as a salad plant and potherb.

	WT	CAL	CHO	FAT	TRY	LEU	LYS	MET	Na	Ca	P	THI	NIA	A
		Proximate			Amino Acids				Minerals			Vitamins		
	g	PRO	FIB	PUFA	PHE	ISO	VAL	THR	K	Mg	Fe	RIB	ASC	D
		g	g	g	mg	mg	mg	mg	mg	mg	mg	mg	mg	iu
CORN, SWEET, WHITE OR YELLOW, RAW	100	96	22.1	1.0	21	385	130	66	tr	3	111	.15	1.7	400*
1 medium ear		3.5	0.7		196	130	220	144	280	48	0.7	.12	12	
CKD, KERNELS ON COB	100	100	21.0	1.0	20	363	122	63	tr	3	89	.12	1.4	400*
1 ear 4 in long		3.3	0.7		185	122	208	135	196		0.6	.10	9	
CND, KERNELS	100	66	15.7	0.6	11	209	70	36	236	4	48	.03	0.9	270
⅖ cup		1.9	0.6		106	70	120	78	97	19	0.4	.05	5	
CND, KERNELS	83	70	16.4	0.7	13	242	81	42	196	4	41	.02	0.7	290
½ cup		2.2	0.7		123	81	14	90	81	17	0.4	.04	4	
FRZN, KERNELS ON COB, CKD	100	94	21.6	1.0	21	385	130	66	1	3	96	.14	1.7	*350
1 medium ear		3.5	0.7		196	130	220	144	231	34	0.8	.08	7	
COWPEAS, IMMATURE SEEDS†	100	127	21.8	0.8	90	675	585	135	2	27	172	.43	1.6	370
⅔ cup		9.0	1.8		468	432	603	351	541		2.3	.13	29	
CKD	80	86	14.5	0.6	65	488	422	98	1	19	117	.02	1.1	280
½ cup		6.5	1.4		338	312	436	254	303		1.7	.09	14	
CND	100	70	12.4	0.3	50	375	325	75	236	18	112	.09	0.5	60
½ cup		5.0	0.7		260	240	335	195	352		1.5	.05	3	
FRZN, CKD	100	130	23.5	0.4	89	668	578	134	39	25	168	.40	1.4	170
3½ oz		8.9	1.5		463	427	596	347	337	55	2.8	.11	9	
COWPEAS, MATURE SEEDS, DRY, RAW	100	343	61.7	1.5	228	1710	1482	342	35	74	426	1.05	2.2	30
½ cup		22.8	4.4		1186	1094	1528	889	1024	230	5.8	.21		
CKD	100	76	13.8	0.3	51	382	332	76	8	17	95	.16	0.4	10
3½ oz		5.1	1.0		265	244	342	199	229		1.3	.04		
COWPEAS, YOUNG GREEN PODS, CKD	100	34	7.0	0.3					3	55	49	.09	0.8	1400
1 cup raw		2.2	1.7						196		0.7	.09	17	
CRESS, GARDEN, RAW	10	3	0.6	0.1	5	23	16	2	1	8	8	.01	0.1	930
5 to 8 sprigs		0.3	0.1		11	14	15	15	61		0.1	.03	7	
CKD, SMALL AMT WATER	100	23	3.8	0.6	30	146	101	11	8	61	48	.06	0.8	7700
3½ oz		1.9	0.9		68	86	93	93	353		0.8	.16	34	
CUCUMBER, RAW, NOT PARED	50	8	1.7	0.1	4	23	22	5	3	13	14	.01	0.1	125
½ medium		0.5	0.3		12	16	17	14	80	6	0.6	.02	6	
PARED	50	7	1.6	0.1	2	13	13	3	3	9	9	.01	0.1	tr
½ medium		0.3	0.2		7	9	10	8	80	5	0.2	.02	6	
DANDELION GREENS, RAW	100	45	9.2	0.7					76	187	66	.19		14000
½ cup after ckd		2.7	1.6						397	36	3.1	.26	35	
CKD	100	33	6.4	0.6					44	140	42	.13		11700
½ cup		2.0	1.3						232		1.8	.16	18	
DASHEEN (JAPANESE TARO), RAW	100	98	23.7	0.2					7	28	61	.13	1.1	20
1⅓ corms		1.9	0.8						514		1.0	.04	4	
DOCK OR SORREL, RAW	100	28	5.6	0.3					5	66	41	.09	0.5	12900
½ cup after ckd		2.1	0.8						338		1.6	.22	119	
CKD, DRAINED	100	19	3.9	0.2					3	55	26	.06	0.4	10800
½ cup		1.6	0.7						198		0.9	.13	54	
EGGPLANT, RAW	100	25	5.8	0.2	11	74	32	6	2	12	26	.05	0.6	10
2 slices; ½ cup diced		1.2	0.9		53	61	71	42	214	16	0.7	.05	5	
CKD, DRAINED	100	19	4.1	0.2	9	62	27	5	1	11	21	.05	0.5	10
½ cup diced		1.0	0.9		44	51	59	35	150		0.6	.04	3	
ENDIVE, RAW	100	20	4.1	0.1					14	81	54	.07	0.5	3300
20 long/40 small leaves		1.7	0.9						294	10	1.7	.14	10	
ESCAROLE, RAW	20	4	0.8	tr					3	16	11	.01	0.1	660
7 small inner leaves		0.3	0.2						59		0.3	.03	2	
RAW	100	20	4.1	0.1					14	81	54	.07	0.5	3300
4 large leaves		1.7	0.9						294		1.7	.14	10	
FENNEL, COMMON, LEAVES, RAW	100	28	5.1	0.4						100	51			3500
3½ oz		2.8	0.5						397		2.7		31	
GINGER ROOT, FRESH	100	49	9.5	1.0					6	23	36	.02	0.7	10
3½ oz		1.4	1.1						264		2.1	.04	4	
INDIAN SPINACH (BASELLA), RAW	100	19	3.4	0.3						109	52	.05	0.5	8000
3½ oz		1.8	0.7								1.2		102	
JERUSALEM ARTICHOKE, RAW	100	‡	16.7	0.1						14	78	.20	1.3	20
4 sm (1½ in dm)		2.3	0.8							11	3.4	.06	4	

*Vitamin A values are for yellow varieties; white varieties contain only a trace.

†Including blackeye peas.

‡Caloric values range from 7 per 100 grams for freshly harvested to 75 after long storage.

	WT	CAL	CHO	FAT	TRY	LEU	LYS	MET	Na	Ca	P	THI	NIA	A
		PRO	FIB	PUFA	PHE	ISO	VAL	THR	K	Mg	Fe	RIB	ASC	D
	g	g	g	g	mg	mg	mg	mg	mg	mg	mg	mg	mg	iu
KALE, RAW, W/STEMS	100	38	6.0	0.8	46	273	130	38	75	179	73			8900
¾ cup when ckd		4.2	1.3		172	143	197	151	318	37	2.2		125	
CKD, DRAINED, W/STEMS	100	28	4.0	0.7	35	208	99	29	43	134	46			7400
¾ cup		3.2	1.1		131	109	150	115	221		1.2		62	
CKD, DRAINED, W/O STEMS	100	39	6.1	0.7	50	292	139	40	43	187	58	.10	1.6	8300
¾ cup		4.5			184	153	212	162	221		1.6	.18	93	
FRZN; CKD	100	31	5.4	0.5	33	195	93	27	21	121	48	.06	0.7	8200
3½ oz		3.0	0.9		123	102	141	108	193	31	1.0	.15	38	
KOHLRABI, RAW	100	29	6.6	0.1					8	41	51	.06	0.3	20
⅔ cups, diced		2.0	1.0						372	37	0.5	.04	66	
CKD, DRAINED	100	24	5.3	0.1					6	33	41	.06	0.2	20
⅔ cup		1.7	1.0						260		0.3	.03	43	
LAMBSQUARTERS OR PIGWEED, RAW	100	43	7.3	0.8						309	72	.16	1.2	11600
3½ oz		4.2	2.1								1.2	.44	80	
CKD, DRAINED	100	32	5.0	0.7						258	45	.10	0.9	9700
½ cup		3.2	1.8								0.7	.26	37	
LEEKS, RAW	100	52	11.2	0.3					5	52	50	.11	0.5	40
3 to 4 (5 in long)		2.2	1.3						347	23	1.1	.06	17	
LENTILS, DRY, WHOLE, RAW	100	340	60.1	1.1	222	1754	1507	73	30	79	377	.37	2.0	60
½ cup dry		24.7	3.9		1136	1309	1334	864	790	80	6.8	.22		
CKD, DRAINED	100	106	19.3	tr	70	554	476	55		25	119	.07	0.6	20
⅔ cup		7.8	1.2		359	413	421	273	249		2.1	.06	0	
LENTILS, DRY, SPLIT, RAW	100	345	61.8	0.9	222	1754	1507	173		46	260	.37	2.0	60
½ cup dry		24.7	1.7		1136	1309	1334	864		80	6.8	.22		
LETTUCE, BUTTERHEAD	100	14	2.5	0.2	12		70	4	9	35	26	.06	0.3	970
3½ oz		1.2	0.5						264	11	2.0	.06	8	
COS, ROMAINE	100	18	3.5	0.3	13		75	4	9	68	25	.05	0.4	1900
3½ oz		1.3	0.7						264	11	1.4	.08	18	
CRISPHEAD, ICEBERG	100	14	2.5	0.2	12		70	4	9	35	26	.06	0.3	970
3½ oz		1.2	0.5						264	11	2.0	.06	8	
LETTUCE, BUNCHED, LEAF	100	18	3.5	0.3	13		75	4	9	68	25	.05	0.4	1900
3½ oz		1.3	0.7						264		1.4	.08	18	
LOTUS ROOT	100	69	15.7	0.1						30	103	.14	0.3	0
⅔ avg segment		2.8									0.6	.01	75	
MIX VEG, CND, DRAINED	161	85	17.1	0.5					504	50	68	.08	0.8	10124
1 cup		4.2	2.3						205	35	2.3	.06	8	
MIX VEG, FRZN	182	116	24.4	0.5	34	322	254	50	96	46	115	.22	2.0	9010
1 cup		5.8	1.5		170	238	254	136	348	44	2.4	.13	15	
MIX VEG W/ ONION SCE, FRZN, BIRDS	100	133	15.6	6.9					412	65	87	.11	0.9	2700
EYE 1 cup		3.6	0.9						218	23	1.0	.15	9	
MUSHROOMS, FRESH, RAW	100	28	4.4	0.5	8	316		189	15	6	116	.10	4.2	tr
(AGARICUS) 10 small/4 large		2.7	0.8			597	424		414	13	0.8	.46	3	
SAUTEED OR FRIED (LACTARIUS)	70	78	2.8	7.4	3	82	51	122		8	81	.05	2.9	173
4 medium		1.7	0.7		10	117	68	92			0.7	.27	tr	
CND, SOLIDS & LIQUIDS	100	17	2.4	0.1	6	222		133	400	6	68	.02	2.0	tr
½ cup		1.9	0.6			420	298		197	8	0.5	.25	2	
CND, DRAINED	90	17	2.3	0.2	3	129		77		7	81	.02	1.8	0
⅓ cup		1.1				243	173			7	0.7	.22		
MUSHROOMS W/BUTTER, FRZN	100	55	3.0	3.5					210	6	75	.05	1.9	200
3½ oz		1.5	0.5						251		0.4	.16	3	
MUSTARD GREENS, RAW	100	31	5.6	0.5	48	81	147	30	32	183	50	.11	0.8	7000
3½ oz		3.0	1.1		96	99	141	78	377	27	3.0	.22	97	
CKD	100	23	4.0	0.4	35	59	108	22	18	138	32	.08	0.6	5800
½ cup		2.2	0.9		70	73	103	57	220		1.8	.14	48	
FRZN	100	20	3.1	0.4	35	59	108	22	10	104	43	.03	0.4	6000
½ cup		2.2	1.0		70	73	103	57	157	23	1.5	.10	20	
MUSTARD SPINACH, RAW	100	22	3.9	0.3						210	28			9900
3½ oz		2.2	1.0								1.5		130	
CKD	100	16	2.8	0.2						158	18			8200
3½ oz		1.7	0.8								0.8		65	
NEW ZEALAND SPINACH, RAW	100	19	3.1	0.3					159	58	46	.04	0.6	4300
3½ oz		2.2	0.7						795	40	2.6	.17	30	
CKD	100	13	2.1	0.2					92	48	28	.03	0.5	3600
3½ oz		1.7	0.6						463		1.5	.10		

	WT	Proximate			Amino Acids				Minerals			Vitamins		
		CAL	CHO	FAT	TRY	LEU	LYS	MET	Na	Ca	P	THI	NIA	A
		PRO	FIB	PUFA	PHE	ISO	VAL	THR	K	Mg	Fe	RIB	ASC	D
	g	g	g	g	mg	mg	mg	mg	mg	mg	mg	mg	mg	iu
OKRA, RAW	100	36	7.6	0.3	24	134	101	29	3	92	51	.17	1.0	520
8 to 9 pods (3 in long)		2.4	1.0		86	91	120	89	249	41	0.6	.21	31	
CKD	100	29	6.0	0.3	20	112	84	24	2	92	41	.13	0.9	490
8 to 9 pods		2.0	1.0		72	76	100	74	174		0.5	.18	20	
FRZN	100	38		8.8	22	123	92	26	2	94	43	.14	1.0	480
3½ oz		2.2	1.0		79	84	110	81	164	53	0.5	.17	12	
ONIONS, MATURE, RAW	100	38	8.7	0.1	22	39	69	14	10	27	36	.03	0.2	*40
1 onion (2¼ in. dm)		1.5	0.6		42	22	33	24	157	12	0.5	.04	10	
CHOPPED, RAW	10	4	0.9	tr	3	5	9	2	1	3	4	tr	tr	4
1 T		0.2	0.1		6	3	4	3	16	1	0.1	tr	1	
CKD	100	29	6.5	0.1	18	31	55	11	7	24	29	.03	0.2	*40
½ cup		1.2	0.6		34	18	26	19	110		0.4	.03	7	
DEHYDRATED FLAKES	100	350	82.1	1.3	130	226	400	78	88	166	273	.25	1.4	200
3½ oz		8.7	4.4		244	130	191	139	1383	106	2.9	.18	35	
ONIONS, CND	100	29	5.9	0.0					240	24		.03	0.2	40
3½ oz		1.2							110		0.4	.03	7	
ONIONS, FRZN	100	38	7.9	0.2					6	14	36	.03	0.1	100
3½ oz		1.2	1.3						147		0.2	.02	8	
ONIONS IN CREAM SCE, FRZN, BIRDS EYE 3½ oz	100	138	13.1	8.6					437	8	62	.05	0.1	360
		2.1	0.6						161	16	0.2	.13	8	
ONIONS IN CREAM SCE, FRZN, GREEN GIANT 3½ oz	100	45	7.0	1.0					326	65	60	.03	0.5	18
		2.0	0.7						160		0.3	.09	10	
ONIONS, YOUNG (SCALLIONS)	100	45	10.5	0.2	16	29	51	10	5	40	39	.05	0.4	tr
5 (5 in long, ½ in dm)		1.1	1.0		31	16	24	18	231		0.6	.04	25	
PARSLEY, RAW	100	44	8.5	0.6	72		230	18	45	203	63	.12	1.2	8500
3½ oz		3.6	1.5						727	41	6.2	.26	172	
CHOPPED, RAW	10	4	0.8	tr	7		23	2	4	20	6	.01	0.1	850
1 T		0.4	tr						73	4	0.6	.02	17	
PARSNIPS, RAW	100	76	17.5	0.5					12	50	77	.08	0.2	30
½ large		1.7	2.0						541	32	0.7	.09	†16	
CKD	100	66	14.9	0.5					8	45	62	.07	0.1	30
½ cup diced		1.5	2.0						379		0.6	.08	10	
PEAS, RAW	100	84	14.4	0.4	50	397	296	44	2	26	116	.35	2.9	640
¾ cup shelled		6.3	2.0		246	290	258	101	316	35	1.9	.14	27	
CKD	100	71	12.1	0.4	43	340	254	38	1	23	99	.28	2.3	540
⅔ cup		5.4	2.0		211	248	221	86	196		1.8	.11	20	
CND, SOLIDS & LIQUID	100	66	12.5	0.3	26	208	155	23	236	20	66	.09	0.9	450
⅔ cup		3.3	1.5		129	152	135	53	96	(20)	1.7	.05	9	
CND, DRAINED SOLIDS	100	88	16.8	0.4	38	296	221	33	236	26	76	.09	0.8	690
¾ cup		4.7	2.3		183	216	193	75	96	20	1.9	.06	8	
FRZN	100	68	11.8	0.3	41	321	240	36	115	19	86	.27	1.7	600
3½ oz		5.1	1.9		199	235	209	82	135	24	1.9	.09	13	
PEAS IN CREAM SCE, FRZN, BIRDS EYE	100	156	18.1	9.1					557	53	85	.23	1.6	880
3½ oz		5.6	1.5						256	21	0.7	.25	16	
PEAS IN CREAM SCE, FRZN, GREEN GIANT 3½ oz	100	54	8.0	1.0					250	50	65	.12	0.5	2078
		3.5	1.7						140		0.6	.10	10	
PEAS IN GOLDEN SCE, FRZN, GREEN GIANT 1 cup	150	123	15.8	5.3					731	36	84	.12	1.5	780
		4.7	1.5						125		1.4	.12	26	
PEAS & CARROTS, FRZN	160	85	16.2	0.5					134	40	91	.30	2.1	14880
1 cup		5.1							251		1.8	.11	13	
PEAS & CARROTS, CND	167	100	19.5	0.7					534	45	82	.12	1.3	10711
1 cup		5.0	2.7						177	35	2.2	.14	10	
PEAS & CAR IN CREAM SCE, FRZN GREEN GIANT 3½ oz	100	55	9.0	1.0					250	55	70	.10	0.5	2000
		3.0							140		0.8	0.6	10	
PEAS & CAULIFLOWER IN CREAM SCE, BIRDS EYE 3½ oz	100	108	12.7	5.4					389	42	68	.20	1.2	620
		4.7	1.5						220	19	0.7	.14	20	
PEAS & ONIONS, FRZN, GREEN GIANT 3½ oz	100	80	8.6	3.0					400	20	55	.15	0.8	574
		3.4	1.8						105		0.7	.08	15	
PEAS & PEARL ONIONS, FRZN, BIRDS EYE 3½ oz	100	64	12.5	0.3					329	16	62	.25	1.9	670
		4.8	1.8						181	19	0.9	.08	20	

*Vitamin A value based on yellow-fleshed varieties; white varieties contain only a trace.

†Ascorbic acid values vary with season and storage, averaging 24 mg per 100 grams during first 3 months of storage in the fall and dropping to less than 12 after 6 months storage.

	WT	CAL	CHO	FAT	TRY	LEU	LYS	MET	Na	Ca	P	THI	NIA	A
		PRO	FIB	PUFA	PHE	ISO	VAL	THR	K	Mg	Fe	RIB	ASC	D
	g	g	g	g	mg	mg	mg	mg	mg	mg	mg	mg	mg	iu
PEAS W/ POT IN CREAM SCE, FRZN,	100	165	21.2	9.3					608	53	80	.20	1.4	680
BIRDS EYE 3½ oz		5.0	1.3						306	19	0.5	.17	17	
PEAS W/ SLICED MUSHROOMS, FRZN,	100	70	11.4	0.5					257	18	76	.26	2.2	730
BIRDS EYE 3½ oz		4.9	0.1						189	22	1.2	.11	22	
PEAS, DRIED SEEDS, WHOLE	30	102	18.0	0.4	79	598	526	86	10	19	102	.22	0.9	36
½ cup when ckd		7.2	1.5		360	403	403	281	301	54	1.5	.09		
WHOLE	100	340	60.3	1.3	265	2000	1759	289	35	64	340	.74	3.0	120
½ cup		24.1	4.9		1205	1350	1350	940	1005	180	1.5	.29		
SPLIT	30	104	18.8	0.3	80	606	532	87	12	10	81	.22	0.9	36
½ cup when ckd		7.3	0.4		365	409	409	285	268		1.5	.09		
SPLIT	100	348	62.7	1.0	266	2008	1766	290	40	33	268	.74	3.0	120
½ cup		24.2	1.2		1210	1356	1356	944	895		5.1	.29		
CKD	100	115	20.8	0.3	88	664	584	96	13	11	89	.15	0.9	40
3½ oz		8.0	0.4		400	448	448	312	296		1.7	.09		
PEPPERS, HOT CHILI, MATURE RED,	100	65	15.8	0.4					25	16	49	.10	2.9	21600
PODS, RAW 3½ oz		2.3	2.3						564		1.4	.20	369	
PEPPERS, HOT CHILI, MATURE RED,	50	160	29.9	4.5					186	65	120	.11	5.2	38500
PODS, DRIED 1¾ oz		6.4	3.1						600		3.9	.66	6	
PEPPERS, HOT CHILI, MATURE RED,	100	93	18.1	2.3					9	29	78	.22	4.4	21600
PODS & SEEDS, RAW 3½ oz		3.7	9.0						420	27	1.2	.36	369	
PEPPERS, GREEN, RAW	100	22	4.8	0.2	8	46	50	16	13	9	22	.08	0.5	420
1 large shell		1.2	1.4		55	46	32	50	213	18	0.7	.08	128	
CKD, DRAINED	100	18	3.8	0.2	7	38	42	13	9	9	16	.06	0.5	420
3½ oz		1.0	1.4		46	38	27	42	149		0.5	.07	96	
BAKED	65	17	3.9	0.1	6	30	34	10		7	16	.03	0.3	481
1 shell, no filling		0.8	1.0		37	30	22	34			0.3	.05	64	
PIGEON PEAS, IMMATURE, RAW	100	117	21.3	0.6					5	42	127	.40	2.2	140
½ cup		7.2	3.3						552		1.6	.17	.39	
MATURE SEEDS, RAW	100	342	63.9	1.4	102	1591	1469	245	26	107	316	.32	3.0	80
3½ oz		20.4	7.0		1754	1265	1081	775	981	121	8.0	.16		
PIMIENTOS, CND	40	11	2.3	0.2						*3	7	.01	0.2	920
1 medium		0.4	0.2								0.6	.02	38	
CND	100	27	5.8	0.5						*7	17	.02	0.4	2300
3 medium		0.9	0.6								1.5	.06	95	
POI, TWO FINGER	100	67	16.0	0.1						11	22	.04	0.3	0
⅔ cup		0.6									0.4	.01	†5	
POTATO, RAW	100	76	17.1	0.1	21	105	111	25	3	7	53	.10	1.5	tr
1-2¼ in. dia		2.1	0.5		92	92	111	86	407	34	0.6	.04	‡20	
BAKED (W/O SKIN)	100	95	21.1	0.1	26	130	138	31	4	9	65	.10	1.7	tr
1-2½ in. dm		2.6	0.6		114	114	138	107	503		0.7	.04	20	
BAKED (W/O SKIN)	150	139	31.7	0.2	39	195	207	47	6	14	98	.15	2.6	tr
1-3¼ in. dm		3.9	0.9		172	172	207	160	755		1.0	.06	30	
POTATO, BAKED W/CHEESE, FRZN,	142	250	30.0	12.5					703	48	107	.04	1.4	63
GREEN GIANT 1 potato		4.5	0.7						727		1.7	.11	7	
POTATO, BAKED W/SOUR CREAM,	142	243	32.0	10.7					540	43	105	.03	1.7	95
FRZN, GREEN GIANT 1 potato		4.7	0.7						721		2.0	.13	12	
BOILED (PARED BEFORE CKD)	100	65	14.5	0.1	19	95	101	23	2	6	42	.09	1.2	tr
1-2¼ in. dm		1.9	0.5		84	84	101	78	285		0.5	.03	16	
BOILED IN SKIN	100	76	17.1	0.1	21	105	111	25	3	7	53	.09	1.5	tr
1 medium		2.1	0.5		92	92	111	86	407		0.6	.04	16	
CND, SOLIDS & LIQUID	100	44	9.8	0.2	11	55	58	13	1	4	30	.04	0.6	tr
⅔ cup		1.1	0.2		48	48	58	45	250		(0.3)	.02	13	
DEHYDRATED FLAKES	100	364	84.0	0.6	72	360	382	86	80	35	173	.23	5.4	tr
3½ oz		7.2	1.6		317	317	382	295	1600	100	1.7	.06	32	
FRENCH FRIED	50		18.0	6.6	21	105	111	25	3	8	56	.06	1.6	tr
10 pieces (½ × ½ × 2 in.)		2.1	0.5		92	92	111	86	427		0.6	.04	10	
FRENCH FRIED, FRZN	100	220	33.7	8.4	36	180	191	43	4	9	86	.14	2.6	tr
3½ oz		3.6	0.7		159	159	191	149	652	25	1.8	.02	21	
FRIED FROM RAW	85	228	27.8	12.1	34	170	180	41	190	13	86	.10	2.4	tr
½ cup		3.4	0.8		150	150	180	139	658		0.9	.06	16	

*Calcium salts may be added as a firming agent, not to exceed 26 mg per 100 grams of finished product (Federal standard).
†The Hawaii Agricultural Experiment Station notes that commercial poi is subjected to long cooking and may contain little or no ascorbic acid.

‡Ascorbic acid values range from about 26 mg per 100 grams in recently dug potatoes to about 12, after 3 months' storage, and about 8 after 6 months storage.

		Proximate			Amino Acids				Minerals			Vitamins		
	WT	CAL	CHO	FAT	TRY	LEU	LYS	MET	Na	Ca	P	THI	NIA	A
		PRO	FIB	PUFA	PHE	ISO	VAL	THR	K	Mg	Fe	RIB	ASC	D
	g	g	g	g	mg	mg	mg	mg	mg	mg	mg	mg	mg	iu
HASHED BROWN	100	229	29.1	11.7	31	155	164	37	288	12	79	.08	2.1	tr
½ cup		3.1	0.8		136	136	164	127	475		0.9	.05	9	
MASHED, MILK & MARGARINE	100	94	12.3	4.3	21	105	111	25	331	24	48	.08	1.0	170
½ cup		2.1	0.4		92	92	111	86	250		0.4	.05	9	
POTATO, MASHED, FROM DEHYD	210	166	27.5	4.6					491	65	92	.06	1.7	190
1 cup		4.2							704		1.3	.11	6	
POTATO, MASHED, FRZN	195	181	30.6	5.5					700	49	82	.12	1.4	273
1 cup		3.5	0.8						419	23	1.2	.08	8	
POTATO, SCALLOPED	122	127	17.9	4.8					433	66	90	.07	1.2	195
½ cup		3.7	0.4						399		0.5	.11	13	
POTATO, SCALLOPED, FRZN, STOUFFERS 3½ oz	100	108	11.6	5.0					430	96		.06	0.6	200
		3.5									0.5	.11	5	
POTATO, SCALLOPED, W/CHEESE	122	177	16.6	9.6					545	155	149	.07	1.1	390
½ cup		6.5	0.4						373		0.6	.15	12	
POTATO, SCALLOPED W/HAM, HORMEL	100	90	8.6	3.7					576	48	116	.18	1.0	50
		5.7	0.3							13	0.5	.07	6	
POTATO, TATOR TOTS, FRZN, ORE-IDA 3½ oz	100	200	23.5	11.8					545	11	40	.11	1.7	100
		2.4	0.5						395		0.5	.02	8	
POTATO, TINY TATERS, FRZN, BIRDS EYE 3½ oz	100	271	27.3	16.4					565	9	64	.10	0.1	
		3.6	0.6						457	18	0.7	.08	10	
POTATO & PEAS, CREAMED	100	75	12.9	1.4					72	35	58	.84	0.9	208
3½ oz		2.9	0.8	0.1					211	8	0.7	.07	11	9
POTATO AU GRATIN, GRN GIANT	100	95	12.7	3.1					529	85	178	.06	0.7	26
3½ oz		3.5	0.5						209		0.4	.11	7	
POTATO AU GRATIN, FRZN, STOUFFERS 3½ oz	100	123	11.0	6.7					380	96		.07	0.9	160
		3.8									0.5	.08	4	
POTATO DELMONICO, FRZN, STOUFFERS 3½ oz	100	116	10.7	6.7					410	68		.07	0.7	375
		2.8									0.5	.09	7	
POTATO MAISON, FRZN, GREEN GIANT 3½ oz	100	89	11.9	3.5					476					
		2.3							212					
POTATO O'BRIEN	100	92	13.4	3.6					520	10	39	.07	1.4	87
3½ oz		1.8	0.5						263		0.5	.04	21	
POTATO CHIPS	20	113	10.0	8.0	11	55	58	13		8	28	.04	1.0	tr
10 pieces (2 in dm)		1.1	0.3		48	48	58	45	226		0.4	.01	3	
POTATO CHIPS	100	568	50.0	39.8	53	265	281	64		40	139	.21	4.8	tr
3½ oz		5.3	1.6		233	233	281	217	1130		1.8	.07	16	
POTATO CHIPS, BBQ FLAVOR, WONDER 1 oz	28	152	13.7	10.5					197	13	41	.09	1.4	0
		1.5									0.5	.03	4	
POT CHIPS, CHIPSTERS, NABISCO	28	130	18.0	6.0					420	9	31	.10	0.9	
1 oz		1.0							31		0.3	.11		
POT CHIPS, MUNCHOS	28	157	14.4	10.5					230	17	49	.02	1.1	24
1 oz		1.5	0.4	1.1					260	17	0.4	.03	5	
POT CHIPS, PRINGLES	28	146	15.4	9.5					252	10	39	.04	1.4	4
1 oz		1.4		0.8					196	20	0.3	.01	25	
POT CHIPS, SOUR CREAM & ONION FLAVOR, FRITO-LAY 1 oz	28	153	14.5	9.7					114	6	9	.01	0.3	14
		2.0	0.7						127	4	0.1	.02	1	
POTATO FRIES, FRITO-LAY	28	158	14.8	10.1										
1 oz		1.5	0.4											
POTATO STICKS	28	152	14.2	10.2					280	12	39	.06	1.3	
1 oz		1.8	0.4						316		0.5	.02	11	
PRICKLYPEAR, RAW	100	42	10.9	0.1	4	21	20	4	2	20	28	.01	0.4	60
3½ oz		0.5	1.6		27	20	18	24	166		0.3	.03	22	
PUMPKIN, RAW	100	26	6.5	0.1	13	79	46	19	1	21	44	.05	0.6	1600
3½ oz		1.0	1.1		56	56	54	30	340	12	0.8	.11	9	
CND	100	33	7.9	0.3	13	79	46	19	2	25	26	.03	0.6	6400
⅖ cup		1.0	1.3		56	56	54	30	240		0.4	.05	5	
CND	115	38	9.1	0.3	16	63	58	11	2	29	30	.03	0.7	7400
½ cup		1.2	1.5		31	44	44	28	276		0.5	.06	6	
PURSLANE, RAW	100	21	3.8	0.4						103	39	.03	0.5	2500
3½ oz		1.7	0.9								3.5	.10	25	
CKD	100	15	2.8	0.3						86	24	.02	0.4	2100
3½ oz		1.2	0.8								1.2	.06	12	
RADISH, RED, RAW	100	17	3.6	0.1	4		28	2	18	30	31	.03	0.3	10
10 small (1 in dm)		1.0	0.7				25	49	322	15	1.0	.03	26	

	WT g	CAL / PRO	CHO / FIB	FAT / PUFA	TRY / PHE mg	LEU / ISO mg	LYS / VAL mg	MET / THR mg	Na / K mg	Ca / Mg mg	P / Fe mg	THI / RIB mg	NIA / ASC mg	A / D iu
RADISH, ORIENTAL, RAW	100	19	4.2	0.1						35	26	.03	0.4	10
3½ oz		0.9	0.7						180		0.6	.02	32	
RHUBARB, RAW	100	16	3.7	0.1					2	96	18	.03	0.3	100
3½ oz		0.6	0.7						251	16	0.8	.07	9	
CKD, SWEETENED	100	141	36.0	0.1					2	78	15	.02	0.3	80
⅜ cup		0.5	0.6						203	13	0.6	.05	6	
FRZN, SWEETENED	100	143	36.2	0.2					3	78	12	.02	0.2	70
⅜ cup		0.5	0.8						176	12	0.7	.04	6	
RUTABAGA, RAW	100	46	11.0	0.1					5	66	39	.07	1.1	580
3½ oz		1.1	1.1						239	15	0.4	.07	43	
CKD	100	35	8.2	0.1					4	59	31	.06	0.8	550
½ cup cubed		0.9	1.1						167		0.3	.06	26	
SALSIFY, RAW	100	*	18.0	0.6						47	66	.04	0.3	10
⅔ cup when ckd		2.9	1.0						380		1.5	.04	11	
SALSIFY, BOILED, DRAINED	135	16-94*												
1 cup														
SAUERKRAUT, CND	100	18	4.0	0.2					747	36	18	.03	0.2	50
⅔ cup drained solids		1.0	0.7						140		0.5	.04	14	
SEAWEED														
AGAR, RAW	100			0.3						567	22			
3½ oz			0.7								6.3			
DULSE, RAW	100			3.2					2085	296	267			
3½ oz			1.2						8060					
IRISHMOSS, RAW	100			1.8					2892	885	157			
3½ oz			2.1						2844		8.9			
KELP, RAW	100			1.1					3007	1093	240			
3½ oz			6.8						5273					
LAVER, RAW	100			0.6										
3½ oz			3.5											
SHALLOT BULBS, RAW	50	36	8.4	0.1					6	18	30	.03	0.1	tr
1¾ oz		1.2	0.4						167		0.6	.01	4	
SOYBEANS, IMMATURE, RAW	100	134	13.2	5.1	164	926	752	164		67	225	.44	1.4	690
3½ oz		10.9	1.4		589	643	632	469			2.8	.16	29	
CKD	100	118	10.1	5.1	147	833	676	147		60	191	.31	1.2	660
⅔ cup		9.8	1.4		529	578	568	421			2.5	.13	17	
CND, DRAINED SOLIDS	100	103	7.4	5.0	135	765	621	135	236	67	114	.06		340
⅔ cup		9.0	1.4		486	531	522	387			2.8		2	
SOYBEANS, MATURE SEEDS, RAW	100	403	33.5	17.7	512	2898	2353	512	5	226	554	1.10	.22	80
½ cup		34.1	4.9		1841	2012	1978	1466	1677	265	8.4	.31		
CKD	100	130	10.8	5.7	165	935	759	165	2	73	179	.21	5.6	30
½ cup		11.0	1.6		594	649	638	423	540		2.7	.09	0	
SOYBEAN CURDS (TOFU)	100	72	2.4	4.2				94	7	128	126	.06	0.1	0
3½ oz		7.8	0.1						42	111	1.9	.03	0	
SOYBEANS, FERMENTED, MISO	100	171	23.5	4.6					2950	68	309	.06	0.3	40
3½ oz		10.5	2.3						334		1.7	.10	0	
SOYBEANS, FERMENTED, NATTO	100	167	11.5	7.4						103	182	.07	1.1	0
3½ oz		16.9	3.2						249		3.7	.50	0	
SPINACH, RAW	100	26	4.3	0.3	51	246	198	54	71	93	51	.10	0.6	8100
3½ oz		3.2	0.6		138	150	176	141	470	88	3.1	.20	51	
CKD	90	21	3.2	0.3	43	208	167	46	45	83	33	.06	0.5	7300
½ cup		2.7	0.5		116	127	148	119	291		2.0	.13	25	
CND, SOLIDS & LIQUID	116	22	3.5	0.5	37	177	143	39	273	98	30	.02	0.3	6400
½ cup		2.3	0.8		99	108	126	101	290	73	2.4	.12	16	
CND, DRAINED SOLIDS	90	22	3.2	0.5	39	185	149	41	212	106	23	.01	0.3	7200
½ cup		2.4	0.8		103	113	131	105	227	56	2.3	.11	13	
FRZN, CHOPPED, CKD	100	23	3.7	0.3	48	231	186	51	52	113	44	.07	0.4	7900
½ cup		3.0	0.8		129	141	165	132	333	65	2.1	.15	19	
FRZN, LEAF, CKD	100	24	3.9	0.3	46	223	180	49	49	105	44	.08	0.5	8100
½ cup		2.9	0.8		125	136	160	128	362	(65)	2.5	.14	28	

*Salsify—Caloric values range from 16 when prepared from freshly harvested salsify to 94 when prepared from stored salsify. A large portion of carbohydrate may be inulin (which is of doubtful availability) in freshly harvested samples. In stored samples, inulin may have been converted to available sugars.

	WT	Proximate			Amino Acids				Minerals			Vitamins		
		CAL	CHO	FAT	TRY	LEU	LYS	MET	Na	Ca	P	THI	NIA	A
		PRO	FIB	PUFA	PHE	ISO	VAL	THR	K	Mg	Fe	RIB	ASC	D
	g	g	g	g	mg	mg	mg	mg	mg	mg	mg	mg	mg	iu
SPINACH, CREAMED, FRZN, BIRDS	100	60	4.7	3.9					327	64	40	.07	0.3	6500
EYE		3.1	0.9						338	22	0.7	.13	16	
SPINACH IN BUTTER SCE, FRZN,	200	90	4.4	6.0					770	158	80	.12	0.6	6574
GREEN GIANT 1 cup		5.0	1.4						620		3.6	.20	44	
SPINACH IN CREAM SCE, FRZN,	200	120	10.0	7.0					760	182	100	.12	0.6	4357
GREEN GIANT 1 cup		5.2	1.6						550		1.6	.20	30	
SPINACH, DEVILED, FRZN, GREEN	100	70	4.6	3.6					407	120	115	.06	0.3	1794
GIANT		4.9	1.2						305		1.2	.18	20	
SQUASH, SUMMER, RAW	100	19	4.2	0.1	9	50	42	14	1	28	29	.05	1.0	410
½ cup when ckd		1.1	0.6	0.1	30	35	41	25	202	16	0.4	.09	22	
BOILED, DRAINED	100	14	3.1	0.1	7	41	34	12	1	25	25	.05	0.8	390
½ cup		0.9	0.6		24	29	33	21	141		0.4	.08	10	
FRZN, CKD	100	21	4.7	0.1	11	63	53	18	3	14	32	.06	0.4	140
½ cup		1.4	0.6		38	45	52	32	167	16	0.7	.04	8	
SQUASH, WINTER, RAW	100	50	12.4	0.3	11	63	53	18	1	22	38	.05	0.6	3700
½ cup when ckd		1.4	1.4		38	45	52	32	369	17	0.6	.11	13	
BAKED	100	63	15.4	0.4	14	81	68	23	1	28	48	.05	0.7	4200
½ cup		1.8	1.8		49	58	66	41	461		0.8	.13	13	
BOILED, MASHED	100	38	9.2	0.3	9	50	42	14	1	20	32	.04	0.4	3500
⅖ cup		1.1	1.4		30	35	41	25	258		0.5	.10	8	
FRZN, CKD	100	38	9.2	0.3	10	54	46	16	1	25	32	.03	0.5	3900
½ cup		1.2	1.2		32	38	45	28	207	17	1.0	.07	8	
SQUASH, WINTER														
ACORN, RAW	244	97	24.6	0.2					2	68	51	.11	1.3	2640
½ squash		3.3	3.4						843	78	2.0	.24	31	
ACORN, BAKED	195	86	21.8	0.2					2	61	45	.08	1.1	2180
½ squash		3.0	3.5						749		1.7	.20	20	
ACORN, BOILED, MASHED	245	83	20.6	0.2					2	69	49	.10	1.0	2700
1 cup		2.9	3.4						659		2.0	.25	20	
BUTTERNUT, RAW	200	108	28.0	0.2					2	64	116	.10	1.2	1140
1 cup		2.8	2.8						574	33	1.6	.22	18	
BUTTERNUT, BAKED, MASHED	205	139	35.9	0.2					2	82	148	.10	1.4	13120
1 cup		3.7	3.7						1248		2.1	.27	16	
BUTTERNUT, BOILED, MASHED	245	100	25.5	0.2					2	71	120	.10	1.0	13230
1 cup		2.7	3.4						835		1.7	.25	12	
HUBBARD, RAW	200	78	18.8	0.6					2	38	62	.10	1.2	8600
1 cup		2.8	2.8						434	38	1.2	.22	22	
HUBBARD, BAKED, MASHED	205	103	24.0	0.8					2	49	80	.10	1.4	9840
1 cup		3.7	3.7						556		1.6	.27	21	
HUBBARD, BOILED, MASHED	245	74	16.9	0.7					2	42	64	.10	1.0	10050
1 cup		2.7	3.4						372		1.2	.25	15	
SUCCOTASH (CORN & LIMA BEANS),	200	186	41.0	0.8					76	26	17.0	.18	2.6	600
FRZN 1 cup		8.4	1.8						492	70	2.0	.10	12	
SWEET POT FIRM FLESH, RAW	180	165	36.5	1.1	56	185	153	59	16	52	76	.16	1.0	14900
(JERSEY TYPE) 1 potato		2.9	1.6		180	157	243	153	394	56	1.1	.10	37	
SWEET POT, SOFT FLESH, RAW	180	190	44.2	0.5	56	185	153	59	16	52	76	.16	1.0	14090
(PUERTO RICO TYPE) 1 med		2.8	1.3		180	157	243	153	394	56	1.1	.10	32	
SWEET POTATO, RAW	100	114	26.3	0.4	29	97	80	31	10	32	47	.10	0.6	8800*
1 small		1.7	0.7		94	82	128	80	243	31	0.7	.06	21	
BAKED IN SKIN	100	141	32.5	0.5	36	120	99	38	12	40	58	.09	0.7	8100*
1 small		2.1	0.9		116	101	157	99	300		0.9	.07	22	
BAKED IN SKIN	180	254	58.5	0.9	65	217	179	68	22	72	104	.16	1.3	14600*
1 large		3.8	1.6		209	182	285	179	540		1.6	.13	40	
SWEET POT, BOILED IN SKIN	180	172	39.8	0.6					15	48	71	.14	0.9	11940
1 large		2.6	1.3		153				367		1.1**	.09	26	
CND, SYRUP PACK	100	114	27.5	0.2	17	57	47	18	48	13	29	.03	0.6	3000*
1 small, syrup		1.0	0.6		55	48	75	47	120		0.7	.03	8	
CND, SOLID PACK	100	108	24.9	0.2	34	114	94	36	48	25	41	.05	0.6	7800*
1 small		2.0	1.0		110	96	150	94	200		0.8	.04	14	

*Vitamin A value based on freshly harvested squash. During storage, the carotenoid content increases, adding to the Vitamin A value.

	WT g	CAL PRO	CHO FIB	FAT PUFA	TRY PHE	LEU ISO	LYS VAL	MET THR	Na K	Ca Mg	P Fe	THI RIB	NIA ASC	A D
DEHYDRATED FLAKES	100	379	90.0	0.6	71	239	197	76	181	60	80	.06	1.3	47000*
3½ oz		4.2	3.2		231	202	315	197	562	100	2.2	.13	45	
TAMPALA LEAVES, RAW	100	37	0.2	8.4						6	229	1.6		12300
½ cup when ckd		2.1	0.9								3.4		51	
TARO, HAWAIIAN, CORMS, RAW	100	98	23.7	0.2	34	169	110	21	7	28	61	.13	1.1	820
¾ cup, cubed		1.9	0.8		99	99	114	89	514		1.0	.04	4	
LEAVES OR LUAU	100	40	7.4	0.8	54	267	174	33		76	59			
8 leaves		3.0	1.4		156	156	180	141			1.0		31	
TARO, JAPANESE-SEE DASHEEN														
TOMATO, GREEN, RAW	100	24	5.1	0.2					3	13	27	.06	0.5	270
1 small		1.2	0.5						244		0.5	.04	20	
TOMATO, RIPE, RAW	100	22	4.7	0.2	10	45	46	8	3	13	27	.06	0.7	900
1 small		1.1	0.5		31	32	31	36	244	14	0.5		23	
RAW	150	33	7.0	0.3	15	68	69	12	4	20	40	.09	1.0	1350
1 medium		1.6	0.8		46	48	46	54	366	21	0.8	.06	34	
RAW	200	44	9.4	0.4	20	90	92	16	6	26	54	.12	1.4	1800
1 large		2.2	1.0		62	64	62	72	488	28	1.0	.08	46	
BOILED	100	26	5.5	0.2	12	53	55	9	4	15	32	.07	0.8	1000
½ cup		1.3	0.6		36	38	36	43	287		0.6	.05	24	
CND	100	21	4.3	0.2	9	41	42	7	130	6	19	.05	0.7	900
½ cup		1.0	0.4		28	29	28	33	217	12	0.5	.03	17	
TOMATO, SCALLOPED	100	94	12.3	3.9					164	39	45	.08	0.9	652
3½ oz		2.8	0.4	0.4					185	14	0.8	.07	13	0
TOMATO PASTE, CND	100	82	18.6	0.4					38	27	70	.20	3.1	3300
3½ oz		3.4	0.9						888	20	3.5	.12	49	
TOMATO PUREE, CND	100	39	8.9	0.2	15	70	71	12	399	13	34	.09	1.4	1600
6 T		1.7	0.4		48	49	48	56	426	20	1.7	.05	3.3	
CND	249	97	22.2	0.5	38	172	176	29	1000	32	85	.22	35	4000
1 cup		4.2	1.0		118	122	118	139	1060	50	4.2	.12	82	
TOMATO SCE, CND, CONTADINA	124	45	9.0	0.2					654	10	40	.05	1.6	1250
½ cup		1.7							589	20	0.9	.05	9	14
TOMATO SCE, CND, DEL MONTE	125	43	9.8	0.3					656	14	46	.10	1.5	1648
½ cup		1.9	0.4						463	22	1.0	.05	23	
TOMATO SCE W/ MUSHROOMS, CND, DEL MONTE 3½ oz	100	41	9.4	0.2					412	11	35	.08	1.3	968
		1.5							407	20	0.9	.07	16	
TOMATO SCE W/ TIDBITS, CND, DEL MONTE 3½ oz	100	36	8.4	0.2					437	12	29	.07	1.0	918
		1.2							346	18	0.9	.05	8	
TURNIP GREENS, RAW	100	28	5.0	0.3	48	213	135	54		246	58	.21	0.8	7600
½ cup after ckd		3.0	0.8		150	111	153	129		58	1.8	.39	139	
CKD, SMALL AMT WATER	100	20	3.6	0.2	35	156	99	40		184	37	.15	0.6	6300
⅔ cup		2.2	0.7		110	81	112	95			1.1	.24	69	
CKD, LARGE AMT WATER	100	19	3.3	0.2	35	156	99	40		174	34	.10	0.5	5700
⅔ cup		2.2	0.7		110	81	112	95			1.0	.23	47	
CND, SOLIDS & LIQUID	100	18	3.2	0.3	24	106	68	27	236	100	30	.02	0.6	4700
½ cup		1.5	0.7		75	56	76	64	243		1.6	.09	19	
TURNIP ROOT, RAW	100	30	6.6	0.2			52	11	49	39	30	.04	0.6	tr
¾ cup diced		1.0	0.9		18	18			268	20	0.5	.07	36	
CKD	100	23	4.9	0.2			42	9	34	35	24	.04	0.3	tr
⅔ cup diced		.8	0.9		14	14			188		0.4	.05	22	
VEGETABLE COMBINATIONS														
BAVARIAN-STYLE BEANS (SPAETZLE), FRZN, BIRDS EYE 3½ oz	100	53	10.6	1.1					218					
		3.2							154					
BOMBAY, FRZN, GREEN GIANT	100	43	5.9	1.7					284	20	29	.03	0.3	4105
3½ oz		1.1	0.8						156		0.3	.07	15	
CHINESE STYLE, FRZN,	100	21	5.3	0.0					463					148
3½ oz		2.1							144					
CHOW MEIN, CND, CHINA BEAUTY	227	39	7.7	0.2										
1 cup		4.0	1.8											

*Vitamin A values have a wide range: 21,000 to 72,000 iu per 100 grams in dehydrated form, and 5,000 to 18,000 as prepared for serving.

		Proximate			Amino Acids				Minerals			Vitamins		
	WT	CAL	CHO	FAT	TRY	LEU	LYS	MET	Na	Ca	P	THI	NIA	A
		PRO	FIB	PUFA	PHE	ISO	VAL	THR	K	Mg	Fe	RIB	ASC	D
	g	g	g	g	mg	mg	mg	mg	mg	mg	mg	mg	mg	iu
DANISH STYLE, FRZN, BIRDS EYE	100	32	8.5	0.0					335					
3½ oz		2.1							170					
FLORIDA-STYLE, FRZN, GREEN	100	50	4.4	3.0					356	20	30	.02	0.2	3366
GIANT 3½ oz		1.0	0.5						137		0.4	.04	5	
HAWAIIAN STYLE, FRZN, BIRDS EYE	227	219	29.0	0.0					737					
1 cup		2.4							266					
ITALIAN STYLE W/SCE, FRZN, BIRDS	100	99	8.9	6.7					549	26	32	.07	0.5	930
EYE 3½ oz		2.5	1.1						164	16	0.7	.08	26	
JAPANESE STYLE, FRZN, BIRDS EYE	100	97	8.5	6.5					280	33	33	.05	0.5	690
3½ oz		1.6	0.9						30	14	0.4	.08	16	
JUBILEE, FRZN, BIRDS EYE	100	128	18.4	5.6					270	43	84	.10	1.2	1610
3½ oz		3.1	0.6						208	20		.12	10	
MEDITERRANEAN STYLE, FRZN,	100	48	11.7	0.0					479					
BIRDS EYE 3½ oz		2.1							186					
MEXICAN STYLE W/ SCE, FRZN,	100	78	17.6	0.3					265	29	64	.10	1.1	400
BIRDS EYE 3½ oz		4.4	1.4						201	18	0.7	.06	10	
NEW ENGLAND STYLE, FRZN, BIRDS	100	65	12.8	1.2					339	25	45	.10	0.6	730
EYE 3½ oz		3.2	0.8						170	23	0.7	.08	9	
NEW ORLEANS STYLE, FRZN, GREEN	100	80	8.0	4.0					605	23	45	.03	0.3	544
GIANT 3½ oz		3.0	1.0						273		0.8	.08	6	
NEW ORLEANS CREOLE STYLE, FRZN,	100	70	14.2	0.4					378	29	49	.26	0.7	630
BIRDS EYE 3½ oz		2.9	0.4						286	23	0.8	.10	19	
NORTHWEST STYLE, FRZN, GREEN	100	85	9.3	3.5					480	30	65	.05	0.5	570
GIANT 3½ oz		3.7	1.1						242		1.2	.06	9	
PARISIAN STYLE W/ SCE, FRZN,	100	92	6.3	6.8					225	26	29	.03	0.5	7150
BIRDS EYE 3½ oz		1.4	0.9						83	7	0.3	.06	5	
PENNA DUTCH STYLE, FRZN, BIRDS	100	43	7.3	1.0					368	24	28	.05	0.4	6970
EYE 3½ oz		1.7	0.6						171	12	0.3	.16	6	
PENNA DUTCH STYLE, FRZN, GREEN	100	100	15.0	3.0					350	15	40	.03	0.3	4500
GIANT 3½ oz		2.0	1.1						378		0.6	.10	18	
SAN FRANCISCO STYLE, FRZN,	100	48	6.8	0.9					324	24	40	.06	0.5	
BIRDS EYE 3½ oz		2.5	0.4						173	16	0.6	.06	11	
SAN FRANCISCO STYLE, FRZN,	100	35	5.0	1.0					532	20	35	.01	0.3	383
GREEN GIANT 3½ oz		1.6	0.6						135		0.4	.11	22	
SWISS STYLE, FRZN, BIRDS EYE	100	37		0.0					303					
3½ oz		2.1							176					
WISCONSIN STYLE, FRZN, BIRDS	100	43	6.8	1.0					301	47	49	.07	0.4	4780
EYE 3½ oz		2.8	1.0						209	13	0.5	.07	32	
VEGETABLE MARROW, CKD	100	16	3.9	0.1						15	15	.04	0.6	260
½ cup		0.6	0.5								0.4	.07	11	
WATER CHESTNUTS, CHINESE	25	20	4.8	0.1					5	1	16	.04	0.2	0
4 chestnuts		0.4	0.2						125		0.2	.05	1	
CHINESE	100		19.0	0.2					20	4	65	.14	1.0	0
16 chestnuts		1.4	0.8						500	12	0.6	.20	4	
WATERCRESS RAW	10	2	0.3	tr	4	17	12	1	5	15	5	.02	0.1	490
10 sprigs		0.2	0.1		8	10	11	11	28	2	0.2	.01	8	
RAW	100	19	3.0	0.3	35	169	117	13	52	151	54	.08	0.9	4900
3½ oz		2.2	0.7		79	99	108	108	282	20	1.7	.16	79	
YAM, RAW	100	101	23.2	0.2	36		111	34		20	69	.10	0.5	*tr
⅔ cup when ckd		2.1	0.9						600		0.6	.04	9	
CKD IN SKIN	100	105	24.1	0.2	41		127	38		4	50	.09	0.6	*tr
½ cup		2.4	0.9								0.6	.04	9	
CKD IN SKIN	200	210	48.2	0.4	82		254	76		8	100	.18	1.2	*tr
1 cup		4.8	1.8								1.2	.08	18	
YAMS & APPLES, FRZN, STOUFFERS	100	133	27.4	1.8					73	17		.01	0.0	4259
3½ oz		1.7									0.5	.09	27	
YAMBEAN	100	55	12.8	0.2						15	18	.04	0.3	tr
3½ oz		1.4	0.7								0.6	.03	20	
YAUTIA (MELANGA, TANIER)	100	136	31.5	0.4	29		82	19		4	50	.10	0.6	*tr
1 corm		2.1	0.8								0.6	.15	11	

*Vitamin A values vary with carotenoid content; white varieties have only a trace.

	WT	CAL	CHO	FAT	TRY	LEU	LYS	MET	Na	Ca	P	THI	NIA	A
	g	PRO	FIB	PUFA	PHE	ISO	VAL	THR	K	Mg	Fe	RIB	ASC	D
		g	g	g	mg	mg	mg	mg	mg	mg	mg	mg	mg	iu
AMARANTH*, RAW (PURPLE HEART)	10	4	0.7	0.1	4	21	14	3		27	7	.01	0.1	610
		0.4	0.1		10	16	14	6	41		0.4	.02	8	
BAKING PWDR, HOME USE, STRAIGHT PHOSPHATE 1t	3	4	0.9	tr					247	188	283	.00	0.0	0
		tr	0.0						5		0.0	.00	0	
BAKING PWDR, HOME USE, TARTRATE 1t	3	2	0.6						219	0	0	.00	0.0	0
		tr							114		0.0	.00	0	
BAKING PWDR, COMMERCIAL, PYROPHOSPHATE 1t	3	3	0.8						486	27	367			
		tr										.00	0	
BAKING PWDR, COMMERCIAL, LOW-SODIUM 1t	3	5	1.2	tr					tr	145	219	.00	0.0	0
		tr							328			.00	0	
BAKING SODA 1t	3	0	0.0	0.0					821					
		0.0												
CORNSTARCH 1T	10	36	8.7	0.1										
		tr	0.0											
COUGH SYRUP, TRIAMINICOL 1T	5	14												
COUGH SYRUP, TRIAMINIC EXPECTORANT 1t	5	11												
GELATIN, DRY 1 envelope	7	23	0.0	tr					8	0	0	.00	0.0	
		6.0							180		0.4	.00	4	
MALT, DRY 1T	28	103	21.7	0.5								.14	2.5	
		3.7									1.1	.09		
MALT EXTRACT, DRIED	10	37	8.9						8	5	29	.04	1.0	
		0.6							23	14	0.9	.05		
MEAT TENDERIZER, ADOLPHS 1t	5	2	0.5	0.0					1745	372	16			
		0.0							tr		tr			
MEAT TENDERIZER, SEASONED, ADOLPHS 1t	5	2	0.3	tr					1761	63	24			
		0.0							5		tr			
MEAT TENDERIZER, LOW-SODIUM, ADOLPHS 1t	5	0	0.0	0.0					1	40	18			
		0.0							2392		tr			
MEAT TENDERIZER, LOW-SODIUM, SEASONED, ADOLPHS 1t	5	5	1.1	tr					1	84	40			
		tr							1652		0.1			
PECTIN, CERTO 1oz	28	3	1.5						tr					
									67					
PECTIN, SURE-JELL 1oz	28	76	21.1						34					
RENNIN TABLET (SALTS, STARCH RENNIN, ENZYME)	10	11	2.4	0.1					2230	351	20	.00	0.0	0
		tr	0.0									.00	0	
SOYBEAN PROTEIN	10	32	1.5	tr					21	12	67			
		7.5	tr						18				0	
SOYBEAN PROTEINATE	10	31	0.8	tr					120					
		8.1	tr										0	
VINEGAR, CIDER 1T	15	2	0.8						tr	1	2			
		0.0	0.0						15		0.1			0
VINEGAR, CIDER ½ cup	120	14	6.0						1.2	8	12			
		0.0	0.0						120		0.6			0
VINEGAR, DISTILLED 1T	15	2	0.8						tr					
									2	tr				
WHEY, ACID, DRY	10	34	7.3	tr	24	112	101	22	97	205	135	.06	0.1	6
		1.1	0.0	tr	39	58	58	59	229	20	0.1	.21		
WHEY, ACID, FLUID	10	2	0.5	tr	2	7	6	1	5	10	8	tr	tr	1
		0.1	0.0	tr	2	4	4	4	14	1	tr	.01	tr	
WHEY, SWEET, DRY	10	35	7.4	0.1	21	119	103	24	108	80	93	.05	0.1	5
		1.3	0.0		41	73	70	82	208	18	0.1	.22	tr	
WHEY, SWEET, FLUID	10	3	0.5	tr	1	7	7	1	5	5	5	tr	tr	2
		0.1	0.0	tr	2	5	5	5	16	1	tr	.01	tr	
YEAST, BAKERS, COMPRESSED, FORTIFIED 1 cake	12	10	1.3	tr					2	2	47	1.66	17.3	
		1.5							73	7	0.6	.20		
YEAST, BAKERS, COMPRESSED, UNFORTIFIED 1 cake	12	10	1.3	tr					2	2	47	.09	1.3	
		1.5							73	7	0.6	.20		
YEAST, BAKERS, DRY (ACTIVE) 1T	8	23	3.1	0.1	35	318	253	68	4	4	103	.19	2.9	
		3.0			167	182	232	182	160	5	1.3	.43		

*A red acid AZO dye; C_{20}, H_{11}, N_2, Na_3, O_{10}, S_3.

	WT	Proximate			Amino Acids				Minerals			Vitamins		
		CAL	CHO	FAT	TRY	LEU	LYS	MET	Na	Ca	P	THI	NIA	A
		PRO	FIB	PUFA	PHE	ISO	VAL	THR	K	Mg	Fe	RIB	ASC	D
	g	g	g	g	mg	mg	mg	mg	mg	mg	mg	mg	mg	iu
YEAST, BREWERS, DEBITTERED	10	28	3.8	0.1	74	338	345	89	12	21	175	1.52	3.8	
1 T		3.9	0.2		201	252	287	248	189	23	1.7	.43		
YEAST, TORULA	10	28	3.7	0.1					2	42	171	1.40	4.4	
1 T		3.9	0.3						205	17	1.9	.51		

TABLE I   151

Alcoholic Beverages
Kilocalories, Carbohydrate and Alcoholic Content

	MEASURE	WEIGHT	KCAL (g)	CHO (g)	ALCOHOL
DISTILLED LIQUORS					
liqueurs (cordials)					
anisette	1 cordial glass	(20g)	75	7.0	7.0
apricot brandy	1 cordial glass	(20g)	65	6.0	6.0
benedictine	1 cordial glass	(20g)	70	6.6	6.6
creme de menthe	1 cordial glass	(20g)	67	6.0	7.0
curacao	1 cordial glass	(20g)	55	6.0	6.0
brandy, Calif	1 brandy glass	(30g)	73		10.5
brandy, cognac	1 brandy pony	(30g)	73		10.5
cider, fermented	6 oz	(180g)	71	1.8	9.4
gin, rum, vodka,					
whiskey (rye/scotch)					
80-proof	1 jigger, 1½ oz	(45g)	104	0.0	15.0
86-proof	1 jigger, 1½ oz	(45g)	112	0.0	16.2
90-proof	1 jigger, 1½ oz	(45g)	118	0.0	17.1
94-proof	1 jigger, 1½ oz	(45g)	124	0.0	17.9
100-proof	1 jigger, 1½ oz	(45g)	133	0.0	19.1
WINES					
champagne, dom	1 wine glass	(120g)	85	3.0	11.0
dessert (18.8% alcohol					
by vol)	1 wine glass	(100g)	137	7.7	15.3
madeira	1 wine glass	(100g)	105	1.0	15.0
muscatel/port	1 wine glass	(100g)	158	14.0	15.0
red, Calif	1 wine glass	(100g)	85		10.0
sauterne, Calif	1 wine glass	(100g)	85	4.0	10.5
sherry, dry, dom	1 wine glass	(60g)	85	4.8	9.0
table (12.2% alcohol					
by vol)	1 wine glass	(100g)	85	4.2	9.9
vermouth, dry (french)	1 wine glass	(100g)	105	1.0	15.0
vermouth, sweet (italian)	1 wine glass	(100g)	167	12.0	18.0
MALT LIQUORS (american)					
ale, mild	8 oz	(230g)	100	8.0	8.9
ale, mild	12 oz	(345g)	148	12.0	13.1
beer	8 oz	(240g)	114	10.6	8.9
beer	12 oz	(360g)	175	15.8	13.3
beer, Budweiser	12 oz	(360g)	150	14.0	17.6
beer, lite	12 oz	(360g)	96	2.8	12.1
beer, Michelob	12 oz	(360g)	160	16.0	18.0
beer, Natural Light	12 oz	(360g)	100	6.0	14.4
beer, Near	12 oz	(360g)	65		1.3
COCKTAILS					
daiquiri	1 cocktail	(100g)	125	5.2	15.1
eggnog (christmas)	4 oz punch cup	(123g)	335	18.0	15.0
gin rickey	8 oz	(120g)	150	1.3	21.0
highball	8 oz	(240g)	165		24.0
manhattan	1 cocktail	(100g)	165	7.9	19.2
martini	1 cocktail	(100g)	140	0.3	18.5
mint julep	10 oz	(300g)	212	2.7	29.2
old-fashioned	4 oz	(100g)	180	3.5	24.0
planters punch	4 oz	(100g)	175	7.9	21.5
rum sour	4 oz	(100g)	165		21.5
tom collins	10 oz	(300g)	180	9.0	21.5
whiskey sour	1 cocktail	(75g)	138	7.7	

**Sources of Protein, Fat and
Carbohydrate in Commercial Infant Formulas**

FORMULA	PROTEIN SOURCE	FAT SOURCE	CARBOHYDRATE SOURCE
Advance	cow milk & soy protein	soy–corn oils	lactose
Enfamil & Enfamil w/ iron	cow milk	soy–cocnt oils	lactose
Isomil	soy isolate	cocnt–soy oils	corn syrup & sucrose
i-Soyalac	soy isolate	soy oil	sucrose & tapioca starch
MBF (meat-base formula)	beef heart	sesame oil	cane sugar & tapioca starch
M-J Premature formula	cow milk	MCT corn–cocnt oils	lactose & sucrose
Neomullsoy	soy isolate	soy oil	sucrose & tapioca
Nutramigen	casein hydrolysate	corn oil	sucrose, tapioca starch & dextrin
Nursoy	soy isolate	oleo–cocnt–safflower–soy oils	corn syrup solids & sucrose
Pregestimil	casein hydrolysate	MCT–corn oil	dextrose, tapioca starch & corn syrup solids
Pro Sobee	soy isolate	soy oil	sucrose & corn syrup solids
Similac & Similac w/ iron (20, 24 & 27 kcal/oz)	cow milk	cocnt–soy oils	lactose
Similac 24 LBW	cow milk	MCT–cocnt–soy oils	lactose & corn syrup solids
Similac PM 60/40	casein & whey	cocnt–corn oils	lactose
SMA (20, 24 & 27 kcal/oz)	cow milk & whey	oleo–cocnt–safflower–soy oils	lactose
Soyalac	soybean solids	soy oil	dextrins, maltose, dextrose & sucrose

TABLE III 153

NAMES OF BEEF CUTS

NAMES COMMONLY USED	OTHER NAMES USED REGIONALLY
ROUND	Bucket Steak, Top Round, Bottom Round, Eye of Round (boneless), "Full Cut" Round, "Swiss" Steak
HEEL OF ROUND	Pike's Peak, Diamond, Wedge, Gooseneck, Horseshoe, Upper Round, Lower Round, Jew Daube, Denver Pot Roast
BOTTOM (OUTSIDE) ROUND	Silverside, Gooseneck, Silver Tip, "Swiss" Steak
SIRLOIN TIP (KNUCKLE)	Short Sirloin, Top Sirloin, Sirloin Butt, Crescent, Veiny, Bell of Knuckle, Face, Face Rump, Round, Boneless Sirloin, Round Tip, Ball Tip, Loin Tip, Family Steak, Sandwich Steak
BONELESS SIRLOIN STEAK	Top Loin Steak, Hip Steak, Rump Steak, "Top of Iowa" Steak, Top Sirloin Butt Steak, Bottom Sirloin Butt Steak
SIRLOIN (LOIN END)	Hip, Short Hip, Head Loin, Rump, K-Style Butt, Sirloin Butt Bone-in, Sirloin Butt, Sir Butt, Sirloin Butt (boneless), Family Steak
LOIN STRIP STEAK	Top Loin Steak, Sirloin Steak, Boneless Sirloin Steak, New York Steak, Kansas City Steak, Club Steak, Delmonico Steak, Shell Steak, Strip Steak, Boneless Top Sirloin Steak, Boneless Hotel Steak, Boneless Hip Steak, Minute Sirloin Steak, Key Strip Steak
PORTERHOUSE STEAK	T-Bone Steak, Large T-Bone Steak, Tenderloin Steak, King Steak
T-BONE STEAK	Porterhouse Steak, Small T-Bone Steak, Club Steak, Tenderloin Steak
CLUB STEAK	Sirloin Steak, Sirloin Strip Steak, Delmonico Steak, Market Steak, Individual Steak
FLANK STEAK	London Broil, Cube Steak, Minute Steak, Flank Steak Filet, "Swiss" Steak
RIB EYE STEAK	Market Steak, Spencer Steak, Beauty Steak, Delmonico Steak, "Boneless" Delmonico Steak, Center Cut Steak, Boneless Rib Steak, Club Steak, Boneless Club Steak, Boneless Rib Club Steak, Country Club Steak, Regular Roll Steak
TENDERLOIN	Filet Mignon, Petite Filet, Tenderloin Roast, Tenderloin Tips, "Tips"
SKIRT STEAK	Skirt Steak Filets
SHOULDER CLOD	Scalped Shoulder, Shoulder Roast, Boneless Shoulder, Cross Rib, Rolled Cross Rib, Clod "Roast," Boneless Clod "Roast," London Broil
ARM POT ROAST	Cross Rib Roast, Thick Rib Roast, Thick End Roast, Round Bone Roast, Shoulder Roast, Round Shoulder Roast
BLADE POT ROAST	Chuck Roast, Blade Cut Chuck Roast, Square Cut Chuck Roast, English Cut Roast, 7 Roast, 7 Bone Roast, Flat Bone Roast
ENGLISH CUT	Boston Cut, Bread and Butter, Boneless English Cut
CHUCK (SHORT RIBS)	Flanken, Brust Flanken
CHUCK TENDER	Scotch Tender, Jewish Tender, Kosher Filet, Round Muscle, Fish Muscle, Top Eye Pot Roast, "Cat Fish" Pot Roast
SHORT RIBS	Middle Ribs, English Short Ribs
BRISKET	Deckle, Boneless Brisket, Bone-In Brisket, Fresh Boneless Brisket, Beef Breast, Brisket Pot Roast, "Barbecue" Beef Brisket, Corned Beef
FORE SHANK	Shin, Fore Shin, Shank
MECHANICALLY TENDERIZED STEAKS	"Cubed," "Chicken," Minute, Quick Steak, Sandwich Steak

Caffeine Content of Selected Items
(mg/serving)

AVERAGE VALUES*
  coffee beverage
    brewed, ground   85/cup (range 85-200/cup)
    percolated   110/cup (range 97-125/cup)
    dripolated   146/cup (137-153/cup)
    instant   60/cup
    instant, decaffeinated   3/cup

  tea beverage (bagged & loose teas brewed 5 min)
    reg, bagged   46/cup
    reg, loose   40/cup
    green, bagged   31/cup
    green, loose   35/cup
    darjeeling, loose   28/cup
    oolong, bagged   40/cup
    oolong, loose   24/cup
    japanese panfried, loose   21/cup
    japanese green, loose   20/cup
    instant   30/cup

  cocoa/hot choc beverage   13/cup (range 6-42/cup)

  cola beverage   47/12 oz (range 30-90/12 oz)

SPECIFIC BRANDS
  coffee, instant dry powder
    Brim, freeze-dried, decaffeinated   3/t
    Cafe Francais   30/t
    Cafe Vienna   32/t
    Maxim, freeze-dried   61/t
    Mellow Roast (coffee & grain)   56/t
    Minicaf, freeze-dried, decaffeinated
      American Hospital Supply   0/pkg
    Nescafe, Nestlé   59/t
    Orange Cappaccino   33/t
    Sanka   3/t
    Sanka, freeze-dried, decaffeinated   3/t
    Suisse Mocha   29/t
    Tasters Choice, Nestlé   59/t
    Tasters Choice, decaffeinated, Nestlé   5/t
    Yuban   56/t
  coffee, instant beverage
    Hills Brothers   189/cup

SPECIFIC BRANDS
  tea beverage
    Bigelow Constant Commet   31/cup
    Bostons 99½% Caffeine-Free Tea   9/cup
    Brooke Bond Red Rose   54/cup
    Canterbury   54/cup
    Grand Union   48/cup
    Harvest Day   45/cup
    Jacksons of Piccadilly Earl Grey   61/cup
    Lipton, bagged   54/cup
    Lipton, loose   51/cup
    MJB   62/cup
    Nestlé , instant iced w/lemon flavor   42/cup
    Our Own   60/cup
    Pantry Pride   45/cup
    Royal Jewel   44/cup
    Salada   59/cup
    Stewarts   53/cup
    Swee-touch-nee   47/cup
    Tender Leaf   66/cup
    Tetley   61/cup
    Twinings English   61/cup
    White Rose   55/cup

*Wide fluctuations are due to variations in beans, leaves, or cocoa, and brew strength.
Values were most commonly reported per cup of beverage. However, some references
listed "5 oz cups" or "6 oz cups" rather than the standard 8 oz cup.

TABLE IV (Continued) 155

Caffeine Content of Selected Items
(mg/serving)

choc beverage dry mix	
choc powder, instant dry, Hershey	10/T
cocoa, dry, Hershey	10/T
cocoa, instant, Hershey	10/T
cocoa mix, instant	9/T
cocoa mix, instant, Carnation	13/pkg
instant brkfst, choc, Carnation	tr/pkg
milkshake, choc malt, Delmark	tr/oz
choc beverage	
cocoa mix (water), Carnation	14/cup
carbonated beverage	
Coca Cola	65/12 oz
Diet-Rite Cola	33/12 oz
Dr. Pepper	61/12 oz
Dr. Pepper, s-f	54/12 oz
Mountain Dew	55/12 oz
Mr. Pibb	57/12 oz
Pepsi Cola	43/12 oz
Pepsi Cola, s-f	36/12 oz
Pepsi Light	36/12 oz
Royal Crown Cola	34/12 oz
Royal Crown Cola, s-f	33/12 oz
Royal Crown w/a twist	21/12 oz
Tab, s-f	45/12 oz
drugs	
aspirin compound–phenacetin–caffeine	32/pill
Cope, Midol, etc.	32/pill
Excedrin, Anacin	60/pill
Pre-Mens	66/pill
No Doz, Vivarin	100-200/pill
Dristan, Sinarest	30/pill

Margarine and Spreads
(values/T)

MARGARINE OR SPREAD	FORM	KCAL	FAT (g)	SFA (g)	PUFA (g)	PUFA (%)	NA (mg)
Allsweet margarine	cube	100	11		3	27	108
A&P marg	cube	100	11		4	32	96
A&P soft marg	tub	100	11		5	43	102
A&P corn oil table spread	cube	100	11		4	34	119
Blue Bonnet marg	cube	100	11		4	32	89
Blue Bonnet soft marg	tub	100	11		4	35	109
Blue Bonnet soft whip marg	tub	65	7		2	33	55
Blue Bonnet liq marg	plst bot*	100	11				
Blue Bonnet soft diet imitation marg	tub	50	6		2	39	95
Blue Bonnet light tasty spread	tub	80	8		3	32	74
Blue Seal Brand marg	cube	100	11				
Bonnie Hubbard marg	cube	100	11	2	3	27	
Chiffon soft stick marg	cube	100	11		3	29	101
Chiffon soft marg	tub	100	11		5	44	99
Chiffon sweet unsalt marg	tub	100	11		5	43	1
Chiffon whip marg	tub	70	8	1	3	42	60
Coldbrook oleomarg	cube	100	11				
Coldbrook soft marg	tub	100	11		4	34	108
Empress marg	cube	100	11		4	33	95
Empress soft marg	tub	100	11		4	40	86
Fleischmanns marg	cube	100	11	2	4	35	92
Fleischmanns sweet unsalt marg	cube	100	11		4	37	4
Fleischmanns soft marg	tub	100	11	2	5	44	97
Fleischmanns diet imitation marg	tub	50	6	1	2	40	107
Hain marg	cube	100	11				
Holiday marg	cube	100	11	2	3	27	
Hollywood marg	cube	100	11	2	5	45	
Imperial marg	cube	100	11	4	4	33	103
Imperial soft marg	tub	100	11	3	4	36	87
Imperial diet imitation marg	tub	45	5	2	2	40	102
Imperial light blend	cube					19	73
Individual Servings colored vegetable oleomarg	pats	100	11				
Instant Whip marg	tub	100	11	2	3	27	
Kitchen Craft marg	cube	100	11	2	6	55	
Kroger marg	cube	100	11		3	28	83
Lady Lee marg	cube	100	11	2	3	27	
Lady Lee soft marg	tub	100	11	2	4	36	
Land o Lakes marg	cube	100	11		1	10	87
Land o Lakes soft marg	tub	100	11		4	37	96
Mazola marg	cube	100	11	2	4	36	115
Mazola sweet unsalt marg	cube	100	11	2	4	37	3
Mazola diet imitation marg	tub	50	6	1	3	45	130
Miracle Whip marg	tub	65	7		3	38	65
Mothers sweet unsalt marg	cube	100	11		4	33	1
Mrs. Filberts Golden Quarters marg	cube	100	11		2	14	92
Mrs. Filberts soft 100% corn oil marg	cube	100	11	2	5	44	97
Mrs. Filberts soft golden marg	tub	100	11	2	4	32	99

Explanatory Notes

Margarine is usually made from hydrogenated or partially hydrogenated vegetable oils, milk solids, emulsifiers, preservatives, artificial flavor and artificial color (usually carotene). Margarine contains 80% fat, about 16.5% water, no protein or carbohydrate, and usually no cholesterol unless made with some animal fat. Of the brands listed here only one, Blue Seal, contained an animal fat (hydrogenated lard). Vitamin A is added at a level of 15,000 IU/lb or 10% of the USRDA/serving (1T). Vitamin D may be added at a level of 2,000 IU lb or 15% of the USRDA/serving (1T). Margarine is available in cubes, tubs or liquid form. Unsalted, whipped, diet, or imitation varieties are also available.

The caloric value of regular margarine is 100 kcal/T. Diet or imitation margarines and spreads contain less fat, more water, and fewer calories than regular margarine. Diet or imitation margarines usually contain 40% fat, 38-57% water and about 50 kcal/T. Spreads contain about 58% fat and about 74 kcal/T. Whipped margarine has about 65 kcal/T because air has replaced some of the volume.

This table lists various margarines and spreads by brand name with their content of fat, saturated fatty acids, polyunsaturated fatty acids and sodium. For comparative purposes, butter is listed at the end of the table. Butter, like margarine, is 80% fat and has 100 kcal/T. The fat composition of butter is 65% saturated, 31% monounsaturated and 4% polyunsaturated. The cholesterol content of butter is 31 mg/T.

*plst bot = plastic bottle

TABLE V (Continued) **157**

Margarine and Spreads
(values/T)

MARGARINE OR SPREAD	FORM	KCAL	FAT (g)	SFA (g)	PUFA (g)	PUFA (%)	NA (mg)
Mrs. Filberts spread	tub					32	73
Nucoa marg	cube	100	11	3	4	33	174
Nucoa no-burn marg	cube	100	11	2	3	27	160
Nucoa soft marg	tub	100	11	2	4	39	163
Nu-Maid soft marg	tub	100	11		5	43	116
Nuspread veg oil spread	cube	80	8				
Parkay marg	cube	100	11		2	19	96
Parkay soft marg	tub	100	11	2	4	38	106
Parkay soft corn oil marg	tub	100	11		5	44	97
Parkay whip marg	tub	65	7		3	40	64
Parkay Squeeze marg	plst bot*	100	11	2	4	45	100
Parkay soft diet imitation marg	tub	50	6		2	40	85
Promise marg	cube	100	11		5	45	104
Promise marg	tub	100	11	2	10	85	96
Saffola marg	cube	100	11	2	5	42	103
Satin Gold marg	tub	100	11	2	4	36	
Shedds liq marg	plst bot*	100	11	2	6	55	
Southern Belle marg	cube	100	11	2	4	36	111
Southern Belle corn oil marg	cube	100	11	2	3	27	111
Sun Valley marg	cube	100	11	2	3	27	
Supreme Miami marg	cube	100	11	2	3	29	
Table Maid soft marg	tub	90	10				
Weight Watchers imitation marg		50	6		2	38	83
Willow Run marg	cube	100	11	2	3	27	
butter	cube	100	11	7	0	4	116
butter, sweet	cube	100	11	7	0	4	1
butter, whip	cube	81	9	6	0	4	93

*plst bot = plastic bottle

Cholesterol

FOOD	mg/100g	mg/SERVING	
CANDY			
brkfst bar, Carnation	2	1/bar	(43g)
choc, german	8	2/oz	(28g)
milk choc w/almonds	15	4/oz	(28g)
milk choc w/pnts	10	5/bar	(50g)
Mr. Goodbar, Hershey	10	5/bar	(50g)
CHEESE			
american, past, proc	96	27/oz	(28g)
american, pimento, past, proc	96	27/oz	(28g)
blue	75	21/oz	(28g)
brick	90	25/oz	(28g)
brie	100	28/oz	(28g)
camembert (dom)	70	20/oz	(28g)
cheddar	106	30/oz	(28g)
Cheezola	4	1/oz	(28g)
cheshire	103	29/oz	(28g)
colby	96	27/oz	(28g)
cottage, uncreamed (1% fat)	4	9/cup	(226g)
cottage, uncreamed (2% fat)	8	18/cup	(226g)
cottage, creamed	15	34/cup	(226g)
cottage w/fruit	11	25/cup	(225g)
Countdown	5	1/oz	(28g)
cream	120	34/2T	(28g)
edam	89	25/oz	(28g)
feta	89	25/oz	(28g)
fontina	116	32/oz	(28g)
gouda	114	32/oz	(28g)
gruyere	110	31/oz	(28g)
limburger	93	26/oz	(28g)
mozzarella	78	22/oz	(28g)
mozzarella, low-moist	89	25/oz	(28g)
mozzarella, low-moist, part skim	54	15/oz	(28g)
mozzarella, part skim	57	16/oz	(28g)
muenster	96	27/oz	(28g)
neufchatel	76	21/oz	(28g)
parmesan, grated	79	4/T	(5g)
parmesan, hard	68	19/oz	(28g)
port du salut	123	34/oz	(28g)
provolone	69	19/oz	(28g)
ricotta	51	63/½ cup	(124g)
ricotta, part skim	32	40/½ cup	(124g)
romano	104	29/oz	(28g)
roquefort	93	26/oz	(28g)
swiss	93	26/oz	(28g)
swiss, past, proc	86	24/oz	(28g)
tilsit	102	29/oz	(28g)
cheese fondue, homemade	106	68/¼ cup	(64g)
cheese food			
american, past, proc	64	18/oz	(28g)
american, past, proc			
cold pack	64	18/oz	(28g)
swiss, past, proc	82	23/oz	(28g)
cheese sce	74	27/2T	(38g)
cheese spread, american, past,			
proc	57	16/oz	(28g)
COMBINATION FOODS			
beans & ham	5	10/serving	(200g)
beef goulash w/noodles	18	27/serving	(150g)

Explanatory Notes

Cholesterol is found primarily in foods of animal origin (meat, fish, poultry, eggs, and dairy items) with the highest concentration existing in internal organs (brains, liver, etc.), egg yolk and shellfish. Plant products (vegetables, fruits, and grains) apparently do not concentrate this sterol to any great extent. The amount of cholesterol in breads, desserts and combination foods depends upon the inclusion of milk, eggs, and animal fat in the recipe.

A low-cholestrol diet would emphasize the use of vegetables, fruit, grains, skim, or low-fat milks and cheese, lean meats, poultry, fish, and vegetable oils and margarine, rather than animal fats (lard, butter, etc.).

TABLE VI (Continued) **159**

FOOD	mg/100g	mg/SERVING	
beefaroni, cnd, Chef Boyardee	25	50/serving	(200g)
beef potpie, homemade	21	48/pie	(227g)
beef potpie, commercial, frzn	18	41/pie	(227g)
beef-veg stew, cnd	14	33/cup	(235g)
beef-veg stew, homemade	26	64/cup	(245g)
blintzes, cheese	122	34/serving	(28g)
chicken a la king, homemade	76	186/cup	(245g)
chicken & noodles, homemade	40	96/cup	(240g)
chicken & noodles, frzn, Green Giant	12	20/serving	(170g)
chicken fricassee	40	80/serving	(200g)
chicken potpie, homemade	31	36/pie	(116g)
chicken potpie, frzn	20	45/pie	(227g)
chili con carne	15	15/serving	(100g)
chicken chow mein, homemade	31	68/serving	(220g)
chicken chow mein, cnd	6	13/cup	(220g)
chop suey w/meat, homemade	26	65/cup	(250g)
chop suey w/meat, cnd	12	43/cup	(360g)
eggroll, plain	53	53/serving	(100g)
eggroll, chicken, frzn, La Choy	16	16/serving	(100g)
eggroll, lobster, frzn, La Choy	3	3/serving	(100g)
eggroll, meat-shrimp, frzn, La Choy	10	10/serving	(100g)
eggroll, shrimp, frzn, La Choy	12	12/serving	(100g)
fish cakes w/ cnd fish-pot-egg, fried	108	108/serving	(100g)
fish cakes, frzn, fried	108	108/serving	(100g)
fish creole	31	31/serving	(100g)
fritters, clam	129	129/serving	(100g)
fritters, corn	60	60/serving	(100g)
frzn dinner—Salisbury stk, Morton	15	47/dinner	(311g)
ham croquette, panfried	100	65/piece	(65g)
ham-lima bean cass	21	59/serving	(280g)
hash, corned beef, cnd	60	66/½ cup	(110g)
macaroni & cheese, homemade	21	42/cup	(200g)
macaroni & cheese, cnd, Franco American	13	26/cup	(200g)
pepper, grn, stfd w/beef & crumbs	30	56/1 avg	(185g)
pizza, sausage, homemade	59	40/piece	(67g)
pizza burger	51	51/serving	(100g)
pork & sauerkraut	20	37/serving	(185g)
pork, sweet & sour, homemade	21	42/serving	(200g)
souffle, cheese, homemade	167	251/serving	(150g)
spaghetti-tom sce-cheese,			
homemade	11	33/serving	(302g)
cnd	6	15/serving	(250g)
spaghetti-tom sce-meatballs,			
homemade	30	66/serving	(220g)
cnd	9	23/serving	(250g)
spaghetti sce	25	25/serving	(100g)
tamales, cnd	9	14/2 avg	(155g)
tuna & noodles, frzn, Green Giant	13	22/serving	(170g)
turkey a la king	16	29/serving	(180g)
turkey potpie, homemade	31	70/pie	(227g)
turkey potpie, commercial, frzn	9	20/pie	(227g)
veal scallopini	64	74/serving	(116g)
welsh rarebit	31	72/serving	(232g)
CONDIMENTS & CONDIMENT-TYPE SAUCES			
tartar sce	51	10/T	(20g)
white sce, med	13	4/2T	(33g)
white sce, thick	12	4/2T	(33g)
white sce, thin	14	4/2T	(30g)
DESSERTS			
apple brown betty	11	15/½ cup	(140g)
brkfst pastry			
cinn roll	70	39/1 avg	(55g)
sweet roll	70	39/1 avg	(55g)
brownies, butterscotch	84	25/1 avg	(30g)
brownies, choc w/nuts, homemade	83	25/1 avg	(30g)
cake			
plain w/o icing	45	22/1 piece	(50g)
plain w/choc icing	45	22/1 piece	(50g)
banana w/buttercream icing	15	7/1 piece	(50g)
boston cream pie	48	53/serving	(110g)
carrot w/cream cheese icing	31	15/1 piece	(50g)
choc roll, Richs	87	24/oz	(28g)

FOOD	mg/100g	mg/SERVING	
choc (devils food)			
w/o icing	76	38/1 piece	(50g)
w/choc icing	43	21/1 piece	(50g)
w/choc icing, frzn	63	31/1 piece	(50g)
w/choc icing & whip cream			
fill, frzn	73	62/serving	(85g)
fruitcake			
dark loaf	45	18/1 piece	(40g)
light loaf	63	25/1 piece	(40g)
gingerbread	1	tr/1 piece	(50g)
orange w/icing	24	12/1 piece	(50g)
pineapple upside-down, frzn	31	31/serving	(100g)
pound, modified	99	30/1 piece	(30g)
spice w/icing	21	10/1 piece	(50g)
sponge	246	123/1 piece	(50g)
white w/choc icing	2	1/1 piece	(50g)
yellow, w/o icing	45	22/1 piece	(50g)
yellow w/ choc icing, from mix	48	24/1 piece	(50g)
yellow w/choc icing, homemade	44	22/1 piece	(50g)
cobbler, peach	4	4/serving	(100g)
cookies			
arrowroot, Nabisco	91	9/2 avg	(10g)
choc chip, homemade	90	9/1 avg	(11g)
choc chip, commerc	60	6/1 avg	(11g)
ladyfingers	356	50/1 large	(14g)
Mcdonaldland, McDonalds	14	9/box	(63g)
molasses	45	7/1 avg	(15g)
oatmeal w/raisins	52	7/1 avg	(14g)
peanut	55	7/1 avg	(12g)
pnt btr bars	85	42/bar	(50g)
cream puffs w/custard fill	144	151/1 avg	(105g)
cupcakes			
big wheels, Hostess	25	6/1 avg	(25g)
choc (devils food), Hostess	10	5/1 avg	(50g)
ding dongs, Hostess	25	10/1 avg	(39g)
ho hos, Hostess	55	14/1 avg	(25g)
orange, Hostess	15	6/1 avg	(42g)
suzy q's	15	10/1 avg	(64g)
custard			
baked, homemade	105	165/³⁄₅ cup	(157g)
banana, from mix	17	19/½ cup	(112g)
chocolate, from mix	14	16/½ cup	(112g)
coconut, from mix	13	15/½ cup	(112g)
lemon, from mix	21	24/½ cup	(112g)
vanilla, from mix	21	24/½ cup	(112g)
doughnuts			
cake-type	83	27/1 avg	(32g)
plain, Hostess	25	7/1 avg	(27g)
powdered	25	8/1 avg	(30g)
eclair, choc, frzn, Richs	80	62/1 avg	(78g)
ice cream			
reg (10% fat)	40	53/cup	(133g)
reg (12% fat)	60	80/cup	(133g)
rich (16% fat)	57	84/cup	(133g)
french frzn custard	73	97/cup	(133g)
french van, soft-serve	89	154/cup	(173g)
ice milk			
strawberry	20	18/⅙ qt	(90g)
vanilla	14	13/⅙ qt	(90g)
vanilla, soft-serve	7	13/cup	(175g)
junket (whole milk)	11	15/½ cup	(135g)
peach crisp	4	4/serving	(100g)
pie			
apple, frzn	10	16/serving	(160g)
apple, Hostess	35	44/serving	(127g)
blueberry, frzn, Morton	10	10/serving	(100g)
butterscotch cream	9	9/serving	(100g)
cherry, frzn, Morton	10	10/serving	(100g)
cherry, Hostess	35	44/serving	(127g)
choc bavarian cream, frzn, Sara Lee	23	20/serving	(85g)
cocnt custard, frzn, Morton	60	60/serving	(100g)
custard, homemade	105	158/serving	(150g)
lemon, Hostess	35	44/serving	(127g)
lemon chiffon, homemade	169	181/serving	(107g)

TABLE VI (Continued)   **161**

FOOD	mg/100g	mg/SERVING	
lemon meringue, homemade	93	130/serving	(140g)
mince, homemade	12	19/serving	(160g)
mince, frzn, Morton	10	10/serving	(100g)
peach, frzn, Morton	10	10/serving	(100g)
pecan, homemade	48	38/serving	(80g)
pumpkin, homemade	61	92/serving	(150g)
pumpkin, frzn, Morton	40	40/serving	(100g)
piecrust, graham cracker, homemade	73	23/serving	(32g)
pudding			
bread w/raisins	64	106/¾ cup	(165g)
butterscotch (skim milk)	2	3/½ cup	(130g)
butterscotch, cnd, Del Monte	9	9/3½ oz	(100g)
choc, starch base	10	14/½ cup	(144g)
choc (skim milk)	1	1/½ cup	(130g)
choc, cnd	12	12/serving	(100g)
rice w/raisins, homemade	11	16/¾ cup	(145g)
tapioca, homemade	10	11/½ cup	(105g)
tapioca, cnd	53	53/serving	(100g)
tapioca cream	97	126/½ cup	(130g)
vanilla, starch base	14	18/½ cup	(125g)
vanilla (whole milk)	11	16/½ cup	(148g)
vanilla (skim milk)	2	3/½ cup	(148g)
vanilla tapioca	10	15/½ cup	(145g)
sherbet, orange	7	7/serving	(100g)
snoballs, Hostess	25	11/1 avg	(42g)
twinkies, Hostess	50	20/1 avg	(39g)
DESSERT TOPPINGS			
cream, whip, pressurized	85	2/T	(3g)
EGGS, EGG DISHES, EGG SUBSTITUTES			
chicken eggs			
whole, fresh/frzn	550	264/1 med	(48g)
whole, dried, stabilized	2017	141/T	(7g)
whites, fresh/frzn	0	0/1 med	(31g)
yolks, fresh	1602	272/1 med	(17g)
yolks, frzn	1301	221/1 med	(17g)
boiled (hard/soft)	548	263/1 med	(48g)
fried	534	262/1 med	(50g)
omelet, plain	388	241/1 med	(62g)
omelet, frzn, heat & serve	412	412/serving	(100g)
omelet, spanish	268	407/2 eggs	(152g)
poached	545	262/1 med	(48g)
scrambled	388	252/1 med	(65g)
duck, whole	884	654/1 med	(74g)
quail, whole	844	84/1 med	(10g)
turkey, whole	933	746/1 med	(80g)
egg substitutes			
frzn	2	1/¼ cup	(60g)
liquid	tr	tr/1½ fl oz	(47g)
EGG SUBSTITUTES			
powder	570	57/serving	(10g)
Bud Anheuser-Busch	1	1/¼ cup	(60g)
Chono powder, General Mills	0	0/T	(7g)
Chono recon, General Mills	0	0/serving	(50g)
Egg Beaters, Fleishmanns	0	0/¼ cup	(60g)
Eggstra powder, Tillie Lewis	330	23/T	(7g)
Eggstra, recon, Tillie Lewis	115	57/serving	(50g)
Eggtime, Nestle	0	0/serving	(55g)
Second Nature Avoset Foods	0	0/serving	(50g)
Super Scramble, Trans American	2	1/¼ cup	(60g)
FAST FOODS			
Arthur Treachers			
chicken filet	47	64/serving	(136g)
chicken sandwich	21	33/serving	(156g)
chips (french fries)	1	1/serving	(113g)
coleslaw	8	7/serving	(113g)
fish chowder	5	9/serving	(170g)
fish, fried	38	44/serving	(147g)
fish sandwich	27	42/serving	(156g)
krunch pup	44	25/serving	(57g)
shrimp, fried	81	93/serving	(115g)
Dairy Queen			
ice cream–cone, small	15	11/serving	(71g)
ice cream–cone, large	15	32/serving	(213g)

FOOD	mg/100g	mg/SERVING	
Kentucky Fried Chicken			
chicken, original recipe	133	133/3½ oz	(100g)
chicken, crispy recipe	116	116/3½ oz	(100g)
coleslaw	4	4/serving	(90g)
dinner roll	2	tr/1 med	(17g)
mash pot w/gvy	3	3/serving	(113g)
McDonalds			
big mac	40	75/serving	(187g)
egg mcmuffin	145	191/serving	(132g)
eggs, scrambled	390	300/serving	(77g)
english muffin, buttered	19	12/serving	(62g)
fish filet	33	43/serving	(131g)
french fries	14	10/serving	(69g)
hamburger	26	26/serving	(99g)
hamburger w/ cheese	36	41/serving	(114g)
pancakes–btr–syrup	17	35/serving	(206g)
pie, apple	15	14/serving	(90g)
pie, cherry	16	15/serving	(92g)
pork sausage	90	45/serving	(50g)
quarter pounder	42	69/serving	(164g)
quarter pounder–cheese	49	95/serving	(193g)
shake (van, choc, straw)	10	29/serving	(290g)
FATS AND OILS			
bacon fat	6	1/T	(14g)
beef separable fat, raw	75	21/oz	(28g)
chicken fat	65	9/T	(14g)
lamb fat	75	21/oz	(28g)
lard	95	13/T	(14g)
pork fat	70	20/oz	(28g)
salt pork, raw	110	31/oz	(28g)
veg oils & shortenings	0	0/T	(14g)
FISH AND SHELLFISH			
caviar, sturgeon, granular	250	25/t	(10g)
clams, cnd	80	80/½ cup	(100g)
clams, raw	50	50/3½ oz	(100g)
cod, raw	50	50/3½ oz	(100g)
cod, dried, salted	82	82/3½ oz	(100g)
cod–lemon sce	56	56/3½ oz	(100g)
crab, steam	100	100/3½ oz	(100g)
crab, cnd	101	86/½ oz	(85g)
crab, deviled	100	100/serving	(100g)
eel, smoked	70	35/serving	(50g)
fish loaf (bread, egg, tom, onion)	99	99/serving	(100g)
fish sticks, frzn	70	70/4-5 sticks	(100g)
flatfishes, broil	61	61/3½ oz	(100g)
haddock, raw	60	60/3½ oz	(100g)
haddock, fried	60	60/3½ oz	(100g)
halibut, atlantic & pacific, raw	50	50/3½ oz	(100g)
halibut, atlantic & pacific, broil	60	60/3½ oz	(100g)
halibut, calif, raw	50	50/3½ oz	(100g)
halibut, greenland, raw	50	50/3½ oz	(100g)
herring, atlantic, raw	85	85/3½ oz	(100g)
herring, cnd	98	98/3½ oz	(100g)
lobster, northern, raw	200	200/3½ oz	(100g)
mackerel, atlantic, raw	95	95/3½ oz	(100g)
mackerel, atlantic, cnd	94	9/½ cup	(105g)
mussels, pacific, cnd	150	150/3½ oz	(100g)
oysters, raw	200	200/5-8 med	(100g)
oysters, cnd	230	230/5-8 med	(100g)
oysters, fried	230	230/5-8 med	(100g)
roe (salmon, sturgeon, turbot), raw	360	360/3½ oz	(100g)
salmon, broil/baked	47	47/3½ oz	(100g)
salmon, atlantic, raw	35	35/3½ oz	(100g)
salmon, atlantic, cnd	35	35/3½ oz	(100g)
salmon, chinook, cnd	60	60/3½ oz	(100g)
salmon, sockeye, raw	36	36/3½ oz	(100g)
salmon, sockeye, cnd	36	36/3½ oz	(100g)
salmon, smoked	60	60/3½ oz	(100g)
salmon patty	64	64/3½ oz	(100g)
sardines, atlantic, cnd in oil	120	120/8 med	(100g)

TABLE VI (Continued) 163

FOOD	mg/100g	mg/SERVING	
sardines, pacific, cnd in oil	140	140/3½ oz	(100g)
scallops, bay & sea, raw	35	35/3½ oz	(100g)
scallops, bay & sea, steamed	53	53/3½ oz	(100g)
shrimp, raw	150	150/3½ oz	(100g)
shrimp, cnd	150	150/3½ oz	(100g)
shrimp, french fried	120	120/½ cup	(100g)
trout, brook, raw	55	55/3½ oz	(100g)
trout, rainbow/steelhead, raw	55	55/3½ oz	(100g)
tuna, raw	60	60/3½ oz	(100g)
tuna, cnd in oil	63	63/½ oz	(100g)
tuna, cnd in water	63	63/½ oz	(100g)
tuna patty	47	47/3½ oz	(100g)
FORMULAS			
dietene, choc powder, Doyle	24	2/10g	
meritene, plain mix, Doyle	19	2/T	(11g)
meritene–whole milk, Doyle	20	48/8 oz	(240g)
FORMULAS, INFANT			
Advance		10/liter	
Isomil		0/liter	
i-Soyalac		0/liter	
Lofenalac powder	13	1/T	(9g)
MBF (meat-base formula)		710/liter	
Neomullsoy		0/liter	
Pro Sobee		0/liter	
Similac & Similac w/iron (20 kcal/fl oz)		11/liter	
Similac & Similac w/iron (24 kcal/fl oz)		14/liter	
Similac (27 kcal/fl oz)		15/liter	
Similac PM 60/40 (20 kcal/fl oz)		22/liter	
SMA (20 kcal/fl oz)		33/liter	
SMA (24 kcal/fl oz)		40/liter	
SMA (27 kcal/fl oz)		45/liter	
GRAIN PRODUCTS			
biscuits, baking powder, homemade	44	15/1 avg	(35g)
biscuits (cnd dough)	44	15/1 avg	(35g)
biscuits, from mix	6	2/1 avg	(28g)
bread			
cornbread, from mix	69	31/piece	(45g)
cornbread, homemade	46	18/piece	(40g)
cornbread, southern style	70	32/piece	(45g)
cornpone (whole ground cornmeal)	7	3/cake	(45g)
Hillbilly	5	1/slice	(23g)
Hollywood, dark/light	5	1/slice	(20g)
Profile, light	5	1/slice	(24g)
Roman Meal	5	1/slice	(23g)
spoonbread (white whole ground cornmeal)	95	91/serving	(96g)
chips & snacks			
cheese straws	32	8/4 pieces	(24g)
french toast	171	111/slice	(65g)
muffins			
plain	53	21/1 avg	(40g)
blueberry	49	20/1 avg	(40g)
corn	46	21/1 avg	(45g)
noodls, chow mein, cnd	12	6/cup	(50g)
noodles, egg, dry	94	69/cup	(73g)
noodles, egg, ckd	31	50/cup	(160g)
pancakes, homemade	71	32/1 avg	(45g)
pancakes, from mix	74	33/1 avg	(45g)
popovers, homemade	142	71/1 avg	(50g)
stuffing, bread, from mix, moist	23	16/½ cup	(70g)
waffles, homemade	85	64/1 med	(75g)
waffles, from mix	60	45/1 med	(75g)
MEATS			
beef cuts	70	70/3½ oz	(100g)
beef, dried, chip, creamed	27	32/½ oz	(120g)
chicken cuts	60	60/3½ oz	(100g)
duck	70	70/3½ oz	(100g)
frogs legs	50	50/3½ oz	(100g)
goose, roast	75	75/3½ oz	(100g)
organ meats			
kidneys			
beef, braised	375	375/3½ oz	(100g)
calf, raw	375	375/3½ oz	(100g)
hog, raw	375	375/3½ oz	(100g)
lamb, raw	375	375/3½ oz	(100g)

FOOD	mg/100g	mg/SERVING	
liver			
calf, raw	300	300/3½ oz	(100g)
calf, fried	438	438/3½ oz	(100g)
chicken, raw	555	555/3½ oz	(100g)
chicken, simmer	746	746/3½ oz	(100g)
goose, raw	300	300/3½ oz	(100g)
hog, raw	300	300/3½ oz	(100g)
hog, fried	438	438/3½ oz	(100g)
lamb, raw	300	300/3½ oz	(100g)
lamb, broil	438	438/3½ oz	(100g)
turkey, raw	435	435/3½ oz	(100g)
turkey, simmer	599	599/3½ oz	(100g)
sweetbread (thymus)			
calf	466	466/3½ oz	(100g)
lamb	466	466/3½ oz	(100g)
tongue			
beef, med fat, braised	140	140/3½ oz	(100g)
beef, smoked	210	210/3½ oz	(100g)
pork			
bacon fat, raw	70	20/oz	(28g)
bacon, cured, ckd	81	13/2 slices	(16g)
bacon, broil, Oscar Mayer	79	5/slice	(6g)
bacon, canadian, broil	90	19/slice	(21g)
bacon, canadian, Oscar Mayer	47	13/oz	(28g)
cuts	70	70/3½ oz	(100g)
hamloaf	81	81/3½ oz	(100g)
rabbit	65	65/3½ oz	(100g)
lamb cuts	70	70/3½ oz	(100g)
luncheon meats			
bbq loaf, Oscar Mayer	37	10/oz	(28g)
bologna	100	28/oz	(28g)
bologna, all beef, Oscar Mayer	55	15/oz	(28g)
Canadian bacon, Oscar Mayer	47	13/oz	(28g)
chopped ham, Oscar Mayer	49	14/oz	(28g)
ham–cheese loaf, Oscar Mayer	5	1/oz	(28g)
headcheese	100	28/oz	(28g)
honey loaf, Oscar Mayer	30	8/oz	(28g)
liver cheese, Oscar Mayer	175	49/oz	(28g)
liverwurst	125	35/oz	(28g)
old-fashioned loaf, Oscar Mayer	55	15/oz	(28g)
olive loaf, Oscar Mayer	36	10/oz	(28g)
pickle–pimento loaf	40	11/oz	(28g)
salami, beef, cotto, Oscar Mayer	36	10/oz	(28g)
salami, cotto, Oscar Mayer	53	15/oz	(28g)
salami, dry, Oscar Mayer	79	22/oz	(28g)
sandwich spread, Oscar Mayer	36	10/oz	(28g)
spam, Hormel	53	15/oz	(28g)
meatloaf	92	92/3½ oz	(100g)
meat substitutes, Morning Star Farms	0	0/serving	(23-38g)
mutton	65	65/3½ oz	(100g)
organ meats			
brains	2100	2100/3½ oz	(100g)
chitterlings (hog intestine)	119	119/3½ oz	(100g)
giblets, chicken, fried	150	42/oz	(28g)
gizzard, chicken, simmer	195	55/oz	(100g)
gizzard, turkey, simmer	229	64/oz	(28g)
heart			
beef, lean, braised	274	274/3½ oz	(100g)
beef, lean & fat, braised	150	150/3½ oz	(100g)
calf, braised	180	180/3½ oz	(100g)
chicken, raw	170	170/3½ oz	(100g)
chicken, simmer	231	231/3½ oz	(100g)
hog, raw	150	150/3½ oz	(100g)
lamb, braised	150	150/3½ oz	(100g)
turkey, simmer	238	238/3½ oz	(100g)
sausages			
blood/blood pudding	100	60/slice	(60g)
cervelat, dry	100	28/oz	(28g)
frankfurter, raw	65	32/1 avg	(50g)
frankfurter, ckd	70	35/1 avg	(50g)
frankfurter, all beef, Oscar Mayer	51	23/1 avg	(45g)
frankfurter, cnd	100	50/1 avg	(50g)
frankfurter, chicken, Tyson	67	23/1 avg	(34g)
frankfurter, chicken, Weaver	140	63/1 avg	(45g)

TABLE VI (Continued)   **165**

FOOD	mg/100g	mg/SERVING	
italian, Holloway House	27	8/oz	(28g)
pork, links/bulk	100	100/3½ oz	(100g)
scrapple	100	57/slice	(57g)
smokie links, Oscar Mayer	56	24/piece	(43g)
turkey cuts	82	82/3½ oz	(100g)
turkey ham loaf	81	81/3½ oz	(100g)
veal cuts	90	90/3½ oz	(100g)
whale meat	15	15/3½ oz	(100g)
MILK, COW, FLUID			
buttermilk (skim milk)	4	10/cup	(245g)
cond, sweet, cnd	34	7/T	(20g)
evaporated, unsweet, cnd	31	10/T	(32g)
evaporated, skimmed, cnd	4	1/oz	(30g)
evaporated, fill, Dairymate	5	1/T	(15g)
low-fat (1% fat)	4	10/cup	(245g)
low-fat (1% fat) w/nonfat milk solids	4	10/cup	(244g)
low-fat (2% fat)	8	20/cup	(245g)
low-fat (2% fat) w/nonfat milk solids	8	20/cup	(245g)
low-fat (2% fat), protein fortified	8	20/cup	(245g)
skim	2	5/cup	(246g)
skim w/nonfat milk solids	2	5/cup	(246g)
skim, protein-fortified	2	5/cup	(246g)
whole (3.25% fat)	14	34/cup	(244g)
whole (3.3% fat)	14	34/cup	(244g)
whole (3.5% fat)	14	34/cup	(244g)
whole (3.7% fat)	14	34/cup	(244g)
MILK, OTHER, FLUID			
goat	11	27/cup	(244g)
human	14	4/oz	(30g)
indian buffalo	19	46/cup	(244g)
MILK, COW, DRY			
buttermilk (skim milk)	69	4/T	(6g)
skim, calcium reduced	2	1/¼ cup	(28g)
skim solids, instant	18	11/cup	(60g)
skim solids, reg	20	14/cup	(70g)
whole, instant	97	97/cup	(100g)
YOGURT (cows milk)			
low-fat w/nonfat milk solids	6	14/cup	(227g)
low-fat, van–coffee flavored w/nonfat milk solids	5	11/cup	(227g)
low-fat, fruit flavor w/nonfat milk solids	4	10/cup	(227g)
skim w/nonfat milk solids	2	4/cup	(227g)
whole	13	30/cup	(227g)
whole	13	30/cup	(227g)
MILK BEVERAGES (cow's milk)			
choc milk (1% fat)	3	8/cup	(250g)
choc milk (2% fat)	7	18/cup	(250g)
choc milk (whole milk)	13	33/cup	(250g)
choc milk w/malt (whole milk)	13	34/cup	(265g)
choc milk, flavor w/Ovaltine (whole milk)	13	33/cup	(250g)
cocoa (hot choc–whole milk)	14	35/cup	(250g)
eggnog (nonalcoholic)	59	149/cup	(254g)
instant brkfst, van (whole milk)	10	28/cup	(276g)
malted milk	14	37/cup	(265g)
milkshake			
choc, thick	11	32/1 avg	(300g)
van, thick	12	37/1 avg	(313g)
orangenog (mix–whole milk), Delmark	13	31/cup	(140g)
MILK MIXES (cow's milk)			
choc malt-flavored powder	5	1/2T	(21g)
cocoa mix, instant, Carnation	7	2/package	(28g)
instant brkfst mix, choc, Carnation	1	tr/package	(36g)
malt powder, Carnation	19	1/t	(5g)
SALADS			
chicken salad, cnd, Spreadables	28	8/oz	(28g)
ham salad, cnd, Spreadables	21	6/oz	(28g)
macaroni salad–mayonnaise	44	84/cup	(190g)
potato salad–mayonnaise	26	65/cup	(250g)
tuna salad, cnd, Spreadables	33	9/oz	(28g)
tuna salad–mayonnaise–egg, homemade	42	42/½ cup	(100g)

FOOD	mg/ 100g	mg/SERVING	
SANDWICHES			
corned beef–rye bread	46	60/1 avg	(130g)
cream chse–olive–white bread	89	106/1 avg	(119g)
pnt btr–jelly–whole wheat bread	1	1/1 avg	(114g)
sloppy joe	68	136/1 avg	(200g)
tuna–white bread, grilled	172	172/1 avg	(100g)
SOUPS			
beans–ham, homemade	5	10/serving	(200g)
bean–pork, cnd, made w/water	5	10/serving	(207g)
beef-flavored noodle, dehyd, Cup-a-Soup	42	5/package	(12g)
beef noodle, cnd, made w/water	3	6/serving	(198g)
beef noodle veg, dehyd, Lipton	81	14/serving	(17g)
beef rice, homemade	5	6/serving	(113g)
bouillon cubes/powder	2	tr/cube	(4g)
cauliflower, cream of, homemade	34	38/serving	(113g)
chicken, cream of, cnd, made w/water	9	22/serving	(240g)
chicken-flavored, dehyd, Cup-a-Soup	53	4/package	(8g)
chicken noodle, cnd, made w/water	3	6/serving	(198g)
chicken noodle, dehyd, Lipton	136	24/serving	(18g)
clam chowder (manhattan), cnd, made w/water	20	40/serving	(201g)
clam chowder (new england), frzn, made w/water	20	39/serving	(193g)
corn chowder, homemade	5	6/serving	(116g)
mulligatawny, homemade	2	3/serving	(125g)
oyster stew, homemade (1 part oysters to 2 parts milk)	26	26/serving	(100g)
oyster stew, homemade (1 part oysters to 3 parts milk)	24	24/serving	(100g)
split pea, homemade	4	8/serving	(200g)
split pea, cnd, made w/water	15	30/serving	(200g)
shrimp, cream of, frzn, made w/milk	20	39/serving	(193g)
tomato, cnd, made w/water	2	4/serving	(198g)
tomato, dehyd, Cup-a-Soup	4	1/serving	(23g)
tomato veg, dehyd, Lipton	48	36/package	(74g)
veg beef, cnd, made w/water	2	4/serving	(203g)
SPICES AND FLAVORINGS			
meat loaf mix–TVP, Delmark	17	2/serving	(10g)
sloppy joe mix, Delmark	29	3/serving	(10g)
taco mix, Delmark	66	7/serving	(10g)
SPREADS, SALAD DRESSINGS, CREAMS, DIPS, GRAVIES			
butter	220	11/t	(5g)
butter, sweet (unsalt)	260	13/t	(5g)
butter, whip	200	8/t	(4g)
cream, half & half	40	6/T	(15g)
cream, light (coffee/table)	67	10/T	(15g)
cream, med (25% fat)	87	13/T	(15g)
cream, sour, cultured	44	5/T	(12g)
cream, sour, half & half, cultured	40	6/T	(15g)
cream, heavy whip, fluid	133	20/T	(15g)
cream, light whip, fluid	111	17/T	(15g)
creamer, nondairy	0	0/T	(10g)
gravy, brown	10	7/4T	(72g)
brown, chicken	28	8/oz	(28g)
gravy, hamburger	42	12/oz	(28g)
margarine, veg oil	0	0/t	(5g)
mayonnaise	70	10/T	(14g)
mayonnaise, imitation, Lady Lee	36	5/T	(14g)
salad dressing			
blue–roquefort	27	4/T	(14g)
blue cheese–yogurt, Henri	5	1/T	(14g)
green goddess	4	1/T	(14g)
homemade, ckd	74	11/T	(15g)
mayonnaise-type	50	7/T	(14g)
russian	65	10/T	(15g)
russian, low-cal	70	10/T	(14g)
spin blend, Hellmanns	49	8/T	(16g)
thousand island	65	9/T	(14g)
thousand island, low-cal	8	1/T	(14g)
thousand island w/yogurt, Henri	9	1/T	(14g)
sand spread w/chop pickle	28	6/T	(20g)
SWEETS			
icing, caramel	22	2/serving	(10g)

TABLE VI (Continued)   **167**

FOOD	mg/100g	mg/SERVING	
icing, choc	22	2/serving	(10g)
icing, choc fudge	22	2/serving	(10g)
icing, coconut	22	2/serving	(10g)
VEGETABLES			
beans, white–pork–sweet sce, cnd	3	8/cup	(255g)
beans, white–pork–tom sce, cnd	1	2/cup	(255g)
potatoes, french fries	20	20/20 pcs	(100g)
potatoes, mash, milk & fat add	14	30/cup	(210g)
potatoes, scallop	6	7/½ cup	(122g)
potatoes, scallop w/ cheese	15	18/½ cup	(122g)
pot–peas, creamed	4	4/serving	(100g)
MISC			
whey, sweet, dry	6	tr/serving	(10g)
whey, sweet, fluid	2	tr/serving	(10g)

TABLE VII

Fluoride Content of Foods
(mcg/100g EP)

Food	Value	Food	Value
almonds	90	kale	16-300
apples	5-130	lamb	120
apricots	2-22	liver, beef	99
bananas	23	liver, calf	19
beans, string	13	mackerel	2,700
beef	29-200	mackerel, cnd	1,200
beets	20	milk	4-55
beer	15-86	oats	25
butter	150	onions	60
cabbage	15	oranges	7-17
cantaloupe	20	oysters	65
carrots	40	parsley	80-100
celery	14	peaches	21
cheese	160	pears	60
cherries	25	peas	60
chicken	140	pineapple	14
chicken, cnd	63	plums	21
chocolate	50	pork	34-98
chocolate milk	50-200	potatoes	7-640
coca-cola	7	radish	80
codfish	700	rhubarb	40
coffee	2-160	rice	10-67
coffee, instant bev	170	round steak	130
corn	62	salmon, cnd	450-900
cornmeal	22	salt pork	100-3,300
crab meat, cnd	200	sardines, cnd	730-1,600
cucumber	20	shrimp, cnd	440
eggs, whole	120	soybeans	130
eggs, white	150	spinach	20-180
eggs, yolk	59	strawberries	18
eggplant	40	tea infusion	120-6,300
figs	21	tea, instant bev	200
fish	100-160	tomatoes	24
fishmeal	8,000-25,000	tuna, cnd	10
frankfurters	170	turnips	30
gooseberries	11-52	veal	90
grapefruit	36	watercress	100
grapes	16	wheat	70
hazelnuts	30	wheat flour	27-35
herring, smoked	350	wheat germ	88-400
honey	100	wine	0-630

Molybdenum Content of Foods
(mcg/serving)

beans, green, raw	66/cup	(100g)	lamb rib chops, broil	410/2 med	(82g)
beans, wax, raw	43/cup	(100g)	lentils, raw	66/½ cup	(95g)
beef, separable lean, ckd	7/3½ oz	(100g)	lettuce, Boston & Bibb	1/cup	(55g)
bread, white, enr	7/slice	(23g)	lettuce, iceburg	2/cup	(75g)
brussels sprouts, raw	4/10 med	(100g)	noodles, egg, enr, dry	34/cup	(73g)
buckwheat, whole grain, raw	485/cup	(100g)	pork, separable lean, ckd	368/3½ oz	(100g)
butter	0/t	(5g)	rye, whole grain	147/3½ oz	(100g)
cabbage, common, raw	1/cup	(90g)	sauerkraut	138/cup	(235g)
carrots, raw	7/cup	(81g)	shrimp, frzn, breaded	3/3½ oz	(100g)
celery, raw	2/cup	(120g)	spinach, raw	14/cup	(55g)
cereal			squash, winter, butternut	18/cup	(200g)
all bran, Kelloggs	154/cup	(56g)	strawberries, raw	20/cup	(149g)
100% natural, Quaker	633/oz	(28g)	tomato sauce	280/3½ oz	(100g)
puffed rice, Quaker	26/cup	(14g)	wheat flour, all purpose, enr	29/cup	(115g)
cucumber w/skin	1/1 med	(105g)	wheat germ	36/T	(7g)
halibut, broil	4/3½ oz	(100g)	wheat, whole grain	48/3½ oz	(100g)
kidneys, beef, braised	2140/3½ oz	(100g)	yam, raw	118/cup	(200g)
kidneys, hog, raw	110/3½ oz	(100g)			

TABLE IX **169**

Zinc Content of Foods
(mg/serving size)

BEVERAGES			pepper	.02/T	(15g)	
alcoholic			relish	.01/T	(15g)	
beer, dom	.11/12 oz	(360g)	sweet	.14/1 large	(100g)	
beer, cnd, Schlitz	.07/12 oz	(360g)	tomato	tr/T	(15g)	
brandy	.02/serving	(30g)	**DESSERTS**			
wine, red, dom	.10/wine glass	(100g)	brkfst pastry			
carbonated			coffee cake	.52/piece	(60g)	
bottled	.01/12 oz	(367g)	danish roll	.33/1 small	(39g)	
cnd	.30/12 oz	(367g)	cake			
cola	.02/12 oz	(360g)	cheese, frzn	.43/piece	(85g)	
cola, s-f	.01/12 oz	(360g)	choc w/choc icing	.34/piece	(50g)	
root beer	.01/12 oz	(360g)	date/nut	.05/piece	(30g)	
non-carbonated			white w/o icing	.20/piece	(86g)	
coffee, dry, instant	.02/T	(2.5g)	cookies			
coffee, fluid bev	.05/6 oz	(180g)	commerc	.08/1 med	(30g)	
coffee, instant, Decaf	.01/8 oz	(240g)	choc chip	.11/1 avg	(11g)	
tea, fluid bev	.04/6 oz	(177g)	choc fudge	.13/1 avg	(14g)	
tea, instant bev	.14/8 oz	(240g)	choc mint	.11/1 avg	(11g)	
CANDY			lemon creme sandwich	.04/1 avg	(12g)	
almond bar	.11/oz	(38g)	oatmeal	.24/1 avg	(18g)	
candy cane	.00/oz	(28g)	shortbread	.06/2 avg	(14g)	
caramel-filled bar	.11/oz	(28g)	doughnut	.20/1 avg	(42g)	
choc bar	.13/oz	(28g)	gelatin dessert, cherry	.02/½ cup	(120g)	
choc-covered mint	.11/oz	(28g)	ice cream	.60/cup	(133g)	
marshmallow	.01/oz	(28g)	ice cream cone, plain	.11/1 avg	(12g)	
mocha-type filled bar	.06/oz	(28g)	ice cream cone, sugar	.12/1 avg	(12g)	
CEREAL, ready-to-eat			ice milk, choc	.68/⅙ qt	(90g)	
cheerios	.71/oz	(28g)	ice milk, straw	.23/⅙ qt	(90g)	
corn flakes	.08/oz	(28g)	ice milk, van	.19/⅙ qt	(90g)	
granola	.60/oz	(28g)	pie			
life	.38/oz	(28g)	apple	.14/piece	(160g)	
oat cereal, puffed	.80/oz	(28g)	blueberry	.02/piece	(160g)	
rice, puffed	.40/oz	(28g)	cherry	.06/piece	(160g)	
total	.40/oz	(28g)	peach	.16/piece	(165g)	
wheat bran flakes, 40%	1.00/oz	(28g)	piecrust, homemade	.72/crust	(135g)	
wheat flakes	.60/oz	(28g)	pudding, bread	1.01/¾ cup	(165g)	
wheat germ, toasted	.90/T	(6g)	pudding, rice	.40/½ cup	(130g)	
wheat, puffed	.70/oz	(28g)	**DESSERT TOPPINGS**			
wheat, shredded	.80/oz	(28g)	choc syrup	.30/2T	(38g)	
CHEESE			non-dairy whip topping	tr/T	(5g)	
camembert, import	.67/oz	(28g)	sundae topping	.01/2T	(40g)	
cheddar	.50/slice	(13g)	**EGGS**			
cottage, creamed	1.04/cup	(225g)	white	.01/1 large	(33g)	
cream	.07/oz	(28g)	yolk	.50/1 large	(17g)	
farmer	.13/oz	(28g)	whole	.50/1 large	(50g)	
gouda, dom	1.08/oz	(28g)	whole, fried in oil	.58/1 med	(50g)	
mozzarella, dom	1.09/oz	(28g)	**FATS AND OILS**			
muenster, dom	.94/oz	(28g)	lard	.03/T	(13g)	
parmesan, dom	1.54/oz	(28g)	oil, salad/cooking	.40/cup	(218g)	
proc spread, american	.84/oz	(28g)	**FISH AND SHELLFISH**			
99% fat-free	.66/oz	(28g)	clams, soft shell, ckd	1.40/3 oz	(85g)	
COMBINATION FOODS			clams, hard shell, raw	1.10/4-5	(70g)	
beef stew, cnd	3.20/serving	(200g)	clams, hard shell, ckd	1.00/4-5	(62g)	
beef stew, cnd, Swanson	2.44/serving	(200g)	clams, surf, cnd	2.70/can	(220g)	
chili con carne, cnd	2.31/5 oz	(142g)	crab	6.70/cup	(155g)	
crab, deviled, frzn	1.71/serving	(100g)	gefilte	.45/3½ oz	(100g)	
hash, corned beef, cnd	2.19/½ cup	(110g)	lobster, ckd	3.10/cup	(145g)	
lasagna, homemade	3.20/serving	(200g)	lobster, steamed	7.95/3½ oz	(100g)	
lasagna, frzn	1.22/serving	(200g)	oysters			
macaroni–chse, frzn	1.34/cup	(200g)	atlantic, raw, cnd	254.30/can	(340g)	
macaroni–tom sce	1.24/serving	(200g)	atlantic, frzn, cnd	268.90/can	(360g)	
pizza, frzn	1.22/piece	(100g)	pacific, raw, cnd	30.6/can	(340g)	
pizza–mushrooms	1.66/piece	(100g)	pacific, frzn, cnd	32.4/can	(360g)	
ravioli, chse, frzn	1.52/serving	(200g)	salmon, cnd	2.10/can	(220g)	
ravioli, beef, cnd	.96/serving	(200g)	scallops, breaded, fried	1.29/3½ oz	(100g)	
spaghetti–meatballs, cnd	1.92/serving	(213g)	shrimp			
CONDIMENTS			boiled	1.70/6 med	(84g)	
horseradish, prepared	.19/T	(18g)	cnd	2.70/cup	(128g)	
mustard, brown	.01/t	(5g)	trout, breaded, fried	.40/3½ oz	(100g)	
mustard, yellow	.03/t	(5g)	tuna, cnd in oil, drained	1.80/can	(157g)	
olives, black	.06/2 large	(20g)	white varieties, flesh only			
olives, green w/pimento	.01/2 med	(13g)	filet, ckd	.90/3 oz	(85g)	
pickles			smoked	.36/3½ oz	(100g)	
dill	.27/1 large	(100g)	steak, ckd	3.10/cup	(145g)	

FLOUR
cornmeal, white/yellow

bolted, dry	2.10/cup	(122g)
degermed, dry	1.20/cup	(138g)
degermed, ckd	.30/cup	(240g)
wheat, whole	2.90/cup	(120g)
wheat, all-purpose, sifted	.80/cup	(115g)
wheat, bread, sifted	.90/cup	(115g)
wheat, cake, sifted	.30/cup	(96g)

FRUIT

apple, raw	.08/1 med	(180g)
applesauce, unsweet	.30/cup	(244g)
banana, raw	.30/1 med	(119g)
cranberry sauce	.01/5T	(100g)
fruit cocktail, cnd, heavy syrup	.02/½ cup	(100g)
grapefruit sections, white	.05/½ cup	(100g)
mango	.47/½ med	(100g)
orange, raw	.20/1 med	(131g)
peach, raw	.20/1 med	(131g)
peach, cnd	.30/cup	(220g)
pear, cnd, heavy syrup	.26/2 halves	(100g)
pineapple tidbits, cnd, heavy syrup	.15/⅜ cup	(100g)
raisins	.02/T	(10g)
strawberries	.08/10 large	(100g)
strawberries, frzn, sweet	.09/½ cup	(128g)

GRAINS AND CEREALS SERVED HOT

corn grits, dry	.70/cup	(160g)
farina, dry	1.00/cup	(180g)
farina, ckd	.20/cup	(245g)
oatmeal, dry	2.70/cup	(80g)
oatmeal, ckd	1.20/cup	(240g)
popcorn, popped	.20/cup	(6g)
popcorn, popped, oil–salt	.30/cup	(9g)
rice, brown, dry	3.40/cup	(185g)
rice, brown, ckd	1.20/cup	(195g)
rice, white, dry	2.50/cup	(185g)
rice, white, ckd	.80/cup	(205g)
rice, white, parboil, dry	.60/cup	(185g)
rice, white, parboil, ckd	.60/cup	(175g)
rice, white, instant, dry	.70/cup	(95g)
rice, white, instant, ckd	.40/cup	(165g)
rice, white, fried mix	1.07/cup	(155g)
rice, white–wild, frzn	.47/serving	(100g)
wheat cereal, whole grain, dry	4.50/cup	(125g)
wheat cereal, whole grain, ckd	1.20/cup	(245g)
wheat germ	1.67/T	(10g)

GRAIN PRODUCTS

bagel	.53/1 med	(55g)

bread

corn–molasses	.27/slice	(32g)
italian	.19/slice	(20g)
oatmeal	.29/slice	(28g)
pumpernickel	.36/slice	(32g)
rye	.40/slice	(25g)
white	.20/slice	(28g)
walnut	.22/slice	(28g)
whole wheat	.50/slice	(28g)
whole wheat, stoneground	.47/slice	(23g)
breadcrumbs	.33/cup	(88g)
corn chips	.40/oz	(28g)

crackers

graham	.20/2 avg	(14g)
round, thin	.17/oz	(28g)
saltines	.10/10 avg	(28g)
sesame seed	.29/oz	(28g)
french toast	.59/slice	(65g)
macaroni, ckd tender	.70/cup	(140g)
pretzels	.14/1 avg	(13g)
pancakes	.31/1 avg	(45g)

rolls

dinner	.43/1 avg	(35g)
hamburger	.20/1 avg	(40g)
hard	.40/1 avg	(35g)
parkerhouse	.19/1 avg	(23g)

JUICES

apple cider, bottled	.06/6 oz	(185g)
apple jce, bottled	.20/6 oz	(185g)
cranberry-apple jce, Ocean Spray	.62/8 oz	(240g)
fruit punch drink	.05/8 oz	(240g)
lemon juice	tr/T	(15g)
lemonade, frzn, diluted	.02/8 oz	(240g)
orange drink	.10/8 oz	(240g)
orange jce, cnd, unsweet	.20/cup	(249g)
orange jce, fresh or frzn	.05/cup	(248g)
pineapple jce, cnd, Dole	.38/8 oz	(240g)
prune jce, bottled	.01/8 oz	(240g)

MEATS
beef

corned, ckd	1.94/3½ oz	(100g)
ground, ckd	3.80/3 oz	(85g)
london broil	6.77/3½ oz	(100g)
roast w/gravy	5.12/3½ oz	(100g)
separable lean, ckd, dry heat	4.90/3 oz	(85g)
separable lean, ckd, moist heat	5.30/3 oz	(85g)

chicken, broiler-fryer, ckd

breast, meat only	.70/½ breast	(85g)
breast, meat & skin	.90/½ breast	(96g)
leg, thigh, back, meat only	2.40/3 oz	(85g)
drumstick, meat only	1.30/1 med	(45g)
drumstick, meat & skin	1.40/1 med	(54g)

lamb, separable lean

ckd, dry heat	3.70/3 oz	(85g)
ckd, moist heat	4.20/3 oz	(85g)
meatloaf, homemade	2.86/3½ oz	(100g)

organ meats

gizzard, chicken, ckd	6.20/cup	(145g)
gizzard, turkey, ckd	6.00/cup	(145g)
heart, chicken, ckd	6.90/cup	(145g)
heart, turkey, ckd	7.00/cup	(145g)
liver, beef, ckd	2.90/2 oz	(57g)
liver, calf, ckd	3.50/2 oz	(57g)
liver, chicken, chop, ckd	4.70/cup	(140g)
liver, turkey, chop, ckd	4.70/cup	(140g)

pork, separable lean

trimmed lean cuts, ckd	3.20/3 oz	(85g)
boston butt	3.80/3 oz	(85g)
ham/picnic	3.40/3 oz	(85g)
loin	2.60/3 oz	(85g)

sausages & luncheon meats

bologna	.50/oz	(28g)
braunschweiger	.80/oz	(28g)
frankfurters, beef	.90/1 avg	(45g)
frankfurters, beef–pork	.70/1 avg	(45g)
pork	.65/link	(20g)

turkey

light meat, ckd	1.80/3 oz	(85g)
dark meat, ckd	3.70/3 oz	(85g)

veal, separable lean

ckd, dry heat	3.50/3 oz	(85g)
ckd, moist heat	3.60/3 oz	(85g)

MILKS—products, beverages, mixes

cnd, evaporated	1.90/cup	(252g)
dry, nonfat	3.10/cup	(68g)
whole	0.90/cup	(244g)
yogurt, plain	1.09/cup	(227g)

milk bev

choc drink	.80/cup	(250g)
cocoa	2.40/cup	(250g)
instant brkfst w/milk, choc	5.77/cup	(276g)
milkshake, choc	1.47/1 avg	(300g)

milk mixes

cocoa, dry powder	.10/cup	(200g)
cocoa mix, dry powder	.12/T	(8g)
granular mix	.07/T	(8g)

NUTS

almonds, roasted	.72/oz	(28g)
brazil nuts	.76/4 med	(15g)
cashews	.66/6-8	(15g)
filberts	.46/10-12	(15g)
peanuts, roast	.30/T	(9g)
peanut butter	.50/T	(16g)

SALADS

bean (wax, green, kidney), cnd	.26/serving	(100g)

TABLE IX (Continued) **171**

coleslaw, homemade	.29/cup	(120g)
coleslaw–creamy dressing	.19/cup	(120g)
macaroni	.57/cup	(190g)
potato	.60/cup	(250g)
salad tossed–italian dress	.18/serving	(100g)
tuna	.32/½ cup	(100g)

SOUP

bean–bacon, cnd, diluted	.83/serving	(207g)
beef noodle, cnd, cond	1.40/⅓ can	(99g)
clam chowder, cnd, cond	.60/⅓ can	(104g)
mushroom, cream of, cnd, cond	.50/⅓ can	(99g)
shrimp, cream of, cnd, cond	.57/⅓ can	(95g)
tomato, cnd, cond	.07/⅓ can	(99g)
tomato, cream of, cnd, cond	.43/⅓ can	(99g)
vegetable, cnd, cond	.75/⅓ can	(102g)
vegetable beef, cnd, cond	1.64/⅓ can	(104g)
soup, dry, packaged		
chicken broth	.01/T	(12g)
onion	tr/serving	(11g)
chicken noodle	.01/serving	(18g)
lobster bisque	.01/serving	(18g)
split pea	.04/serving	(18g)

SPREADS, SALAD DRESSINGS, CREAMS, GRAVIES

butter	.01/T	(14g)
cream, nondairy	.01/T	(15g)
cream, sour	.04/T	(12g)
cream, table	.04/T	(15g)
creamer, nondairy	.08/T	(15g)
gravy, beef	.40/4T	(72g)
gravy, beef, cnd	.50/oz	(28g)
gravy, chicken, packaged	.03/oz	(28g)
margarine	.03/T	(14g)
mayonnaise	.02/T	(14g)
mayonnaise, imitation	.02/T	(14g)
salad dressing		
commerc	.03/T	(15g)
italian	.02/T	(14g)
roquefort	.04/T	(14g)
russian	.06/T	(14g)
thousand island	.02/T	(14g)
tomato french	.01/T	(14g)

SWEETS

icing, buttercream	.02/serving	(10g)
honey	.02/T	(20g)
jam, peach	.01/T	(20g)
jam, strawberry	tr/T	(20g)
marmalade, orange	tr/T	(20g)
sugar, white granulated	.10/cup	(200g)
syrup, pancake	.01/T	(20g)
tomato ketchup	.04/T	(15g)

VEGETABLES

artichoke, ckd	.35/1 med EP	(100g)
avocado, peeled	.52/½ med	(120g)
beans, common mature		
dry	5.30/cup	(190g)
boiled	1.80/cup	(185g)
beans, lima, mature		
dry	5.00/cup	(180g)
ckd	1.70/cup	(190g)
beans, snap, green		
raw	.40/cup	(110g)
boiled	.40/cup	(125g)
cnd, drained	.40/cup	(135g)
broccoli		
ckd	.15/⅔ cup	(100g)
spears, frzn	.27/3½ oz	(100g)
cabbage, common		
raw	.30/cup	(90g)
boiled	.60/cup	(145g)

carrots		
raw	.30/1 med	(72g)
boiled/cnd	.50/cup	(155g)
chickpeas/garbanzos, mature		
dry	5.40/cup	(200g)
boiled	2.00/cup	(146g)
corn, sweet, yellow		
boiled	.70/cup	(165g)
cnd	.80/cup	(210g)
creamed, cnd	.55/½ cup	(128g)
cowpeas (blackeye)		
raw	4.90/cup	(170g)
boiled	6.70/cup	(250g)
garlic, raw	.04/clove	(3g)
lentils, mature		
dry	5.90/cup	(190g)
boiled	2.00/cup	(200g)
lettuce, head/leaf	.20/cup	(55g)
onions, mature, chopped	.60/cup	(170g)
onion rings, frzn	.35/3½ oz	(100g)
onions, young green, chopped	.30/cup	(100g)
parsley, dry	.25/serving	(10g)
peapods, snow, frzn	.41/3½ oz	(100g)
peas, green, immature		
raw/frzn	1.20/cup	(145g)
boiled	1.20/cup	(160g)
cnd	1.30/cup	(170g)
peas, green, mature		
dry	6.40/cup	(200g)
boiled	2.10/cup	(200g)
pepper, green	.04/1 shell	(65g)
potato		
raw, peeled	.40/1 med	(112g)
baked w/skin	.96/1 med	(160g)
baked w/o skin	.71/1 med	(150g)
boiled, pared before cooking	.30/1 med	(112g)
boiled in skin	.40/1 med	(136g)
chips	.23/oz	(28g)
french fried	.14/10 pieces	(50g)
knishes	.84/serving	(100g)
mashed, instant	.80/cup	(210g)
puffs, frzn	.20/serving	(100g)
sticks, cnd	.07/oz	(28g)
radish	.03/1 small	(10g)
rhubarb		
raw	.37/3½ oz	(100g)
ckd, sugar added	.11/⅜ cup	(100g)
sauerkraut, cnd	.81/⅔ cup	(100g)
spinach		
raw, chopped	.50/cup	(55g)
boiled, drained	1.30/cup	(180g)
cnd, drained	1.60/cup	(205g)
creamed	.61/serving	(100g)
sweet potato, raw	.14/1 large	(180g)
tomato, ripe		
raw	.20/1 med	(123g)
boiled	.50/cup	(241g)
cnd, solids & liquid	.50/cup	(241g)
tomato sauce–meat	.73/½ cup	(125g)
tomato sauce–mushrooms	.68/½ cup	(125g)
vegetables, bavarian style, frzn	.34/serving	(100g)
vegetables, mixed, frzn	.53/cup	(161g)
zucchini, ckd	.18/½ cup	(100g)

MISCELLANEOUS

vinegar	.02/T	(15g)

Choline Content of Foods
(mg/serving)

Food	Amount	Weight	Food	Amount	Weight
avocado, calif	14/½ small	(84g)	Precision Low Residue	10/cup	(240g)
beef rib, choice, separable lean, ckd	82/3½ oz	(100g)	Sustagen (oral)	10/cup	(240g)
beef, round, choice, tot ed (89% lean, 11% fat), raw	68/3½ oz	(100g)	Sustagen (tube)	1/oz	(30g)
beans, green	18/cup	(135g)	liquid formulas for infants & child sensitive to cow milk		
beans, mung, mature seeds, raw	220/½ cup	(105g)	Isomil	84/liter	
bologna	17/oz	(28g)	i-soyalac	90/liter	
butter	1/t	(5g)	Neomullsoy	90/liter	
cabbage, common, raw	16/cup	(70g)	Nursoy	90/liter	
carrots, raw	11/1 med	(81g)	Nutramigen	90/liter	
cashews	37/oz	(28g)	Pregestimil	90/liter	
corn, sweet, white & yellow, raw	61/serving	(100g)	Pro Sobee	90/liter	
cowpeas, young pods–seeds, raw	97/cup	(100g)	Soyalac	110/liter	
cowpeas, mature, raw	218/½ cup	(85g)	margarine	0/t	(5g)
eggs	242/1 med	(48g)	milk, whole	49/cup	(244g)
egg white	1/1 med	(33g)	milk, evaporated	131/liter	
egg yolk	170/1 med	(17g)	milk, dry skim, recon	104/liter	
flour, soybean, full-fat	158/cup	(70g)	milk, dry whole, recon	142/liter	
flour, soybean, low-fat	198/cup	(88g)	molasses	17/T	(20g)
flour, soybean, defatted	225/cup	(100g)	mustard greens, raw	7/cup	(33g)
flour, wheat, all purpose, enr	60/cup	(115g)	oil, corn	0/T	(14g)
frankfurter	29/1 avg	(50g)	rice, brown, raw	112/½ cup	(100g)
instant brkfst mix, choc, Carnation	22/package	(62g)	rice, brown, ckd	218/cup	(195g)
lamb leg, tot ed	84/3½ oz	(100g)	rice, white, enr, raw	58/½ cup	(98g)
lamb loin, tot ed	76/3½ oz	(100g)	rice, white, enr, ckd	89/cup	(150g)
lard	1/T	(14g)	rice, white, enr, long grain, raw	60/½ cup	(62g)
lentils, raw	212/½ cup	(95g)	rice, white, enr, long grain, ckd	147/cup	(150g)
liquid formulas			sausage, pork links, raw	29/piece	(60g)
Advance	2/oz	(30g)	soybeans, immature	315/3½ oz	(100g)
Ensure Osmolite	130/cup	(240g)	soybeans, mature	340/3½ oz	(100g)
Ensure Plus	125/cup	(240g)	soybean protein conc	10/10g	
Flexical	60/cup	(240g)	tomato sauce	20/3½ oz	(100g)
Isocal	62/cup	(240g)	turnip greens, raw	27/3½ oz	(100g)
Meritene (whole milk)	50/cup	(240g)	turnip greens, frzn	45/cup	(165g)
Nutramigen	22/cup	(240g)	wheat bran	13/T	(9g)
Nutri-1000	48/cup	(240g)	wheat germ	28/T	(7g)
Nutri-1000 LF	48/cup	(240g)	wheat, whole, dry	59/½ cup	(63g)
Precision High Nitrogen	10/cup	(240g)	wheat, whole, ckd	230/cup	(245g)
			yeast, Brewers	24/T	(10g)

TABLE XI

Vitamin K Content of Foods
(mcg/serving)

Food	Amount	Weight	Food	Amount	Weight
asparagus, raw	57/5-6 spears	(100g)	Nursoy	110/liter	
beef, ground, lean, raw	59/3 oz	(85g)	Nutramigen	110/liter	
beef, dried	15/3 rnd pieces	(15g)	Pregestimil	110/liter	
broccoli, raw	252/cup	(126g)	Pro Sobee	100/liter	
cabbage, raw	155/cup	(124g)	liver, beef, raw	85/3 oz	(85g)
coffee, bev	76/cup	(200g)	liverwurst	34/1 oz slice	(28g)
lettuce, raw	96/cup	(74g)	milk, cow	60/liter	
liquid formulas			milk, human	15/liter	
Nutri-1000 LR	38/cup		peas, green, boiled	221/½ cup	(85g)
Precision High, Nitrogen	10/cup		spinach, raw	48/cup	(54g)
Precision Isotonic	15/cup		tomato, raw	7/1 med	(148g)
Precision Low Residue	15/cup		turnip greens, raw	351/cup	(54g)
liquid formulas for infants & child			watercress, raw	14/25 sprigs	(25g)
Isomil	150/liter				
Neomullsoy	100/liter				

REFERENCES

1. Adams C: Nutritive Value of American Foods in Common Units, Agriculture Handbook No. 456. Washington, DC, USDA, 1975

2. Amino Acid Content of Foods and Biological Data on Proteins. Rome, Food and Agricultural Organization of the United Nations, 1970

3. Block RJ, Weiss KW: Amino Acid Handbook. Springfield, IL, Thomas Publishing Company, 1956

4. Borgstrom G: Fish as Food, vol 1. Production, Biochemistry, and Microbiology. New York, Academic Press, 1961

5. Boucher RV and Committee: Composition of Concentrate By-Product Feeding Stuffs, Publication 449. Washington DC, National Research Council, 1956

6. Burger M, Hein LW, Teply LJ, Derse PH, Krieger CH: Nutrients in frozen foods. Wisconsin Alumni Research Foundation, Agricultural and Food Chemistry 4:418–425, 1956

7. Chatfield C: Food Composition Tables for International Use. Washington, DC, Food and Agriculture Organization of the United Nations, 1949

8. Chatfield C: Food Composition Tables—Minerals and Vitamins—For International Use. Rome, Italy, Food and Agriculture Organization of the United Nations, 1954

9. Composition of Hawaii Fruit, Bulletin 135. University of Hawaii, Hawaii Agricultural Experiment Station, 1965

10. Davidson CD and Committee: Sodium Restricted Diets, Publication 325. Washington, DC, National Research Council, 1954

11. Diem K (ed): Documented Geigy, Scientific Tables, 6th ed. New York, Geigy Pharmaceuticals, 1962

12. Goddard VR, Goodall L: Fatty Acids in Food Fats, Home Economics Research Report No. 7. USDA, 1959

13. Howard FD, Macgillivary JH, Yamaguchi M: Nutrient Composition of Fresh California-Grown Vegetables, Bulletin 768. University of California, California Agriculture Experiment Station, 1962

14. Kraus B: The Dictionary of Sodium, Fats, and Cholesterol. New York, Grosset and Dunlap, 1974

15. Lee CF: Composition of Cooked Fish Dishes, Fish and Wildlife Service Circular 29. Washington DC, US Department of the Interior, 1959

16. Leung WW, Flores M: Food Composition Table for Use in Latin America. Bethesda, Maryland, International Committee on Nutrition for National Defense, National Institutes of Health, 1961

17. Leung WW, Pecot RK, Watt BK: Composition of Foods Used in Far Eastern Countries, Agriculture Handbook No. 34, USDA, 1952

18. Lowenberg ME, Wilson ED: Nutrients in Frozen Foods. Washington, DC, National Association of Frozen Food Packers, 1959

19. Marsh AC, Moss MK, Murphy EW: Composition of Foods—Spices and Herbs—Raw, Processed, Prepared, Agriculture Handbook 8-2. Washington, DC, USDA, 1977

20. Miller CD, Branthoover B: Nutritive Values of Some Hawaii Foods. Honolulu, Hawaii Agricultural Experiment Station, University of Hawaii, 1960

21. Miller DF and Committee: Composition of Cereal Grains and Forages, Publication 585. Washington, DC, National Research Council, 1958

22. Nutritive Value of Foods, Home and Garden Bulletin No. 72, rev. Washington DC, USDA, 1970

23. Orr ML, Watt BK: Amino Acid Content of Foods, Home Economics Research Report No. 4. Washington, DC, USDA, 1957

24. Paul AA, Southgate DAT: McCance and Widdowson's Composition of Foods, 4th rev ed of MRC Special Report No. 297. London, Her Majesty's Stationery Office, Elsevier/North-Holland Biomedical Press, 1978

25. Posati LP, Orr ML: Composition of Foods—Dairy and Egg Products—Raw, Processed, Prepared. Agriculture Handbook 8-1. Washington, DC, USDA, 1976

26. Stansby ME: Composition of Fish, Fishery Leaflet 116. Washington, DC, Fish and Wildlife Service, 1953

27. Treichler R, Lee CF, Jarvis ND: Chemical Composition of Some Canned Fishery Products, Fishery Leaflet 295. Washington, DC, Fish and Wildlife Service, 1948

28. Turner D: Handbook of Diet Therapy. Chicago, IL, University of Chicago Press, 1965

29. Umbarger BJ: Phenylalanine Content of Foods, rev. Ohio, The Children's Hospital Research Foundation and the University of Cincinnati College of Medicine, 1965

30. Watt BK, Merrill AL: Composition of Foods—Raw, Processed, Prepared. Agriculture Handbook No. 8, rev. Washington, DC, USDA, 1963

Index

A "t" following a page number indicates a table.